Marine Mammal Research in the Southern Hemisphere

Status, Ecology and Medicine

Male Australian Fur Seal *Arctocephalus pusillus doriferus*, at Cape Gantheaume, Kangaroo Island, South Australia. Photo: Simon Goldsworthy.

Marine Mammal Research in the Southern Hemisphere

Volume 1

Status, Ecology and Medicine

Edited by

Mark Hindell and Catherine Kemper

Published by

Surrey Beatty & Sons

The National Library of Australia
Cataloguing-in-Publication entry:

Marine Mammal Research in the Southern Hemisphere.

Bibliography.
Includes index.
ISBN 0 949324 76 0 (v. 1).
ISBN 0 949324 70 1 (set)

1. Marine mammals — Southern Hemisphere. I. Hindell, Mark Andrew. II. Kemper, Catherine Margaret.

599.5

Published February 1997

PRINTED IN AUSTRALIA BY
SURREY BEATTY & SONS PTY LIMITED
43 Rickard Road, Chipping Norton, NSW 2170

Contents

List of colour plates vi

Introduction. By M. A. Hindell and C. M. Kemper vii

List of contributors ix

Acknowledgements x

PART 1: STATUS AND MANAGEMENT

Bottlenose Dolphins *Tursiops truncatus* in south-east Queensland waters: social structure and conservation biology. By P. J. Corkeron 1

Genetic studies of the Southern Elephant Seal *Mirounga leonina*. By R. W. Slade 11

A yacht-based cetacean research cruise to Antarctic waters south of Australia. By P. C. Gill .. 30

Southern Right Whales in southeastern Australia — aerial surveys during 1991–93 and incidental information from 1904. By C. M. Kemper, J. Mole, R. M. Warneke, J. K. Ling, D. J. Needham and J. E. Wapstra 40

A census of fur seals at Montagu Island, New South Wales. By A. Irvine, M. M. Bryden, P. J. Corkeron and R. M. Warneke 56

Field identification of Australian and New Zealand Fur Seals *Arctocephalus* spp., based on external characters. By S. D. Goldsworthy, D. Pemberton and R. M. Warneke 63

PART 2. GENERAL ECOLOGY

The importance of being fat: maternal expenditure in the Southern Elephant Seal *Mirounga leonina*. By M. A. Hindell and D. J. Slip 72

The Australian Sea Lion: a review of an unusual life history. By N. J. Gales and D. P. Costa .. 78

Growth in Subantarctic Fur Seal *Arctocephalus tropicalis* pups as a possible indicator of offshore food availability. By M. N. Bester and A. S. van Jaarsveld 88

Evidence for precocious development in female pups of the Southern Elephant Seal *Mirounga leonina* at Macquarie Island. By C. R. McMahon, J. van den Hoff, H. R. Burton and P. D. Davis 92

PART 3: FORAGING ECOLOGY

Diving behaviour of Antarctic Fur Seals *Arctocephalus gazella* Peters around Heard Island. By K. Green 97

Foraging ecology of Antarctic Fur Seals *Arctocephalus gazella* Peters around Heard Island. By K. Green, R. Williams and H. R. Burton 105

Diving and foraging behaviour of juvenile Southern Elephant Seals from Heard Island. By D. J. Slip 114

Analysis of the diet of New Zealand Fur Seals *Arctocephalus forsteri* in Tasmania. By S. Lake .. 125

Prey of Hooker's Sea Lions *Phocarctos hookeri* based at Otago Peninsula, New Zealand. By C. Lalas 130

The use of satellite telemetry to determine fur seal foraging areas. By R. Harcourt and L. Davis 137

Assessing the use of faecal and regurgitate analysis as a means of determining the diet of New Zealand Fur Seals. By N. Fea and R. Harcourt 143

Diet and diving behaviour of sympatric fur seals *Arctocephalus gazella* and *A. tropicalis* at Macquarie Island. By S. D. Goldsworthy, M. A. Hindell and H. M. Crowley 151

PART 4: MEDICINE HUSBANDRY

The role of stones in the sea lion stomach — investigations using contrast radiography and fluoroscopy. By D. J. Needham 164

Report on the healing of a large wound in a Bottlenose Dolphin *Tursiops truncatus*. By M. B. Orams and R. B. Deakin 170

Preliminary investigations of techniques for killing whales. By D. K. Blackmore, A. Nutman, G. R. G. Barnes, P. Madie, A. S. Davies, M. C. Bowling, M. Donoghue and E. J. Kirk .. 174

On capture, anaesthesia and branding of adult male New Zealand Fur Seals *Arctocephalus forsteri.* By S. Troy, D. Middleton and J. Phelan 179

Index 184

List of Colour Illustrations

FRONTISPIECE

Male Australian Fur Seal *Arctocephalus pusillus doriferus,* at Cape Gantheaume, Kangaroo Island, South Australia. Photo Simon Goldsworthy.

PAGE 39

Above: Killer Whale *Orcinus orca*, Sceale Bay, South Australia, 25th December 1992. Photo: W. Guiddera.

Top right: Southern Right Whale *Eubaleana australis,* Bicheno, Tasmania, 1st June 1993. Adult female. Photo: L. Geeves.

Bottom right: Humpback Whale *Megaptera novaeangliae,* interacting with yacht "Iniquity", Antarctic Waters (63°55′S, 103°38′E), 13th February 1993. Photo: P. Gill.

PAGE 71

Top left: Female Southern Elephant Seal *Mirounga leonina,* with new born pup at Macquarie Island. This female was branded as a pup and the mark was still clear 23 years later. Photo: Mark Hindell.

Top right: Australian Fur Seals *Arctocephalus pusillus doriferus,* at Cape Gantheaume, Kangaroo Island, South Australia. Photo: Simon Goldsworthy.

Bottom left: New Zealand Fur Seal *Artocephalus fosteri,* female and pup, at Cape Gantheaume, Kangaroo Island, South Australia. Photo: Simon Goldsworthy.

Bottom right: Two species of fur seals at Macquarie Island. Left to right: Adult female and adult male Subantarctic Fur Seal *Arctocephalus tropicalis*, and adult female Antarctic Fur Seal *Arctocephalus gazella*. Photo: Simon Goldsworthy.

Right: Adult female Hooker's Sea Lion *Phocarctos hookeri*, Enderby Island. Photo: Mark Hindell.

PAGE 129

Top left: Bottlenose Dolphins *Tursiops truncatus,* 8th November 1985. Photo: P. Corkeron.

Top right: Australian Seal Lion *Neophoca cinerea* female, Dangerous Reef, South Australia, 3rd February 1995. Photo: C. Kemper.

Left: Antarctic Fur Seals *Arctocephalus gazella* at Heard Island, illustrating an adult male, and adult female and her pup. Photo: Simon Goldsworthy.

Bottom left: Female Australian Fur Seals *Arctocephalus pusillus doriferus* at Cape Gantheaume, Kangaroo Island, South Australia. Photo: Simon Goldsworthy.

Bottom centre: Two fur seal pups at Heard Island. The left pup is a Subantarctic Fur Seal *Arctocephalus tropicalis* exhibiting the uniform glossy black coat. The pup on the right is an Antarctic Fur Seal *Arctocephalus gazella* with a more grizzled appearance, pale muzzle and belly. Photo: Simon Goldsworthy.

Bottom right: New Zealand Fur Seal *Arctocephalus forsteri* pup. Photo: Simon Goldsworthy.

Introduction

THE southern hemisphere is home to a rich marine mammal fauna. The majority of the world's 13 otariid pinnipeds (eared seals and sea lions) are found there, as is the most numerous of all phocid pinnipeds, the Crab-eater Seal. The southern hemisphere cetacean fauna is also diverse and includes not only the well-known Great Whales, but also many members of the beaked whale family about which very little is known. Marine mammal conservation issues in the southern hemisphere are growing. New Zealand and Australia have both of the world's rarest sea lions, one of which, Hooker's Sea Lion, is subject to serious incidental mortality in squid trawl fisheries. Gill net fisheries off northern Australia are a potentially serious problem to Dugongs and inshore tropical dolphins. Fundamental marine mammal research, such as is presented in this volume, is vital to helping solve these conservation problems.

Marine mammals are difficult to study because they live in an environment very different from our own. However, technological advances such as electronic data logging, satellite telemetry, photographic matching and DNA genetics have enabled a whole new suite of fundamental questions to be addressed, especially about the lives of animals at sea. Public interest in marine mammals has increased dramatically in the last 20 years and with that has come more funding support for research.

There is a long history of marine mammal research in Australia. Good examples are the pioneering research on seals by Judith King and Basil Marlow and on whales by Herbert Hale and Eric Scott. For some reason, possibly because it has been more closely linked with fisheries science, marine mammalogy has not been a traditionally important component of the Australian Mammal Society's annual meetings. It has only been in the last 10 or so years that there has been an increase in the number of marine mammal papers presented at these meetings and by the early 1990s there were sufficient marine mammal papers to occupy an entire session at the annual meeting.

A symposium on Marine Mammals of Australia and New Zealand was held in Hobart in July 1994. It was different in several important ways from most earlier Australian Mammal Society symposia in that it had a deliberately international list of delegates (so did the American/Australian joint meeting in 1984) and it focused on species, for the most part, not endemic to Australia. Seals and whales show little regard for international boundaries. Some species of fur seals breed in both New Zealand and Australia, while Subantarctic and Antarctic Fur Seals breed on islands with a range of national sovereignties. One of the aims of the symposium was to bring together marine mammalogists studying southern hemisphere species and judging by the attendance of more than 80 delegates from Australia, New Zealand, South Africa and the United States, this was achieved. Forty-two spoken papers and 10 posters were presented at the two-day symposium. For the most part, papers contained in this volume were presented at the conference, but some were offered subsequently.

The proceedings of the Hobart symposium joins two other recent Australian collections of marine mammal research papers. The first was "Studies of Sea Mammals in South Latitudes", edited by John Ling and Michael Bryden (1985, South Australian Museum) and the second was "Marine Mammals of Australasia: Field Biology and Captive Management", edited by Michael Augee (1988 Royal Society of New South Wales). The diversity of subjects (conservation and management, ecology, feeding ecology, veterinary studies) covered at the Hobart symposium showed that marine mammal research is being embraced by scientists from many disciplines.

These proceedings have been divided into four sections. The first deals with conservation and status of marine mammals. Topics range from yacht-based surveys of cetaceans in the Antarctic to studies of population genetics of elephant seals. A mixture of basic survey work and more sophisticated (but no less routine) molecular biology studies are needed to provide the information required for informed management decisions. With the increasing pressure from human fisheries and tourism more and more studies of this nature are required.

The second section deals with general ecology. These papers reflect a growing trend to move away from purely descriptive studies to more integrative and question-directed studies, an encouraging sign that reflects the maturation of the discipline. For example, one study uses estimates of maternal expenditure and pup growth to test predictions from demographic models of elephant seal populations, while another examines the unusual breeding biology (an 18 month, asynchronous breeding cycle) of the Australian Sea Lion in the light of ecosystem processes.

Papers about foraging ecology are presented in the third section. Studies include several on deducing diet from scat analysis, and it is encouraging to note that the authors acknowledge the need to quantify the biases in these methods. At the opposite end of the technological scale to scat collection is the deployment of electronic data logging and telemetric devices. These methods are at the forefront of wildlife ecology studies and have often been pioneered on marine mammal species before being used on terrestrial species. This section is devoted entirely to studies of pinnipeds, either because the types of studies are inappropriate for cetaceans (it is difficult to collect faecal samples from free ranging cetaceans), or because the technology has yet to be widely adapted for cetacean studies.

The final section is devoted to chapters on veterinary aspects of marine mammals. Cetacean and pinniped veterinary science has traditionally been a somewhat specialist area, largely due to the difficult and infrequent access to these animals. In this section there are some valuable contributions, from assessing the most humane method of killing cetaceans, to examining the role of stomach stones in the digestion of sea lions.

Although this volume contains papers from a diverse range of southern hemisphere regions, some species (notably Dugongs) and some areas (notably South America) are not represented. Given the enthusiasm generated by this meeting it will be valuable to have similar meetings on a regular basis, perhaps every four years. Future gatherings will provide an opportunity for expanding the scope of geographic regions and scientific disciplines even beyond those reported in this volume.

Mark A. Hindell

Cath M. Kemper

List of Contributors

BARNES, G. R., Department of Physics and Biophysics, Massey University, Palmerston North, New Zealand.

BESTER, M. N., Department of Zoology and Entomology, University of Pretoria, 0002 South Africa.

BLACKMORE, D. K., 371 Albert Street, Palmerston North, New Zealand.

BOWLING, M. C., Department of Physics and Biophysics, Massey University, Palmerston North, New Zealand.

BRYDEN, M. M., Department of Veterinary Anatomy, The University of Sydney, New South Wales, 2006 Australia.

BURTON, H. R., Australian Antarctic Division, Channel Highway, Kingston, Tasmania, 7050 Australia.

CORKERON, P. J., Department of Veterinary Anatomy, The University of Sydney, New South Wales, 2006 Australia. Present address: Department of Tropical Environment Studies and Geography, James Cook University of North Queensland, Queensland, 4811 Australia.

COSTA, D. P., Institute of Marine Sciences and Department of Biology, University of California, Santa Cruz, CA 95064 USA.

CROWLEY, H. M., Division of Botany and Zoology, Australian National University, Canberra, Australian Capital Territory, 2601 Australia. Present address: Reserve de Berenty, Tolognaro (Fort Dauphin), BP 54, Madagascar.

DAVIES, A. S., Department of Physiology and Anatomy, Massey University, Palmerston North, New Zealand.

DAVIS, L. S., Department of Zoology, University of Otago, PO Box 56, New Zealand.

DAVIS, P. D., Australian Antarctic Division, Channel Highway, Kingston, Tasmania, 7050 Australia.

DEAKIN, R. B., SeaWorld Enterprises, PO Box 190, Surfers Paradise, Queensland, 4217 Australia.

DONOGHUE, M., Department of Conservation, Auckland, New Zealand.

FEA, N., Department of Zoology, University of Otago, PO Box 56, New Zealand.

GALES, N. J., School of Veterinary Studies, Murdoch University, Murdoch, Western Australia, 6150 Australia. Present address: Department of Conservation, PO Box 10-420, Wellington, New Zealand.

GILL, P. C., Oceanic Research Foundation, c/- 148 Copeland Road East, Beecroft, New South Wales, 2119 Australia.

GOLDSWORTHY, S. D., Department of Ecology and Evolutionary Biology, Monash University, Clayton, Victoria, 3168 Australia. Present address: Department of Zoology, University of Tasmania, GPO Box 252C, Hobart, Tasmania, 7001 Australia.

GREEN, K., Australian Antarctic Division, Channel Highway, Kingston, Tasmania, 7050 Australia. Present address: NPWS, Snowy Mountain Region, PMG Cooma, New South Wales, 2630 Australia.

HARCOURT, R., Department of Zoology, University of Otago, PO Box 56, New Zealand.

HINDELL, M. A., Department of Zoology, University of Tasmania, GPO Box 252C, Hobart, Tasmania, 7001 Australia.

IRVINE, A., Department of Veterinary Anatomy, The University of Sydney, New South Wales, 2006 Australia.

KEMPER, C. M., South Australian Museum, North Terrace, Adelaide, South Australia, 5000 Australia.

KIRK, E. J., Department of Physiology and Anatomy, Massey University, Palmerston North, New Zealand.

LAKE, S., Department of Zoology, University of Tasmania, GPO Box 252C, Hobart, Tasmania, 7001 Australia.

LALAS, C., Box 31, Portobello, Dunedin, New Zealand.

LING, J. K., South Australian Museum, North Terrace, Adelaide, South Australia, 5000 Australia.

MADIE, P., Department of Veterinary Pathology and Public Health, Massey University, Palmerston North, New Zealand.

McMAHON, C. R., Australian Antarctic Division, Channel Highway, Kingston, Tasmania, 7050 Australia.

MIDDLETON, D., WildVet, PO Box 626, Lilydale, Victoria, 3140 Australia.

MOLE, J., South Australian Museum, North Terrace, Adelaide, South Australia, 5000 Australia.

NEEDHAM, D. J., 294 Greenhill Road, Glenside, South Australia, 5065 Australia.

NUTMAN, A., Department of Physiology and Anatomy, Massey University, Palmerston North, New Zealand.

ORAMS, M. B., Department of Management Systems, Massey University, Albany, Private Bag 102 904, North Shore MSC, New Zealand.

PEMBERTON, D., Department of Parks, Wildlife and Heritage, PO Box 44A, Hobart, Tasmania, 7001 Australia.

PHELAN, J., Healesville Sanctuary, Healesville, Victoria, 3777 Australia.

SLADE, R. W., Department of Zoology and Centre for Molecular and Cellular Biology, University of Queensland, 4072 Australia. Present address: Queensland Institute of Medical Research, Post Office, Royal Brisbane Hospital, Queensland, 4029 Australia.

SLIP, D. J., Department of Zoology, University of Tasmania, GPO Box 252C, Hobart, Tasmania, 7001 Australia and Australian Antarctic Division, Channel Highway, Kingston, Tasmania, 7050 Australia.

TROY, S. Department of Zoology, University of Melbourne, Parkville, Victoria, 3052 Australia.

VAN DEN HOFF, J., Australian Antarctic Division, Channel Highway, Kingston, Tasmania, 7050 Australia.

VAN JAARSVELD, A. S., Department of Zoology and Entomology, University of Pretoria, 0002 South Africa.

WAPSTRA, J. E., Department of Parks, Wildlife and Heritage, PO Box 44A, Hobart, Tasmania, 7001 Australia.

WARNEKE, R. M., Blackwood Lodge, RSD 273, Mt Hicks Road, Yolla, Tasmania, 7325 Australia.

WILLIAMS, R., Australian Antarctic Division, Channel Highway, Kingston, Tasmania, 7050 Australia.

Acknowledgements

EVERY paper in this book has been reviewed by at least two experts in the relevant field. In order to ensure that each chapter was given the best possible scrutiny we made use of undoubtedly already busy people from all around the world. The quality of science in the following chapters owes an immense debt to these people who devoted considerable time to ensuring that the papers were of a very high standard.

We would also like to thank all the contributors to this volume for their hard work and patience in preparing and revising their manuscripts.

We thank the University of Tasmania which hosted the symposium, and we would particularly like to acknowledge all those who helped organize and run the meeting. The unusually smooth running was largely the result of the efforts of Mike Stoddart, Randy Rose and Sue Jones of the Zoology Department and Stewart Nicol from the Physiology Department. Dianne Moyle, Deborah Theile, Margie Morrice and Cindy Hull all contributed to the success of the meeting in various ways.

The Australian Mammal Society provided valuable financial assistance for the production this book.

Bottlenose Dolphins *Tursiops truncatus* in south-east Queensland waters: social structure and conservation biology

P. J. CORKERON[1,2]

Group sizes and ranges of Bottlenose Dolphins *Tursiops truncatus* which occur in the inshore oceanic waters around Point Lookout, North Stradbroke Island, Queensland were studied in 1992 by photographic identification of individuals from natural marks. Results of the study were compared with data collected previously from Bottlenose Dolphins in an adjacent bay. Group sizes were estimated with two methods; sizes ranged between 1 and 51, with size means estimated by the two methods being 16.3 and 16.18. Three hundred and twenty-one individual dolphins were identified, most on one occasion only, although some were sighted up to 16 times (median 3). Convex polygon ranges were derived for 40 dolphins that were observed more than eight times. Ranges varied between 0.8 km^2 to 15.1 km^2 (mean 4.33 km^2). Mean size of the groups in Moreton Bay was significantly smaller than those from Point Lookout, and the mean of the observed ranges in the bay was significantly larger. Information available on dolphins' predators and prey in both areas as reviewed. There appear to be indications that the food supply for Bottlenose Dolphins in Moreton Bay has been altered, or depleted, to the point where dolphins tend to rely on anthropogenic food sources (commercial prawn trawlers). Obtaining food from trawlers may increase likelihood that dolphins come into contact with predatory sharks, although the available data are insufficient to be definitive about this. In the waters off Point Lookout, fish stocks are such that dolphins do not need "handouts", range over a much smaller area to find sufficient food and may be less prone to shark attack.

Key words: Bottlenose Dolphins, *Tursiops truncatus*, Social structure, Ranging, Conservation biology.

INTRODUCTION

THE overhunting of great whales has been the major issue in cetacean conservation (Brownell *et al.* 1989; Gambell 1993) and the illegal trade in endangered whales, which took place in the recent past (Yablokov 1994) remains a problem (Baker and Palumbi 1994). In recent years, the deaths of small cetaceans in fisheries bycatches (Waring *et al.* 1990; Wade 1993; Ross *et al.* 1994), the impact of pollutants, particularly on inshore populations (Cockcroft *et al.* 1989; Kuehl *et al.* 1991), mass mortalities due to epizootics (e.g., Forcada *et al.* 1994) and the hunting of small cetaceans (Read *et al.* 1988; Lal Mohan 1989) have also become major problems for cetacean conservation. The future of river dolphins and some inshore delphinoid species are of particular concern (Perrin *et al.* 1989; Slooten and Dawson 1992). However, few studies can document what threats such activities pose before they constitute a major problem for a population or species.

Bottlenose Dolphins *Tursiops truncatus*, are cosmopolitan, absent only from polar seas, and occur in coastal and pelagic waters (Leatherwood and Reeves 1983). They are among the best-studied of the cetaceans (e.g., Leatherwood and Reeves 1990) and occur along the entire Australian coastline (Ross and Cockcroft 1990). The threatening processes affecting Bottlenose Dolphins in Australia waters are less severe than in some other areas and Bottlenose Dolphins are not at risk of extinction. However, as they are relatively common and occur in areas close to urban and built-up areas, Bottlenose Dolphins provide a good model for assessing the impact of some threatening processes on small cetaceans.

In this paper, I compare group sizes, ranges, and aspects of predation and feeding ecology of the Bottlenose Dolphins which occur in the waters around Point Lookout, North Stradbroke Island, with data from Bottlenose Dolphins in Moreton Bay. These are adjacent areas which have differing levels of anthropogenic impacts.

Bottlenose Dolphins in south-east Queensland waters

In 1977, the distribution and relative abundance of Bottlenose Dolphins in south-east Queensland waters was assessed with aerial surveys (Lear and Bryden 1980). Bottlenose Dolphins were common in Moreton Bay and the inshore oceanic waters off Moreton and North Stradbroke Islands, particularly around the rocky headlands of Cape Moreton and Point Lookout (Fig. 1), and did not undergo annual migrations in the area.

[1]Department of Veterinary Anatomy, The University of Sydney, New South Wales, Australia 2006.
[2]Present address: Department of Tropical Environment Studies and Geography, James Cook University of North Queensland, Townsville, Queensland, Australia 4811.

Pages 1–10 *in* MARINE MAMMAL RESEARCH IN THE SOUTHERN HEMISPHERE VOLUME 1: STATUS, ECOLOGY AND MEDICINE ed by M. Hindell and C. Kemper. Surrey Beatty & Sons, Chipping Norton. 1997.

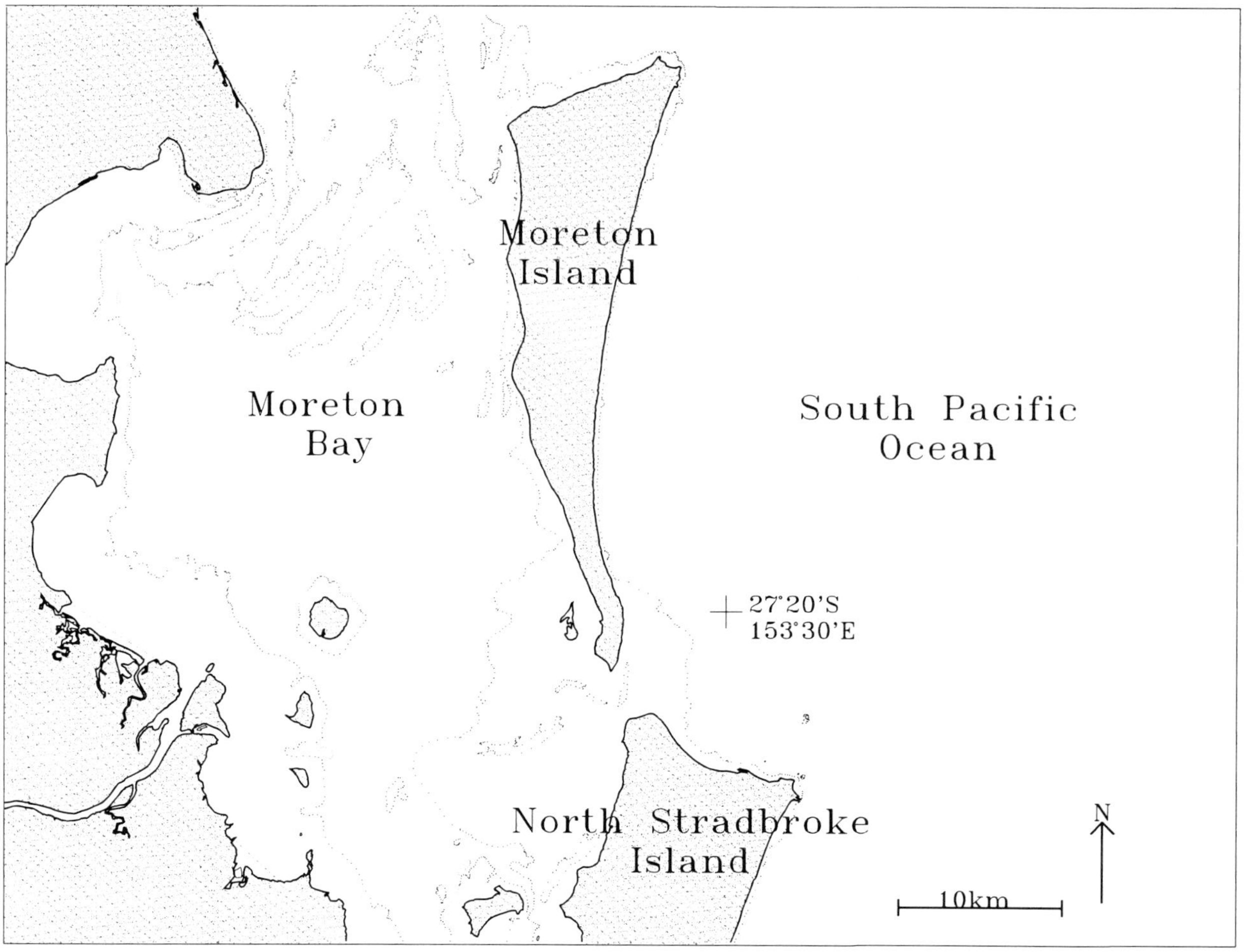

Fig. 1. The study area in south-east Queensland. The dashed lines indicate the 5 m depth contour.

Between 1984 and 1987, over 330 individual Bottlenose Dolphins were identified in Moreton Bay from photographs of natural marks (Corkeron 1990). Information of relevance to this paper include data on group sizes, evidence of shark attack and feeding behaviour of these dolphins. Thirty-seven per cent of the Bottlenose Dolphins in the bay showed definite evidence of shark attack (Corkeron *et al.* 1987a). There was no significant difference between male and female animals in the occurrence of scars from previous attacks, although females with new calves appeared more likely to have fresh wounds (Corkeron *et al.* 1987a). Dolphins healed rapidly from these wounds (Corkeron *et al.* 1987b). Dolphins in the bay feed on discarded catch from prawn trawlers (Corkeron *et al.* 1990; Wassenberg and Hill 1990), demonstrate preferences for particular fish species, and social factors influence the availability of food to individuals of different age and sex classes (Corkeron *et al.* 1990).

Over the same period, occasional surveys were run in the waters off Moreton and North Stradbroke Islands, and 138 individual Bottlenose Dolphins were identified. Moreton Bay and the waters off Point Lookout are adjoining, but no dolphin identified in the bay was observed in oceanic waters (Corkeron 1989). Photo-identification work in the bay in later years confirmed the continuing presence of identified individual dolphins within Moreton Bay (Green and Corkeron 1991). Elsewhere, Bottlenose Dolphins living in inshore waters exhibit similar patterns of site fidelity and there is no migration (e.g., Wells and Scott 1990; Williams *et al.* 1993; Smolker *et al.* 1993), although this is not always the case (Hansen and Defran 1990).

METHODS

Field methods

Fieldwork took place within 15 km of Point Lookout, North Stradbroke Island, although occasionally waters both further offshore and south along the coast of North Stradbroke Island were surveyed. Observations were made from a rigid-hulled inflatable boat and a crew of two or three people. Random searches were carried out for dolphin groups, with the pattern of movement of the vessel dictated by weather and sea conditions. Three species of dolphins were most commonly sighted; Bottlenose Dolphins, Indo-Pacific Humpback Dolphins *Sousa chinensis,*

Long-beaked Common Dolphins *Delphinus capensis*. False Killer Whales *Pseudorca crassidens* and Pantropical Spotted Dolphins *Stenella attenuata* were sighted on rare occasions. Data for this study and a concurrent study of dolphins' acoustic behaviour were collected whenever a group of dolphins was sighted and conditions were suitable. Data recorded for each sighting included; species sighted, date and time of sighting, estimated group size at the time of sighting, position (recorded either with a GPS satellite navigator or by taking bearings on known landmarks), general behaviour of the group and details of the film used.

Photographs were taken to identify individual dolphins using a motordriven Nikon FM2 and a 300 mm Nikkor lens on to slide film. All dolphins' dorsal fins were photographed, except under the following conditions; (1) in poor weather and sea conditions, (2) when there was a clash with the requirements of the concurrent acoustic study of dolphins and (3) when the behaviour of dolphins was to either avoid the vessel, in which case observations were stopped, or to surface irregularly and quickly, thus being difficult to photograph.

Analyses

Individual animals were identified by the distinctive patterns of scars on their dorsal surface, particularly on the caudal edge of their dorsal fin (Würsig and Würsig 1977; Würsig 1978; Würsig and Jefferson 1990). All photographs of dolphins were compared with the existing catalogue to determine whether an individual had been identified previously. When a new identification was made, the animal was given an identification number and representative transparencies were entered into the catalogue. A record of sightings of individual animals was transcribed on to a computerized database. Positions of dolphins were plotted on to a hydrographic chart which had been digitized into AutoCAD (Autodesk 1990). Statistical analyses were carried out using Statistica (Statsoft 1991).

Definitions

A **group** of dolphins comprised either a solitary animal, or any aggregation of dolphins where a member of the aggregation was within 10 m of any other member in the first 10 minutes that we observed the group (definition derived from Smolker *et al.* 1993).

Group sizes were estimated in two ways. In the first instance this was by counting the number of animals in a group while the group was under observation. This is referred to as **counted group size.** Where a range was estimated for counted group size (e.g., 18–20 animals in a group), the lower estimate was accepted. Secondly, group size was estimated by photographing most of the identifiable animals in a group. After analysis of all photo-identifications, the number of identified animals present in each group was summed. This is referred to as **identified group size.**

RESULTS

Bottlenose Dolphins off North Stradbroke Island

Data were collected on 110 groups of Bottlenose Dolphins, between 28 March and 22 November 1992. Positions where dolphins were observed are shown in Figure 2.

Suitable data for both **counted group size** and **identified group size** were collected for 61 groups of inshore Bottlenose Dolphins. Counted group size ranged between 1 and 50 (mean = 16.3, 95% confidence limits = 13.2–19.4, median = 12). Identified group size ranged between 2 and 51 (mean = 16.2, 95% confidence limits = 13.3–19.1, median = 13). The means did not differ significantly ($t_{120} = 0.054$, $P = 0.957$), and there was a strong correlation between counted and identified group size ($r^2 = 0.843$, $P < 0.001$, Fig. 3).

The number of individual inshore Bottlenose Dolphins identified was 321. Most (129) were identified on one occasion only, although some were sighted up to 16 times on 13 separate days (Fig. 4). The mean number of sightings was 3.7 (95% confidence intervals = 3.3–4.0), although the data were highly left-skewed (skewness = 0.96). The median number of times that individual animals were sighted was three.

Convex polygon ranges (Southwood 1966) were constructed on the digitized chart for each of the 40 dolphins that were observed more than eight times. Polygons were derived by joining the outermost sightings of each animal, and joining the northwestern and southeastern sighting with the coastline. If lines through the polygon crossed over land, an extra point was added to the polygon, moving the line to a point on the 5 m depth contour, so the polygon did not include land. Areas of the polygons were estimated with the AREA function in AutoCAD. Ranges varied in size from 0.8 km^2 to 15.1 km^2, with a mean of 4.33 km^2 (95% confidence limits = 3.12–5.54 km^2). There was a significant positive correlation between the number of sightings of animals and the size of their estimated range ($r^2 = 0.16$, $P < 0.05$). Since the correlation explained so little of the variation in the data, the estimated ranges of the animals sighted most often were not necessarily a better representation of their real ranging behaviour than were the ranges of those animals sighted less often. There was no indication that the relationship between number of sightings and range size reached an asymptote with the number of observations of dolphins made during this study.

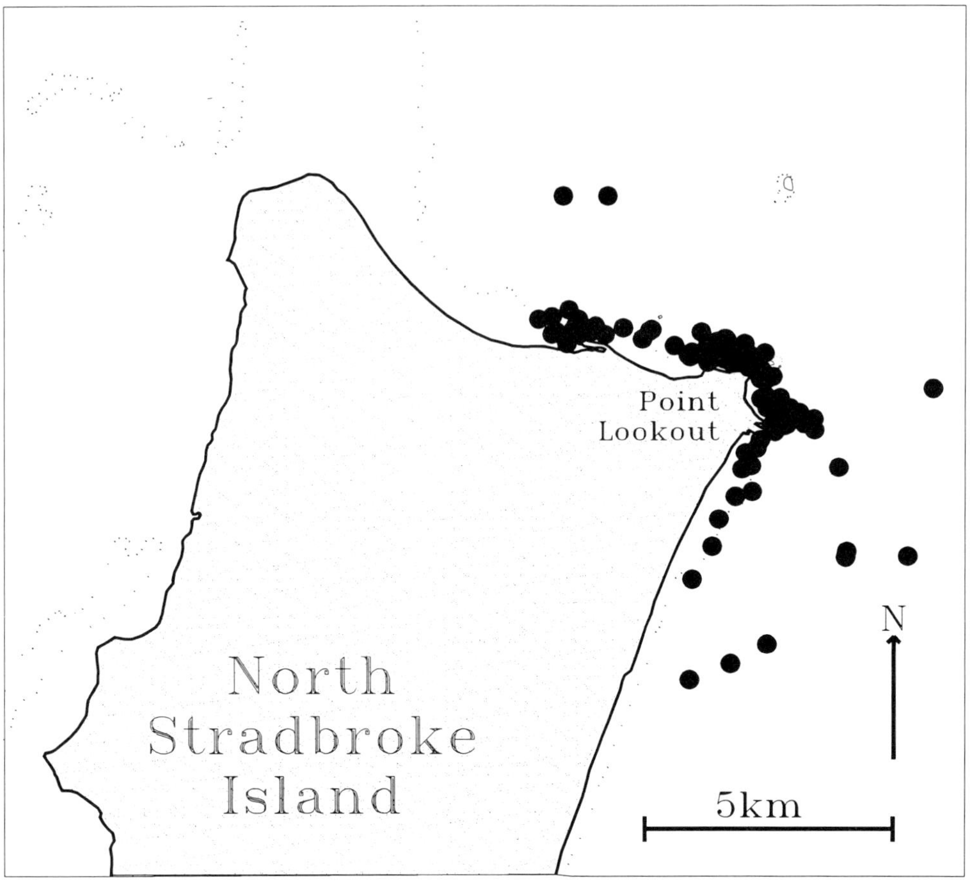

Fig. 2. Sightings of Bottlenose Dolphins in the waters off Point Lookout, North Stradbroke Island. Solid circles indicate positions where Bottlenose Dolphins were observed in this study. Dashed lines represent the 5 m depth contour.

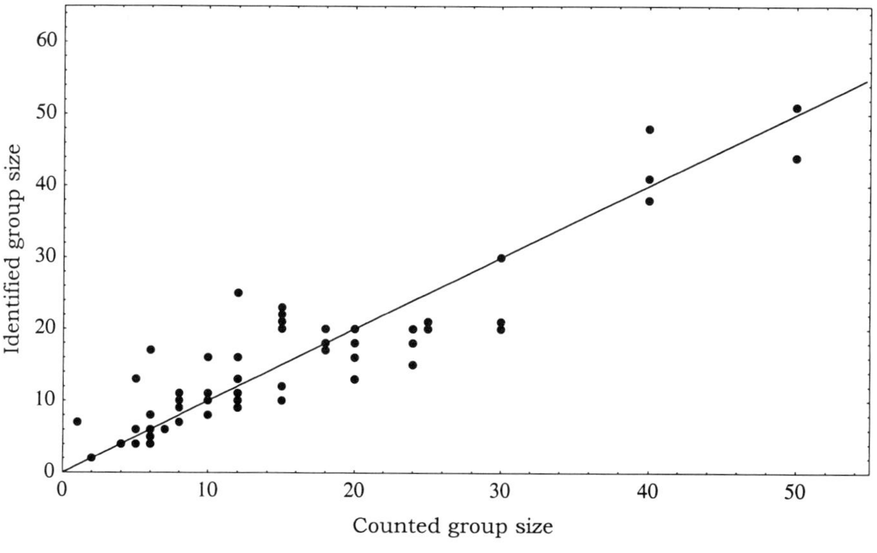

Fig. 3. The relationship between counted and identified group size of Bottlenose Dolphins. The line indicates where counted group size equals identified group size.

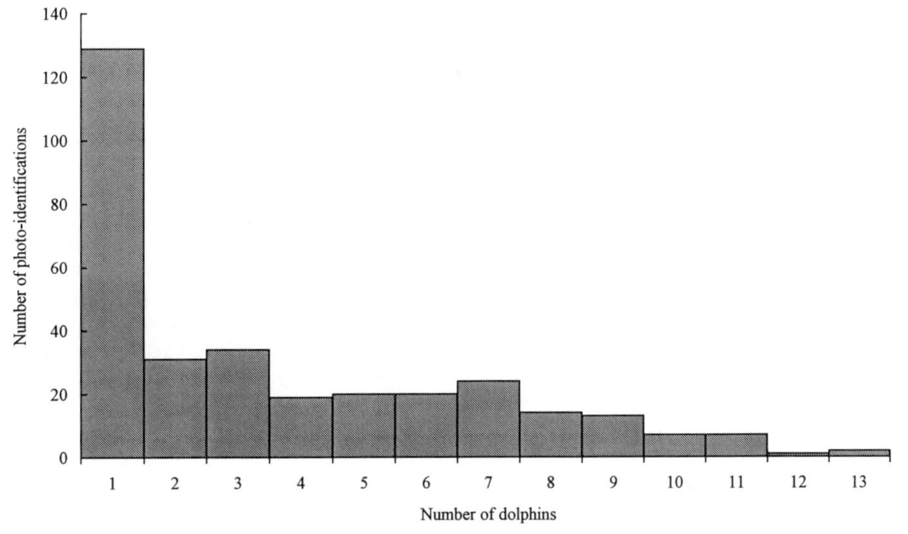

Fig. 4. The number of times that individual Bottlenose Dolphins were photographically identified in the waters off Point Lookout, North Stradbroke Island, during the 1992 study.

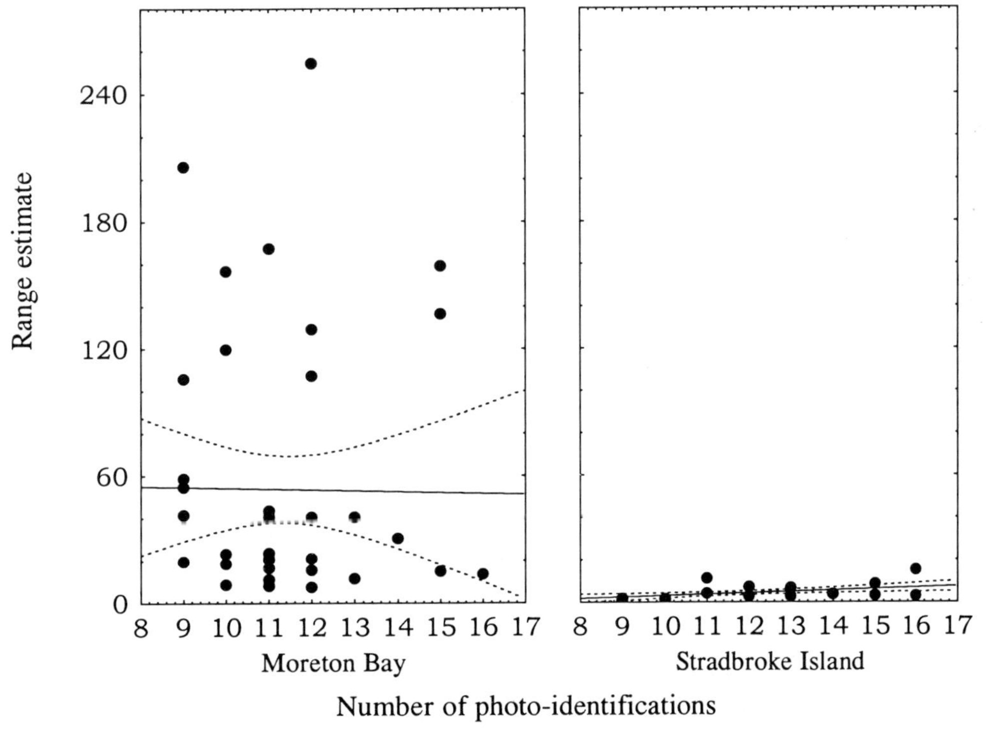

Fig. 5. The relationship between the number of times Bottlenose Dolphins were observed and their estimated range, for Moreton Bay (1984–1987) and the waters off Point Lookout, North Stradbroke Island (1992). Solid lines indicate the fit of linear regressions, dashed lines indicate 95% confidence intervals of these regression lines.

Twenty-seven of the animals photographed had been identified during previous research off North Stradbroke Island. The earliest sighting of any animal was in September 1984. The longest period between first and last sighting of an animal was between September 1984 and November 1992, or eight years and two months. Positions of previous sightings were not used to estimate ranges.

Comparisons between dolphins off North Stradbroke Island and Moreton Bay

The size of groups of Bottlenose Dolphin in Moreton Bay (279 sightings of dolphin groups made from three different types of vessels between May 1984 and February 1987 (Corkeron 1990)) was compared with those from the waters off Point Lookout (98 sightings for which identified group sizes were available). The mean size of the groups in Moreton Bay (10.6) was significantly smaller than that (16.7) from Point Lookout ($t_{375} = 6.205$, $P < 0.001$).

Estimated ranges of dolphins off North Stradbroke Island were compared with previously determined ranges of individual dolphins in Moreton Bay. In the latter case, only the 43 animals sighted between 9 and 16 times were included in analysis, in order to generate data comparable with the point Lookout study. The mean of the observed ranges in Moreton Bay (53.62 km^2) was significantly larger than that (4.33 km^2), off Point Lookout ($t_{71} = 4.389$, $P < 0.001$). Also, the relationship between the number of sightings of animals and their observed range varied between the two studies (ANCOVA, $F_{(1,70)} = 18.703$, $P < 0.001$, Fig. 5).

DISCUSSION

Bottlenose Dolphins in the waters off Point Lookout

Bottlenose Dolphins live in complex fission/fusion societies (Smolker *et al.* 1993). However, the size of instantaneously sampled groups appear to provide a good measure of difference in social structure between areas (e.g., Shane *et al.* 1986). The sizes of groups observed off Point Lookout are within the range of those reported for Bottlenose Dolphins elsewhere, although they are larger than many reported (e.g., Shane *et al.* 1986).

Determining the size of groups of dolphins can be difficult, particularly for pelagic dolphins, which may occur in groups numbering in the thousands (e.g., Holt and Sexton 1989). The size of groups of Bottlenose Dolphins has been determined in the past by both the methods used in this paper. Initially, photo-identification was suggested as a method by which precise and accurate estimates of group size could be derived (Würsig and Würsig 1977; Würsig 1978). Ballance (1990) suggested that obtaining four photographs of each individually identifiable animal will ensure that all identifiable members of a group are photographed. This method cannot account for individuals lacking identifying natural marks (Corkeron 1990; Williams *et al.* 1993) and young animals can lack such markings. The proportion of identifiable individuals in a population (based on the proportion of unidentifiable individuals in photographs) has been estimated as 71.4% (Corkeron 1984) and 65.5% (Williams *et al.* 1993). Therefore, identified group size can underestimate the number of animals in a group.

In this study, counted group sizes were often given as ranges and the lower value of the range was accepted. From comparison with identified groups it would seem more appropriate to accept the upper estimates. However, these values are likely to be inaccurate as there is a tendency for counted group sizes to cluster around particular values (e.g., 15, 18, 24, 30). Identified group sizes could not be used to derive precise and accurate estimates of group sizes of Bottlenose Dolphins, where the groups comprise more than approximately a dozen animals. For counted group sizes it may be possible to use high, low, "best" or the mean of the high and low values, as has been done for Long-finned Pilot Whales *Globicephala melas* (Buckland *et al.* 1994). Recordings of dolphin groups with high resolution video photography may be the best option for deriving precise and accurate estimates of group size.

There have been few estimates of home ranges of Bottlenose Dolphins. Lockyer and Morris (1986) gave an approximate estimate of the home range of a solitary dolphin as 77 km^2. Würsig and Würsig (1979) found that their study site (with an area of approximately 50 km^2, Shane *et al.* 1986) did not encompass the entire range of the 53 dolphins that comprised their study population. A community of approximately 100 dolphins in Sarasota Bay, Florida use a small, shallow bay with an area of 85 km^2, although individual animals may not use the entire bay, and there are age and sex difference in ranging patterns (Wells *et al.* 1980; Scott *et al.* 1990). However, individual Bottlenose Dolphins in inshore waters off the oceanic coast of southern California can travel over far more than 32 km of coastline (Hanson and Defran 1993).

The ranges reported here for the dolphins off Point Lookout are small, and may underestimate the true ranging patterns of these animals. However, in this study, few sightings were made of dolphins at the extremities of the study area (Fig. 2). Also, Bottlenose Dolphins were observed within 300 m of one site on Point Lookout for 36% of the time during land based surveys for Humpback Whales *Megaptera novaeangliae* in the winter of 1993 (Bryden *et al.* 1994). The area within which dolphins were recorded as being present was approximately 0.2 km^2. Although these data are obviously seasonally biased, they add weight to the data presented above that dolphins which have Point Lookout as a core area have relatively small ranges. Finally, aerial surveys reported extremely low densities of dolphins in areas immediately north-west of Point Lookout (Lear and Bryden 1980). Densities of dolphins on the ocean beach immediately south of Point Lookout were higher, although no distinction was made between the ocean beach close to the point and the beach at the southern end of North Stradbroke Island (Lear and Bryden 1980). It is likely that the high density of dolphins in this area was due to sightings of dolphins in the vicinity of

Point Lookout, but locations of sightings were not presented in the paper. The available data indicate that the small ranges reported here are not simply due to sampling bias.

Comparisons between Bottlenose Dolphins in Moreton Bay and off Point Lookout

Bottlenose Dolphins in Moreton Bay occur in smaller groups than those off Point Lookout. This appears to confirm the trend described by Norris and Dohl (1980) that dolphins in deeper waters, further from shore, occur in larger groups than those close to shore. However, the region where most sightings of Bottlenose Dolphins were made in Moreton Bay is deeper (30–40 m) than where most sightings were made off Point Lookout (generally 10–25 m). Proximate influences on group size in mammals include responses to predation (both detection and dilution effects), responses to the density and patchiness of food resources, and social responses to group sizes generated by the previous two (Wrangham and Rubenstein 1986).

It has been suggested that predation is a major influence on grouping behaviour in dolphins (Johnson and Norris 1986; Norris and Schilt 1988) and that the risk of attack from large sharks and the need for vigilance from attack is greater in deeper waters (Norris and Dohl 1980; Johnson and Norris 1986). Little is known of the patterns of attack by dolphin predators, but it is clear that large sharks attack marine mammals in shallow water. Great White Sharks *Carcharodon carcharias* are known to attack at least four different species of seals in shallow water (Klimley *et al.* 1992; Klimley 1994) and most records of white sharks in South Australia have been in shallow water (Bruce 1992). Tiger Sharks *Galaeocerdo cuvier* occur in shallow water at night (Randall 1992) and have been caught regularly in gill nets and on baited hooks set in shallow water off Townsville (Simpendorfer 1992).

Great White and Tiger Sharks occur in shallow waters in Moreton Bay, and large Whaler Sharks *Carcharhinus* spp. have been observed in the vicinity of working trawlers (Grant 1982; Corkeron *et al.* 1987a). There have been no studies of the shark fauna in the waters off Point Lookout but Great White, Tiger and large Whaler Sharks occur in the area (pers. obs.; P. Spinner, pers. comm.). Sharks are observed regularly in shallow water close to shore during shore-based surveys of Humpback Whales (pers. obs.; M. Brown, pers. comm.). Although there are no quantitative studies of the shark fauna off south-east Queensland, it seems unlikely that the density of sharks which prey on dolphins is greater in Moreton Bay than off Point Lookout. In both areas, such sharks are known to occur in shallow water.

Twenty-one fresh wounds caused by sharks were observed on the dolphins in Moreton Bay over three years (Corkeron *et al.* 1987b). The only observed shark attack was while the dolphin was following a working trawler. No fresh wounds have been observed on the dolphins off Point Lookout during the 1992 study. As large wounds can be visible on animals for several months (Corkeron *et al.* 1987b), this suggests that the dolphins off Point Lookout were less likely to suffer a shark attack than were the animals in Moreton Bay. The incidence of healed wounds on the Bottlenose Dolphins off Point Lookout appears to be far lower than on those in Moreton Bay. However, the observation technique of photo-identification from the surface was less likely to detect wounds than was underwater observation made at times in Moreton Bay.

The discussion above suggests that Bottlenose Dolphins off Point Lookout may be less vulnerable to predation than those in Moreton Bay. If this is so, it implies that the group sizes of dolphins in the bay are suboptimal for predator detection and defence. If group sizes in the bay are suboptimal, why have selective pressures not encouraged the formation of larger groups?

It is difficult to disentangle the cause-effect relationship between group size and predation in delphinids. For birds and other mammals, it is difficult to establish that vigilance varies with group size (Elgar 1989), and there are populations where vigilance does not covary with group size (e.g., Island Silvereyes *Zosterops lateralis*, Catterall *et al.* 1992). Also, the interaction between the vigilance and dilution effects of group size can be complex in mammals (e.g., Dehn 1990). We lack an understanding of how echolocation functions in group vigilance, as individuals may be capable of "sharing" received echoes (e.g., Jerison 1986). Further, we have few data on the way that dolphins interact with potential predators, although co-operative defence against sharks by dolphins (not necessarily Bottlenose Dolphins) has been observed or inferred (Wood *et al.* 1970; Saayman and Tayler 1979; Corkeron *et al.* 1987a).

To conclude, it may be that predation pressure affects group sizes of Bottlenose Dolphins in Moreton Bay and in the waters off North Stradbroke Island. However, there are insufficient data available on predator distribution, abundance and behaviour, and on the behavioural mechanisms by which dolphins respond to predator presssure to move beyond conjecture. The problem extends beyond this study and represents a major difficulty in developing an understanding of delphinind behavioural ecology.

There are no data on the standing stocks of fish or crustaceans exploited by commercial and amateur fishers in Moreton Bay and it is not known whether fisheries in the Moreton Bay region are sustainable (Williams 1992). However, catches of Tailor *Pomatomus saltatrix*, Summer Whiting *Sillago ciliata* and *S. analis* and Yellow Bream *Acanthopagrus australis* are declining, and there is anecdotal evidence for declines in Schnapper *Chrysophrys auratus*, Mullet *Mugil* spp. and Flathead *Platycephalus* spp. (Quinn 1992). Most of these fish are consumed by Bottlenose Dolphins in south-east Queensland waters (Lear and Bryden 1980; Corkeron *et al.* 1990).

Another problem when comparing the impact of fisheries on the dolphins in Moreton Bay and off Point Lookout is that both areas are considered part of the one region in published analyses of catches (e.g., Quinn 1992; Williams 1992). The major commercial fishery in the region is the Moreton Bay prawn trawling industry and trawlers in the bay are subject to different regulations to those working in offshore waters such as off Point Lookout and Cape Moreton (Quinn 1992). Off Point Lookout, beach seining for Mullet and Tailor, line fishing (predominantly for Mackerel *Scomberombus* spp. and Schnapper) and offshore prawn trawling (generally further than 15 km from shore) are the main commercial fisheries (Quinn 1992).

In Moreton Bay, Bottlenose Dolphins are observed regularly feeding in association with prawn trawlers (Blaber and Wassenberg 1989; Corkeron *et al.* 1990). Trawlers probably discard sufficient material to feed at least 100 dolphins each night (Wassenberg and Hill 1990). Dolphins feed in the vicinity of nets while fishing is underway, apparently feeding on fish and invertebrates disturbed by trawling, and also taking animals through the nets' meshes (also described by Leatherwood 1975). As dolphins follow trawlers for hours to obtain food in this way (Corkeron 1989), it is likely that the quantity of food obtained in this manner is at least comparable to that obtained from discards. Commercial trawlers are legally prevented from working in Moreton Bay on the weekends (Quinn 1992). When research trawls were conducted on weekends, dolphins spent more time in the vicinity of the research boat than at other times, as this was then the only source of discards (Green and Corkeron 1991). The available evidence suggests that trawlers represent a major source of food for the Bottlenose Dolphins in Moreton Bay.

Social factors (i.e., dominance relationships) and the size of the discard patch regulate the number dolphins present during the disposal of fish. For example, males may tolerate the presence of females in their vicinity but will drive off other males (Green and Corkeron 1991). It appears that the size of food patches created by trawlers is an important determinant for group size of Bottlenose Dolphins in Moreton Bay.

There are few data on the feeding patterns of Bottlenose Dolphins off Point Lookout. In this study, dolphins were never observed feeding in the vicinity of trawlers but were observed feeding on small fish at the surface, and apparently feeding over rock reefs. Dolphins have also been observed feeding on schools of small bait fish, and on seasonally abundant schools of Tailor and Mullet (P. Spinner, pers. comm.). It appears that the feeding habits of Bottlenose Dolphins in the waters off Point Lookout are less affected by human activities than are those in Moreton Bay.

CONCLUSIONS

The available data indicate that the food supply for Bottlenose Dolphins in Moreton Bay has been altered, or perhaps depleted, to the point where dolphins tend to rely on trawlers for food. Obtaining food from trawlers may increase the likelihood that dolphins come into contact with predatory sharks, although the available data are insufficient to be definitive about this. In the waters off Point Lookout, fish stocks are such that dolphins do not need "handouts", range over a much smaller area to find sufficient food and may be less prone to shark attack.

A problem common to papers discussing the ecological influences on delphinid social organization is that a small raw data set becomes swamped with anecdote and conjecture (e.g., Norris and Dohl 1980; Connor and Norris 1982; Johnson and Norris 1986), because good data on predator and prey densities and behaviour are almost impossible to come by. Although this paper suffers from the same problem, I suggest that the hypothesis raised here could be tested by further comparative work in Moreton Bay and off Point Lookout. Further comparative work on ranging using radio telemetry and theodolites (Würsig *et al.* 1991), on the time budgets of dolphins in the two areas (i.e., do dolphins in Moreton Bay spend more time seeking food and feeding?) and on the manner in which dolphins in the two areas associate would be a suitable starting points. A repeat of the aerial survey of Moreton Bay to compare with the data from 1977 (Lear and Bryden 1980) would be useful to assess changes in the relative size of the population of dolphins in the area. Good data on abundance of fish stocks and level of the harvest would also be informative.

ACKNOWLEDGEMENTS

This research was supported by a grant from the Sea World Research and Rescue Foundation Inc., and carried out under permits from the Queensland Fish Management Authority and The Australian Nature Conservation Agency. I thank Miranda Brown, Ken Schultz, Mairi Macleod, Richard Arnett and many volunteers for assistance in the field. Michael Bryden and Miranda Brown provided helpful comments on this paper. Bernd Würsig and Janet Mann pulled me back into line with their excellent refereeing.

REFERENCES

Autodesk, 1990. The AutoCAD crafting package reference manual. Autodesk Inc.: Sausolito, California.

Baker, C. S. and Palumbi, S. R., 1994. Which whales are hunted? A molecular genetic approach to monitoring whaling. *Science* **265:** 1538–539.

Ballance, L. T., 1990. Residence patterns, group organization and surfacing associations of Bottlenose Dolphins in Kino Bay, Gulf of California, Mexico. Pp. 267–84 *in* The bottlenose dolphin, ed by S. Leatherwood and R. R. Reeves. Academic Press: San Diego.

Blaber, S. J. and Wassenberg, T. J., 1989. The feeding ecology of piscivorous birds, *Phalacrocorax varius*, *P. melanoleucos* and *Sterna bergii* in Moreton Bay, Australia: diet and dependence on trawler discards. *Mar. Biol.* **101:** 1–10.

Brownell, R. L., Ralls, K. and Perrin, W. F., 1989. The plight of the "forgotten" whales. *Oceanus* **32:** 5–11.

Bruce, B. D., 1992. Preliminary observations on the biology of the white shark, *Carcharodon carcharias*, in South Australian waters. *Aust. J. Mar. and Fresh. Res.* **43:** 1–11.

Bryden, M. M., Brown, M. R. and Corkeron, P. J., 1994. Survey of humpback whales *(Megaptera novaeangliae)* off eastern Australia, 1993. Report to the Australian Nature Conservation Agency. 40 pp.

Buckland, S. T., Block, D., Cattanach, K. L., Gunnlaugsson, Th., Hoydal, K., Lens, S. and Sigurjónsson, J., 1994. Distribution and abundance of long-finned pilot whales in the North Atlantic, estimated from NASS-87 and NASS-89 data. *Rep. Int. Whal. Commn. (Special Issue 14)* 33–49.

Catterall, C. P., Elgar, M. A. and Kikkawa, J., 1992. Vigilance does not covary with group size in an island population of silvereyes *(Zosterops lateralis)*. *Behav. Ecol.* **3:** 207–10.

Cockcroft, V. G., De Kock, A. C., Lord, D. A. and Ross, G. J. B., 1989. Organochlorines in bottlenose dolphins, *Tursiops truncatus*, from the east coast of South Africa. *Sth Afr. J. Mar. Sci.* **8:** 207–17.

Connor, R. C. and Norris, K. S., 1982. Are dolphins reciprocal altruists? *Amer. Nat.* **119:** 358–74.

Corkeron, P. J., 1984. Studies of inshore dolphins, *Tursiops* and *Sousa*, in the Moreton Bay region. MSc Qualifying thesis, University of Queensland, Brisbane.

Corkeron, P. J., 1989. Studies of inshore dolphins, *Tursiops* and *Sousa*, in the Moreton Bay region. PhD thesis, University of Queensland, Brisbane.

Corkeron, P. J., 1990. Aspects of the behavioural ecology of inshore dolphins *Tursiops truncatus* and *Sousa chinensis* in Moreton Bay, Australia. Pp. 285–94 *in* The bottlenose dolphin, ed by S. Leatherwood and R. R. Reeves. Academic Press: San Diego.

Corkeron, P. J., Morris, R. J. and Bryden, M. M., 1987a. Interactions between bottlenose dolphins and sharks in Moreton Bay, Queensland. *Aqu. Mam.* **13:** 111–14.

Corkeron, P. J., Morris, R. J. and Bryden, M. M., 1987b. A note on healing of large wounds in bottlenose dolphins, *Tursiops truncatus*. *Aqu. Mam.* **13:** 96–98.

Dehn, M. M., 1990. Vigilance for predators: detection and dilution effects. *Behav. Ecol. Sociobiol.* **26:** 337–42.

Elgar, M. A., 1989. Predator vigilance and group size in mammals and birds: a critical review of the empirical evidence. *Biol. Rev.* **64:** 13–33.

Forcada, J., Aguilar, A., Hammond, P., Pastor, X. and Aguilar, R., 1994. Distribution and numbers of striped dolphins in the western Mediterranean Sea after the 1990 epizootic outbreak. *Mar. Mam. Sci.* **10:** 137–50.

Gambell, R., 1993. International management of whales and whaling: an historical review of the regulation of commercial and aboriginal subsistence whaling. *Arctic* **46:** 97–107.

Grant, E. M., 1982. "Guide to Fishes". Queensland Department of Harbours and Marine: Brisbane.

Green, A. and Corkeron, P. J., 1991. An attempt to establish a dolphin feeding station at Moreton Island, Queensland. *Aqu. Mam.* **17:** 125–29.

Hansen, L. J. and Defran, R. H., 1990. A comparison of photo-identification studies of California coastal Bottlenose Dolphins. *Rep. Int. Whal. Commn. (Special Issue 12)* 101–04.

Hanson, M. T. and Defran, R. H., 1993. The behaviour and feeding ecology of the Pacific coast bottlenose dolphin, *Tursiops truncatus*. *Aqu. Mam.* **19:** 127–42.

Holt, R. S. and Sexton, S. N., 1989. Monitoring trends in dolphin abundance in the eastern tropical Pacific using research vessels over a long sampling period: analyses of 1987 data. *Rep. Int. Whal. Commn. (Special Issue 39)* 347–51.

Jerison, H. J., 1986. The perceptual worlds of dolphins. Pp. 141–66 *in* Dolphin Cognition and Behavior: a Comparative Approach, ed by R. J. Schusterman, J. A. Thomas and F. G. Wood. Lawrence Erlbaum and Associates: Hillsdale, New Jersey.

Johnson, C. M. and Norris, K. S., 1986. Delphinid social organization and social behavior. Pp. 355–66 *in* Dolphin Cognition and Behavior: a Comparative Approach, ed by R. J. Schusterman, J. A. Thomas and F. G. Wood. Lawrence Erlbaum and Associates: Hillsdale, New Jersey.

Klimley, A. P., 1994. The predatory behavior of the white shark. *Amer. Sci.* **82:** 124–33.

Klimley, A. P., Anderson, S. D., Pyle, P. and Henderson, R. P., 1992. Spatiotemporal patterns of white shark *(Carcharodon carcharias)* predation at the South Farallon Islands, California. *Copeia* **1992:** 680–90.

Kuehl, D. W., Haebler, R. and Potter, C., 1991. Chemical residues in dolphins from the U.S. Atlantic coast including Atlantic bottlenose obtained during the 1987/88 mass mortality. *Chemosphere* **22:** 1071–084.

Lal Mohan, R. S., 1989. Conservation and management of the Ganges river dolphin, *Platanista gangetica*, in India. Pp. 64–69 *in* Biology and Conservation of the River Dolphins, ed by W. F. Perrin, R. L. Brownell Jnr., Z. Kaiya and L. Jiankang. Occasional papers of the IUCN species survival commission (SSC). No 3.

Lear, R. J. and Bryden, M. M., 1980. A study of the bottlenose dolphin *Tursiops truncatus* in eastern Australian waters. Australian National Parks and Wildlife Service Occasional Paper 4.

Leatherwood, S., 1975. Some observations of the feeding behavior of Bottlenose Dolphins *Tursiops truncatus* in the northern Gulf of Mexico and (*Tursiops* cf *gilli*) off southern California. *Mar. Fish. Rev.* **37:** 10–16.

Leatherwood, S., Reeves, R. R. and Foster, L., 1983. The Sierra Club Handbook of Whales and Dolphins. Sierra Club Books: San Francisco.

Leatherwood, S. and Reeves, R. R., 1990. The Bottlenose Dolphin. Academic Press: San Diego.

Lockyer, C. and Morris, R. J., 1986. The history and behaviour of a wild, sociable Bottlenose Dolphin *(Tursiops truncatus)* off the north coast of Cornwall. *Aqu. Mam.* **12:** 3–16.

Norris, K. S. and Dohl, T. P., 1980. The structure and function of cetacean schools. Pp. 211–61 *in* Cetacean Behavior: Mechanisms and Processes, ed by L. M. Herman. John Wiley and Sons: New York.

Norris, K. S. and Schilt, C. R., 1988. Cooperative societies in three-dimensional space: on the origins of aggregations, flocks and schools, with special reference to dolphins and fish. *Ethol. Sociobiol.* **9:** 149–79.

Perrin, W. F., Brownell Jnr., R. L., Zhou Kaiya and Liu Jiankang (eds), 1989. Biology and conservation of the river dolphins. Occasional papers of the IUCN species survival commission (SSC) No 3.

Quinn, R. H., 1992. Fisheries Resources of Moreton Bay. Queensland Fish Management Authority: Brisbane.

Randall, J. E., 1992. Review of the biology of the tiger shark *(Galaeocerdo cuvier). Aust. J. Mar. Fresh. Res.* **43:** 21–31.

Read, A. J., Van Waerbeek, K., Reyes, J. C., McKinnon, J. S. and Lehmann, L. C., 1988. The exploitation of small cetaceans in coastal Peru. *Biol. Cons.* **46:** 53–70.

Ross, G. J. B and Cockcroft, V. G., 1990. Comments on Australian bottlenose dolphins and the taxonomic status of *Tursiops aduncus* (Ehrenberg, 1832). Pp. 101–28 *in* The Bottlenose Dolphin, ed by S. Leatherwood and R. R. Reeves. Academic Press: San Diego.

Ross, G. J. B, Heinsohn, G. E. and Cockcroft, V. G., 1994. Humpback dolphins *Sousa chinensis* (Osbeck 1765), *Sousa plumbea* (G. Cuvier 1829) and *Sousa teuszii* (Kukenthal 1892). Pp. 23–42 *in* Handbook of Marine Mammals Volume 5, ed by S. H. Ridgway and R. J. Harrison. Academic Press: London.

Saayman, G. S. and Tayler, C. K., 1979. The socioecology of humpback dolphins (*Sousa* sp.) Pp. 165–226 *in* Behavior of Marine Animals Volume 3, Cetaceans, ed by H. E. Winn and B. L. Olla. Plenum Press: New York.

Scott, M. D., Wells, R. S. and Irvine, A. B., 1990. A long-term study of bottlenose dolphins on the west coasts of Florida. Pp. 235–44 *in* The Bottlenose Dolphin, ed by S. Leatherwood and R. R. Reeves. Academic Press: San Diego.

Shane, S., Wells, R. S. and Würsig, B., 1986. Ecology, behavior and social organization of the Bottlenose Dolphin: a review. *Mar. Mam. Sci.* **2:** 34–63.

Simpendorfer, C., 1992. Biology of tiger sharks *(Galaeocerdo cuvier)* caught by the Queensland Shark Meshing Program off Townsville. *Aust. J. Mar. Fresh. Res.* **43:** 33–43.

Slooten, E. and Dawson, S. M., 1992. Survival rates of photographically identified Hector's dolphins from 1984 to 1988. *Mar. Mam. Sci.* **8:** 327–43.

Smolker, R. A., Richards, A. F., Connor, R. C. and Pepper, J. W., 1993. Sex differences in patterns of association among Indian Ocean Bottlenose Dolphins. *Behaviour* **123:** 38–69.

Southwood, T. R. E., 1966. Ecological Methods. Methuen and Co: London.

Statsoft Inc., 1991. Statistical Reference Manual. Statsoft Inc: Tulsa, Oklahoma.

Wade, P. R., 1993. Estimation of the historical population size of the eastern spinner dolphin *(Stenella longirostris orientalis). Fish. Bull. USNMFS.* **91:** 775–87.

Wassenberg, T. and Hill, B. J., 1990. Partitioning of material discarded from prawn trawlers in Moreton Bay. *Aust. J. Mar. and Fresh. Res.* **41:** 26–36.

Waring, G. T., Gerrior, P., Wayne, P. M., Parry, B. L. and Nicolas, J. R., 1990. Incidental take of marine mammals in foreign fishery activities off the northeast United States, 1977–88. *Fish. Bull., USNMFS* **88:** 347–60.

Williams, J. A., Dawson, S. M. and Slooten, E., 1993. The abundance and distribution of bottlenosed dolphins in Doubtful Sound, New Zealand. *Can. J. Zool.* **71:** 2080–088.

Williams, L. E., 1992. Moreton Bay fisheries. Pp. 71–79 *in* Moreton Bay in the Balance, ed by O. Crimp. The Australian Littoral Society Inc: Brisbane.

Wells, R. S., Irvine, A. B. and Scott, M. D., 1980. The social ecology of inshore odontocetes. Pp. 263–317 *in* Cetacean Behavior: Mechanisms and Processes, ed by L. M. Herman, John Wiley and Sons: New York.

Wells, R. S. and Scott, M. D., 1990. Estimating Bottlenose Dolphin population parameters from individual identification and capture-release techniques. *Rep. Int. Whal. Commn. (Special Issue 12)* 407–15.

Wood, F. G. Jnr, Caldwell, D. K. and Caldwell, M. C., 1970. Behavioral interactions between porpoises and sharks. Pp. 264–77 *in* Investigations on Cetacea. Volume II, ed by G. Pilleri. Brain Anatomy Institute, University of Berne: Berne, Switzerland.

Wrangham, R. W. and Rubenstein, D. I., 1986. Social evolution in birds and mammals. Pp. 452–70 *in* Ecological Aspects of Social Evolution: Birds and Mammals, ed by D. I. Rubenstein and R. W. Wrangham. Princeton University Press: Princeton, New Jersey.

Würsig, B., Cipriano, F. and Würsig, M., 1991. Dolphin movement patterns. Information from radio and theodolite tracking studies. Pp. 79–111 *in* Dolphin Societies. Discoveries and Puzzles, ed by K. Pryor and K. S. Norris. University of California Press: Berkeley.

Würsig, B., 1978. Occurrence and group organization of Atlantic bottlenose porpoises *(Tursiops truncatus)* in an Argentine Bay. *Biol. Bull.* **154:** 348–59.

Würsig, B. and Jefferson, T. A., 1990. Methods of photo-identification for small cetaceans. *Rep. Int. Whal. Commn. (Special Issue 12)* 43–52.

Würsig, B. and Würsig, M., 1977. The photographic determination of group size, composition and stability of coastal porpoises *(Tursiops truncatus). Science* **198:** 755–56.

Würsig, B. and Würsig, M., 1979. Behavior and ecology of the Bottlenose Dolphin, *Tursiops truncatus*, in the south Atlantic. *Fish. Bull., USNMFS* **77:** 399–412.

Yablokov, A. V., 1994. Validity of whaling data. *Nature* **367:** 108.

Genetic studies of the Southern Elephant Seal *Mirounga leonina*

R. W. SLADE[1,2]

Marine mammals have a high capacity for dispersal and inhabit an environment with few geographic barriers. Global population structure has been investigated in a number of cetacean species, but in few pinnipeds. This paper summarizes a study of the population genetic structure of the Southern Elephant Seal, using mitochondrial DNA (mtDNA) and nuclear DNA (nDNA) markers. On the global scale, samples were drawn from subantarctic islands representing the three main oceanic populations; South Georgia (SG) in the south Atlantic Ocean, Heard Island (HI) in the south Indian Ocean and Macquarie Island (MI) in the south Pacific Ocean. To investigate population structure on a finer scale, published mtDNA data from a second south Atlantic Ocean population, Península Valdés (PV) on the Argentinian coast, were incorporated.

Analysis of mtDNA variation showed that geographic structure in the Southern Elephant Seal was high. All populations were genetically distinct for distribution of lineages, distribution of genetic variance and haplotype frequency. The genetic distance between SG and HI was at least an order of magnitude lower than that between other populations. SG and HI share the most recent common ancestry, due either to ongoing gene flow of 3–4 females per generation, or, assuming no gene flow, recent population divergence during the last ice-age some 20 000 years ago. The nDNA loci, although less informative were consistent with the mtDNA results. There was, however, evidence of male-biased gene flow, consistent with studies on individual movements. The proportion of migrants exchanged between populations was negligible; all populations are demographically isolated. This lends support to the hypothesis that the observed declines in the MI and HI populations were due to intrinsic population fluctuations. Sub-populations within the same oceanic region cannot be considered as part of the main population in the same oceanic region, as PV was as close to HI in the south Indian Ocean as it was to SG. In summary, geographic structure in the Southern Elephant Seal was strong and not correlated with dispersal capacity.

The pattern and level of variation from interspecific comparisons of orthologous genes was used to estimate the mutation rate of mitochondrial DNA (mtDNA) and nuclear DNA (nDNA) in pinnipeds. Combined with data on levels of intraspecific variation, this allowed an assessment of the relative contribution of the forces of mutation, drift, migration and selection that may have influenced the level and distribution of mtDNA and nDNA variation. The female effective population size (N_f), and the total effective population size (N_e) were roughly equal ($\sim 4 \times 10^4$), suggesting a 2-fold greater rate of drift for mtDNA. This equality of N_f and N_e can be attributed to the estimated effective breeding ratio of 4–5 males per female. The mutation rate was roughly 50 times greater in the mtDNA sequence (75×10^{-9} substitutions per site per year) compared with the nDNA sequences (1.23×10^{-9}). The difference between mtDNA and nDNA in their ability to detect geographic structure was principally a factor of mutation rate and sex-biased gene flow, rather than effective population size.

INTRODUCTION

It is probably fair to say that more is known of the biology of *Mirounga leonina* than of any other species of seal (Ling and Bryden 1992). The attraction of the Southern Elephant Seal to researchers may be that it exhibits many extreme characteristics. It is the largest of pinnipeds (adult males are >3 000 kg) and the most sexually dimorphic of mammals (adult males are 10 times the weight of adult females), the latter due most likely to sexual selection in a polygynous breeding system. Southern Elephant Seals grow faster, mature earlier and die younger than other pinnipeds of the family Phocidae. Along with its congener, the Northern Elephant Seal *M. angustirostris*, it dives deeper, for longer and more continuously than any other pinniped (Le Boeuf *et al.* 1989; Hindell *et al.* 1991a). Not surprisingly, therefore, its blood oxygen capacity is greater than that recorded for any other marine mammal (Bryden and Lim 1969). Paradoxically, given that they are so highly adapted in this respect to aquatic existence, they have retained the ancestral terrestrial characteristics of breeding and moulting more completely than any other pinniped (Carrick *et al.* 1962a).

There have been several intra-population genetic studies. Seal *et al.* (1971) detected no polymorphism in six proteins from 18 animals of unknown origin. McDermid *et al.* (1972) screened 42 Macquarie Island animals and detected variation in 5 of 19 blood proteins; these data were used by Bonnell and Selander (1974) to estimate heterozygosity for that population as 0.028, although this figure is

[1]Department of Zoology and Centre for Molecular and Cellular Biology, University of Queensland, Queensland, Australia 4072.
[2]Present address: Queensland Institute of Medical Research, Post Office, Royal Brisbane Hospital, Queensland, Australia 4079.
Pp. 11–29 *in* MARINE MAMMAL RESEARCH IN THE SOUTHERN HEMISPHERE VOLUME 1: STATUS, ECOLOGY AND MEDICINE ed by N. Hindell and C. Kemper. Surrey Beatty & Sons, Chipping Norton. 1997.

invalid due to the inclusion of related individuals (Gales *et al.* 1989). Shaughnessy (1974) screened 10–97 Macquarie Island animals for five blood proteins (nine loci) and three milk proteins (three loci) and detected variation in two whey proteins giving a heterozygosity of 0.075. However, variation at whey proteins may not be genetically controlled and the estimate of heterozygosity is questionable *(ibid).* At the DNA level, average nucleotide diversity of 2.3% in the mitochondrial DNA (mtDNA) control region I has been reported from 48 samples from two localities in the south Atlantic Ocean, South Georgia and Península Valdés (Hoelzel *et al.* 1993).

Inter-population genetic studies began with Lydekker (1909) who analysed skull characteristics and proposed that three subspecies be recognized — *falclandicus, crosetensis* and *macquariensis* — corresponding to type localities in the south Atlantic, south Indian and south Pacific oceanic regions respectively. However, Bryden (1968) claimed that these and other morphometric differences were environmentally determined. The single study of inter-population differences utilising molecular markers assessed the distribution of genetic variation between the Macquarie and Heard Island populations by screening 96 and 100 pups, respectively, for variation at 35 allozyme loci (Gales *et al.* 1989). Four loci were polymorphic. The heterozygosity values for each island of 0.029 and 0.034 respectively are among the highest recorded for pinnipeds. There were significant differences in allele frequency between the two populations at three of the four polymorphic loci. The authors concluded that the two populations have diverged genetically and could be considered distinct populations, but could not rule out other possibilities, such as substructuring within islands, to explain the results.

In this paper, I review some results from a recent molecular population and evolutionary genetics study of the Southern Elephant Seal. Specific aims of the study included estimates of the level of genetic diversity, rate of molecular evolution, effective population size and effective breeding ratio, geographic distribution of genetic variation, rate of gene flow and population divergence time. These results are discussed in terms of several broad issues: (i) the population genetic structure of a marine species with high dispersal capacity; (ii) genetic versus demographic structure; and (iii) the decline in numbers of south Pacific and south Indian Ocean populations. The populations studied were the three principal ones; Macquarie Island (MI) in the south Pacific Ocean, Heard and Kerguelen Islands in the south Indian Ocean and represented by samples from Heard Island (HI) and South Georgia (SG) in the south Atlantic Ocean (Fig. 1). Published data on mtDNA control region I sequence variation (Hoelzel *et al.* 1993) from the complete set of SG samples and from the smaller population at Península Valdés (PV), Argentina, were also included for some analyses. The detailed presentation of experimental and analytical procedures have either been previously published (Slade *et al.* 1993; Slade *et al.* 1994) or will be published elsewhere (Slade, Moritz, Burton and Hoelzel, in prep.). Here the emphasis will be on the relevance of the results to the biology and management of the Southern Elephant Seal.

ABBREVIATIONS USED IN THE TEXT

mtDNA = mitochondrial DNA, nDNA = single-copy nuclear DNA, bp = base pairs, π = nucleotide diversity, π_n = nucleotide diversity for nDNA, π_{mt} = nucleotide diversity for mtDNA, N_e = effective population size, N_f = effective population size of females, N_m = effective population size of males, μ = mutation rate per nucleotide site per year, g = generation time in years, μg = mutation rate per nucleotide site per generation, μ_n = mutation rate in nDNA per nucleotide site per year, μ_{mt} = mutation rate in mtDNA per nucleotide site per year, θ = neutral parameter, m = migration rate (the proportion of individuals exchanged between populations per generation), m_f = migration rate of females per generation.

LEVEL OF GENETIC DIVERSITY

MtDNA and nDNA genes were sequenced to obtain estimates of nucleotide diversity (Table 1). The level of nucleotide diversity (π) in 444 base pairs (bp) of mtDNA sequence, containing mainly control region and also some tRNA genes, was 1.95% (i.e., an average of 1.95 differences per 100 bp between any two sequences). This level of mtDNA control region diversity in the Southern Elephant Seal is among the higher levels so far observed among vertebrate species (Table 2), although the precise region sequenced differs between studies.

All else being equal, the amount of genetic diversity is proportional to population size. The Northern Elephant Seal has a level of control region diversity 1/4 to 1/5 that of the Southern Elephant Seal. When the same region of sequence is compared (i.e., 300 bp of control region I only), and when the estimates are corrected for multiple substitutions, the difference increases to roughly 1/10. This difference is still small considering the difference in history of exploitation between the two species. The global Northern Elephant Seal population was apparently reduced to less than 100 individuals by the late 1800s (Bartholomew and Hubbs 1960;

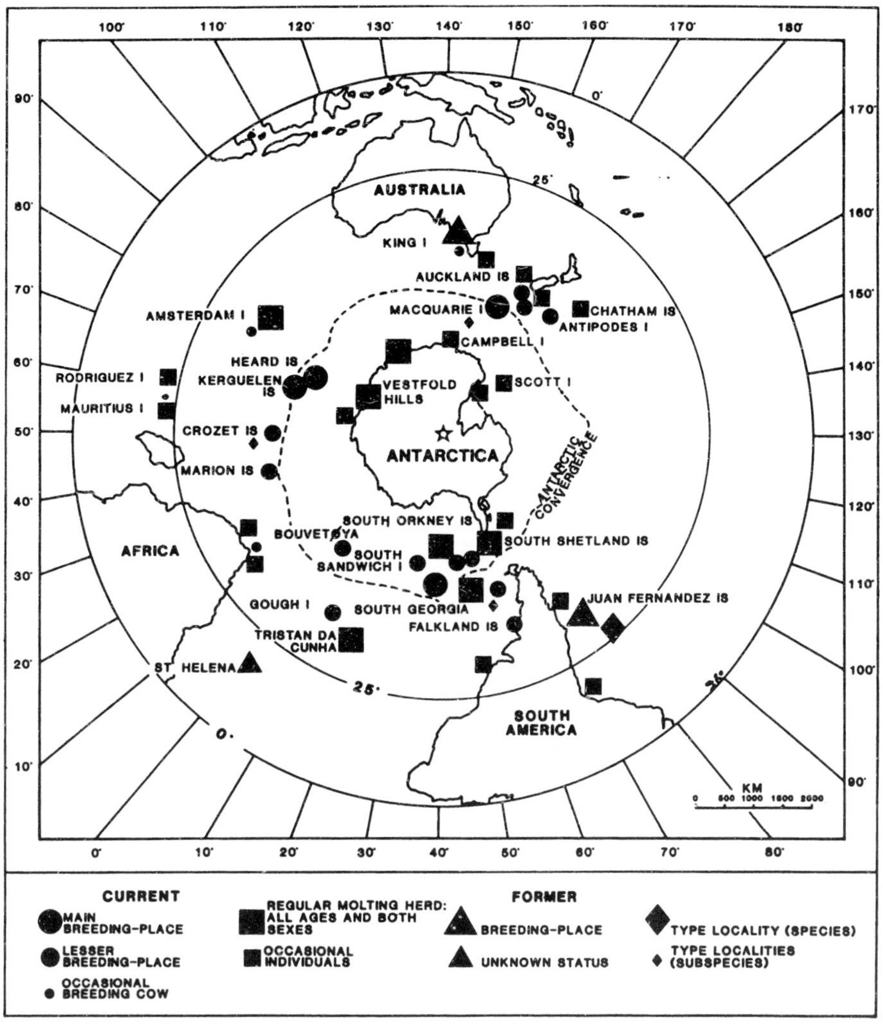

Fig. 1. Map showing past and present distribution of the Southern Elephant Seal *Mirounga leonina* and type localities of species and subspecies (from Ling and Bryden 1992 with permission).

Hoelzel *et al.* 1993), whereas that of the Southern Elephant Seal was probably never reduced below 100 000 individuals. There are several possible explanations for the discrepancy. It may partly reflect stochastic error in the estimates of nucleotide diversity (Tajima 1983). It may also partly reflect the theoretical explanation that little genetic diversity is lost if the bottleneck period is short and population size increases rapidly (Nei *et al.* 1975; Tajima 1993), as was the apparent situation for the Northern Elephant Seal (Bartholomew and Hubbs 1960; Hoelzel *et al.* 1993). In either case, it indicates a potential lack of sensitivity of genetic parameters to recent demographic processes (discussed further below). Alternatively, we should consider the possibility that the global Southern Elephant Seal population may have been historically much smaller than recent records indicate (see effective population size below).

For nDNA, sequencing of 3 594 bp from each of several individuals from each of the three main populations (Table 1 but excluding 519 bp of *β-globin* genes) resulted in π for silent sites (i.e., those not causing an amino acid change) of 0.09%. Less comparative data are available for nDNA compared with mtDNA. Diversity in Southern Elephant Seal nDNA is similar to that in humans of 0.11% (Li and Sadler 1991) and lower than that in *Drosophila pseudoobscura, D. simulans, D. ananassae* or *D. melanogaster* (2.2%, 1.4%, 1.0%, 0.4% respectively; Aquadro 1992). The level of polymorphism (3.2%) in the protein products of 35 nDNA genes (Gales *et al.* 1989) is similar to the mammalian average of 4.1% (Nevo *et al.* 1984). In summary, levels of genetic diversity within the Southern Elephant Seal suggest that it has not experienced any recent severe population bottlenecks.

RATE OF MOLECULAR EVOLUTION

Sequence data were obtained for over 400 bp of mtDNA and 1 200 bp of silent sites in nDNA from a total of six species of seal and a dog

Table 1. Genes sequenced[a].

Gene and region	No. base pairs	No. variant sites	Marker	Sequenced in other species[e]
Lysozyme exon 1/intron 1	303	0	—	—
Lysozyme intron 1/exon 2	311	2	—	—
Corticotropin-releasing factor 5′ UTR[b]	209	0	—	—
β-globin exon 2/intron 2[c]	168	4	—	—
β-globin intron 2/exon 3[c]	351	0	microsatellite	—
H2AF exon 2/intron 2	270	0	—	—
H2AF intron 3/exon 4	293	0	—	—
H2AF exon 4/intron 4	338	1	—	yes
ψH2AF exon 2–5	274	0	—	yes
ALD-A exon 5–6*	478	4	*Nsp*I site	yes
ALD-C exon 5–6	272	0	—	yes
Mhc-DQA exon 2*	169	1	—	yes
Mhc-DQA exon 3*	230	1	—	yes
Mhc-DQA intron 2[d]	447	0	6 bp deletion	
mtDNA tRNA^{thr}, tRNA^{pro}, control region I*	444	28	*Fok*I, *Rsa*I, *Bsl*I sites	yes

[a]Sequenced in one or five(*) individuals from each of MQ, HD and SG populations of the Southern Elephant Seal. For the mtDNA data, 16 sequences were obtained from 15 individuals because one individual from HD contained two mtDNA haplotypes (i.e., it was heteroplasmic).

[b]The 5′ untranslated region.

[c]The *β-globin* sequences were not included in the estimates of nucleotide diversity as they were not single-copy.

[d]*Mhc-DQA* intron 2 was sequenced from only two individuals from HD.

[e]Northern Elephant Seal *M. angustirostris*, Leopard Seal *Hydrurga leptonyx*, Weddell Seal *Leptonychotes weddelli*, Harbour Seal *Phoca vitulina*, New Zealand Fur Seal *Arctocephalus forsteri* and Dog *Canis familiaris*. These species were selected to represent a range of divergence times [<1 million years ago (Mya) to 40–50 Mya] and levels of relationship, including within genus *Mirounga* (Southern and Northern Elephant Seal), within subfamily Monachinae (+ Weddell and Leopard Seals), within family Phocidae (+ Harbour Seal), within pinnipeds (+ Fur Seal) and within order Carnivora (+ Dog). Additional sequence of the mtDNA control region was available for the Grey Seal of the family Phocidae (*Halichoerus grypus*; Árnason *et al.* 1993).

(Table 1; Slade *et al.* 1994). When combined with information from the fossil record on dates of divergence the distance between sequences (after correction for multiple substitutions using a specific model of evolution) allows estimation of the mutation rate (Slade *et al.* 1994; Slade, unpubl. data). The rate of evolution for pinniped nDNA (using a Jukes-Cantor model of evolution; Jukes and Cantor 1969) averaged 1.23 substitutions per nucleotide site per 10^9 years (i.e., a mutation rate of 1.23×10^{-9}); this is at the low end of rates among mammals, i.e., 6.5×10^{-9} in rodents, 3×10^{-9} in primates and artiodactyls and [illegible] $\times 10^{-9}$ in humans (Li *et al.* 1987). A slower rate of nDNA evolution in carnivores compared with other mammals has also been indicated by immunological distances (Sarich 1985). However, these estimates should be treated with caution because they are based on the fossil record and it is clear that reinterpretation of parts of the mammalian fossil record is necessary (Wilson *et al.* 1987; Easteal 1990).

The rate of evolution for pinniped mtDNA was estimated for 300 bp of control region I and was calibrated against only one fossil record date of 4.5 million years divergence between *Mirounga* species and other Antarctic monachines (Ray 1976); other calibrations were not possible due to the high mutation rate. The rate of mtDNA evolution (using a Tamura-Nei model of evolution; Tamura and Nei 1993) was estimated as approximately 75×10^{-9}. This is at least an order of magnitude higher than that for nDNA and it may be close to two orders of magnitude higher. In terms of per cent change, the nDNA is evolving at roughly 0.1% per million years (/Myr) along a lineage, while the mtDNA control region I is evolving at 5–10%/Myr. The rate of substitution averaged over the entire mtDNA molecule in other animals was estimated as 1%/Myr along a lineage (which is 2%/Myr divergence time; Wilson *et al.* 1985), while the silent rate for the ND4 and ND5 mtDNA genes in primates was estimated to be around 5%/Myr along a lineage (Brown *et al.* 1982). More specifically, recent studies have estimated a rate of evolution of 103×10^{-9} and 74×10^{-9} for control regions I and II respectively in great apes (Horai *et al.* 1995), 75×10^{-9} for the entire control region in great apes (Tamura and Nei 1993), 114×10^{-9} for the entire control region in humans (Stoneking *et al.* 1992) and $20–40 \times 10^{-9}$ for control region I in the horse (Ishida *et al.* 1995). Therefore, the estimate presented here of 75×10^{-9} for control

Table 2. Per cent diversity (π) in vertebrate mtDNA control regions.

Species	π	Region sequenced (No. bp)	No. populations	n[a]	Ref.[b]
Mirounga leonina Southern Elephant Seal	1.95	$tRNA^{thr}$, $tRNA^{pro}$, control region I (444)	3	16	1
M. angustirostris Northern Elephant Seal	0.43	control region I (300)	1	40	2
Zalophus californianus California Sea Lion	0.405–1.708	$tRNA^{pro}$, control region I (360)	4	3–17	3
Homo sapiens Human	1.49	control region I (360)	Amerindian	63	4
H. sapiens	2.08	control regions I and II (610)	African	21	5
H. sapiens	1.75	"	Asian	24	5
H. sapiens	1.08	"	Caucasian	15	5
Megaptera novaeangliae Humpback Whale	2.57	control region I (283)	6	90	6
Mus domesticus European House Mouse	0.91	control regions I and II	28	56	7
Macroderma gigas Ghost Bat	4.7	control region I (330)	9	61	8
Macropus rufus Red Kangaroo	3.6	control region I (560)	10	35	9
Dipodomys panamintinus mohavensis Kangaroo Rat	1.66	$tRNA^{thr}$, $tRNA^{pro}$, control region I (225)	1	34	10
D. p. caudatus	0.33	"	1	40	10
D. p. panamintinus	0.79	"	1	32	10
D. panamintinus	2.2	"	3	106	10
Pomatostomus temporalis temporalis Grey-crowned Babbler	3.9	control region I (400)	5	69	11
P. t. rubeculus	2.1	"	7	94	11
Calidris alpina Dunlin	0.37	control region I (295)	1	14	12
"	1.83	"	1	3	12
"	0.96	"	1	14	12
"	0.95	"	1	23	12
"	0.37	"	1	16	12
Acipenser transmontanus White Sturgeon	0.75	control region I (462)	1	11	13
"	1.13	"	1	16	13
Melanochromis auratus Cichlid	0.08	control region I (445)	8	68	14
M. heterochromis	0.5	"	6	42	14

[a]Number of sequences.

[b]1. This study. 2. Hoelzel *et al.* 1993. 3. Maldonado *et al.* 1995. 4. Ward *et al.* 1991. 5. Vigilant *et al.* 1991. 6 Baker *et al.* 1993. 7. Nachman *et al.* 1994b. 8. J. Worthington Wilmer *et al.*, in prep. 9. S. Clegg *et al.*, in prep. 10. Thomas *et al.* 1990. 11. Edwards 1993. 12. Wenink *et al.* 1993. 13. Brown *et al.* 1993. 14. Bowers *et al.* 1994.

region I in the Southern Elephant Seal is consistent with other studies.

EFFECTIVE POPULATION SIZE AND EFFECTIVE BREEDING RATIO

The data on nucleotide diversity and mutation rates for nDNA and mtDNA allow estimation of effective population size *(N_e)* and effective female population size *(N_f)* respectively. The effective population size is the number of individuals in an idealized population where each individual has an equal expectation of progeny (Wright 1931; Kimura and Crow 1963). Estimates of N_e and N_f can be obtained from the formulae linking the neutral parameter (θ) with mutation rate per generation (μg, where μ = mutation rate per site per year and g = generation time in years) and effective population size. For autosomal nDNA loci in a sexually reproducing diploid species $\theta = 4N_e\mu_{n}g$ while for haploid loci such as mtDNA $\theta = 2N_f\, \mu_{mt}g$. The parameter θ can be calculated a number of different ways (Tajima 1993; Fu 1994). Here I use π calculated after application of the Tamura-Nei model to the pairwise sequence comparisons within the Southern Elephant Seal. The generation time *(g)* of approximately eight years for the Southern Elephant Seal was calculated from the life table for females at South Georgia, which has had a constant population size (McCann 1985), as being $\frac{\Sigma x l_x m_x}{\Sigma l_x m_x}$ where x is age in years, l_x is the proportion of females surviving to age x, and m_x is the number of females produced per female at age x (Begon *et al.* 1986). The values of N_e and N_f are given in Table 3.

The Southern Elephant Seal has a female effective population size of roughly $N_f = 4 \times 10^4$, while each of the three main populations has an N_f of approximately $2–3 \times 10^4$. The PV population is an order of magnitude smaller than the other Southern Elephant Seal populations and the effective number of Northern Elephant Seals is an order of magnitude smaller than that for the Southern Elephant Seal. The size of N_e in the Southern Elephant Seal of 4×10^4 is similar to that in humans which is approximately 10^4 to 10^5, depending on the time scale over which the estimates are calculated, while N_f is smaller in humans at approximately 4×10^3 (Takahata

Table 3. Total (N_e) and female (N_f) effective population sizes[a].

Population/species (n)	N_e	N_f
MQ (5)	—	18 000
HD (6)	—	22 000
SG (28)	—	30 000
PV (21)	—	3 000
Sth Elephant (60)	40 000	40 000
Nth Elephant (41)	—	5 000

[a]Calculated from the Jukes-Cantor adjusted nDNA data for N_e, and the Tamura-Nei adjusted mtDNA control region I data for N_f including the data from Hoelzel *et al.* 1993 for SG and PV. There was too little nDNA variation to estimate N_e within populations.

1993). If the worldwide population size of Southern Elephant Seals is 750 000 (McCann 1985) then the ratio of effective to current population size (N_e/N) would be 0.05. N_e/N ratios are highly variable among species; ratios of nine species of mammal ranged from 0.08 to 0.76 (Frankham 1994). The relationship between N_e and current total population size (N) is influenced by a variety of factors including breeding ratio, variance in progeny number, overlapping generations, fluctuation in population size, extinction and recolonization of subpopulations and population structure. When calculated from nucleotide diversity the value of N_e reflects the influence of those factors over the coalescence time of the genes used for the estimate of diversity (roughly $4N_e$ and $2N_f$ generations for N_e and N_f respectively). The low ratio for the Southern Elephant Seal would, in part, reflect its highly polygynous breeding system. It may also reflect a current population size larger than historical population sizes.

It may seem surprising that the effective female population (N_f) is as large as that for females and males combined (N_e). This probably reflects a breeding sex ratio biased towards females where, if the female to male breeding ratio is greater than 3:1, then N_f is greater than N_e (Birky *et al.* 1983). This is a likely scenario for the polygynous Southern Elephant Seal. It is possible to estimate the ratio of N_f to N_m in a sexual population using information on mutation rates for mtDNA (μ_{mt}) and nuclear DNA (μ_n), and information on nucleotide diversity in mtDNA (π_{mt}) and nuclear DNA (π_n). The mutation rates do not depend on effective population size, while diversity does and so it can be shown that $N_f = N_m(\lambda 8-1)$, where $\lambda = \frac{\pi_{mt}\mu_n}{\pi_n\mu_{mt}}$ (Slade, unpubl. data). From this formula the estimated breeding ratio is roughly 4–5 females to one male.

This ratio is perhaps lower than expected considering that, in one study, the average number of cows per bull in a harem varied between 28 and 53 (Carrick *et al.* 1962b). However, not only are estimates of copulatory success often over-estimated by behavioural observations (e.g., Harris *et al.* 1991; Pemberton *et al.* 1993; Lambert *et al.* 1994) but also, while males may sire a large number of offspring in the one or two years they are a beachmaster, their lifetime reproductive success would be much lower than observation during beachmaster years would indicate. A female at South Georgia that survives to breeding age produces, on average, five offspring in her lifetime. [This figure was derived from the female life tables in McCann (1985) and from the estimated fecundity rate of 0.391 females born per female per year (Hindell 1991)]. Considering this, then the above data suggest that the average breeding male would sire 20 offspring in his lifetime, although there would likely be a much larger variance associated with this figure than that for the average female.

THE GEOGRAPHIC DISTRIBUTION OF GENETIC VARIATION

The presence of significant geographic structure was tested in three ways; (i) phylogeographic analysis (Avise *et al.* 1987; Maddison and Slatkin 1991) of mtDNA sequence variation from each of the MI, HI, SG and PV populations, (ii) analysis of the amount of sequence variation within and between those four populations (Excoffier *et al.* 1992) and (iii) analysis of frequency differences (Roff and Bentzen 1989) of mtDNA haplotypes and nDNA alleles between the three main populations of MI, HI and SG. Phylogeographic analysis is concerned with detecting geographic structure from the quantitative and qualitative relationships between alleles, while analysis of allele frequencies detects structure from differences in proportions of alleles with no consideration given to relationships between alleles. Analysis of molecular variation uses information on both allele frequency and relationships. It is important to use several methods to analyse geographic structure, not only because the power of the methods differ, but also because different methods emphasize different aspects of the data and therefore provide different insights into the geographic structure of genetic variation. (For these analyses, the results were essentially the same whether all the SG sequences, or only the subset from this study, were used.)

Phylogeographic analysis

A minimum-evolution tree (Rzhetsky and Nei 1992) was constructed from the 16 Southern Elephant Seal mtDNA sequences obtained in this study with Weddell and Northern Elephant Seal sequences as outgroups (Fig. 2). Sequences PV1–PV3, NTH2 and NTH3 were taken from Hoelzel *et al.* (1993). The MI and PV haplotypes form separate and distinct monophyletic lineages that are strongly supported by the branch length test (significance levels of 89% and 99% respectively). Conversely, the SG and HI populations share haplotype lineages and are polyphyletic.

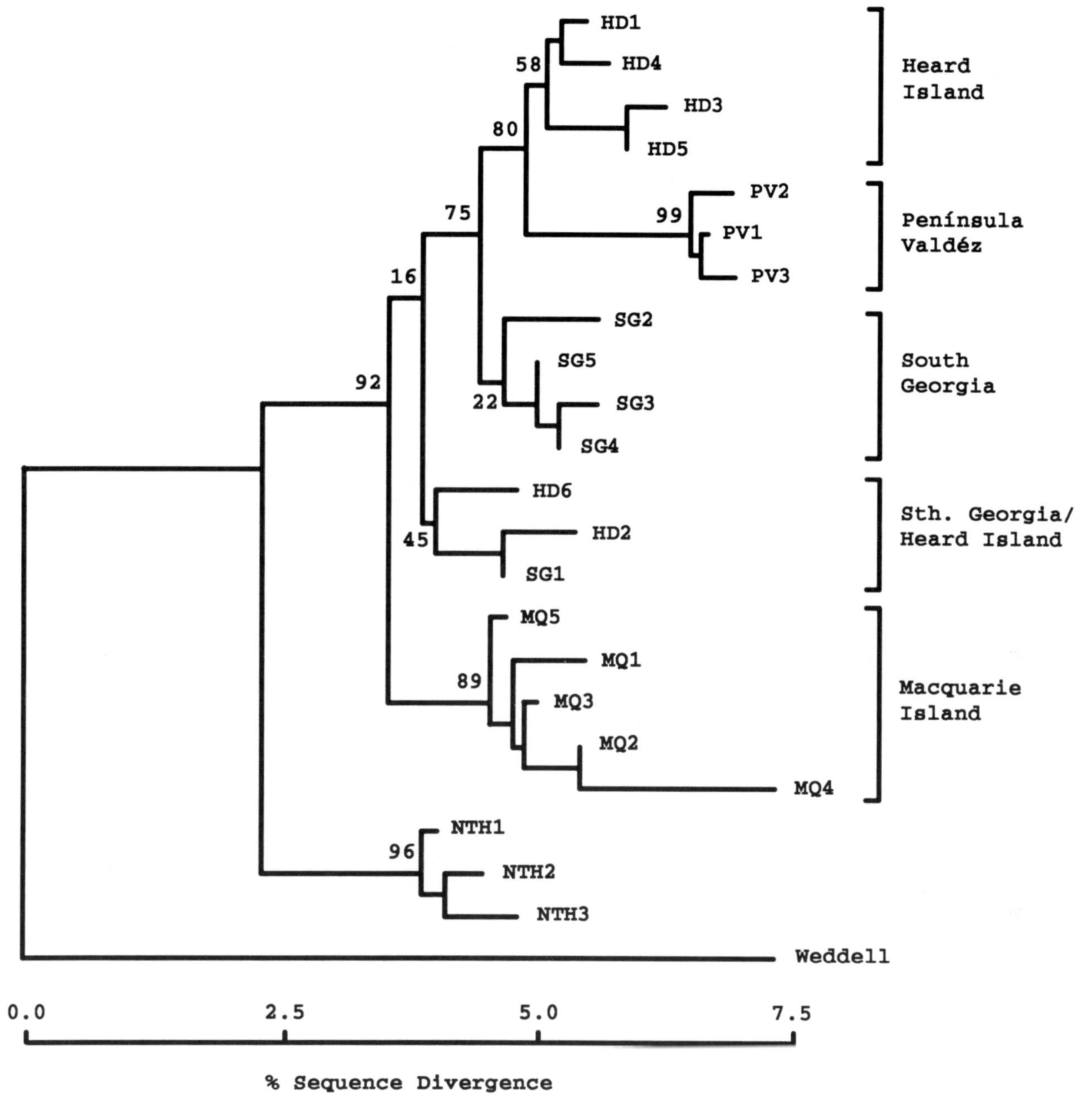

Fig. 2. Minimum-evolution tree of mtDNA control region I haplotypes in the Southern and Northern Elephant Seals, with the Weddell Seal as outgroup. The per cent significance levels of branch-length positiveness are shown. Sample size n = 5 for MI, n = 6 for HI, n = 5 for SG, and n = 21 for PV. For MI, HI, and SG every sequence was unique. For the PV samples n = 8 for PV1, n = 6 for PV2, and n = 7 for PV3 (Hoelzel *et al.* 1993).

The presence of geographic structure within the tree was tested by comparing the branching pattern of the tree with that of 1 000 trees constructed by resampling the data (Maddison and Maddison 1992). The result showed that there is significant geographic structure within the tree ($P < 0.001$), most of which can be attributed to the monophyletic clustering of the MI and PV haplotypes. However, although the SG and HI haplotypes are polyphyletic, their distribution may still be non-random, reflecting some geographic structure. This was tested specifically by repeating the randomization procedure with a minimum-evolution tree of only the SG and HI haplotypes and the result showed significant geographic structure between these two populations ($P < 0.05$)

The low diversity within the nDNA sequences meant that a phylogeographic approach was not applicable. For the nDNA gene with the highest diversity, aldolase A (*ALD-A*), there were five alleles within the Southern Elephant Seal. The most common allele (allele *a*, $p = 0.5$) was found at equal frequency in the three main populations and was 100% homologous to the Northern Elephant Seal sequence (Fig. 3). The four other alleles (*b, c, d, e*) differed from *a* and the Northern Elephant Seal sequence

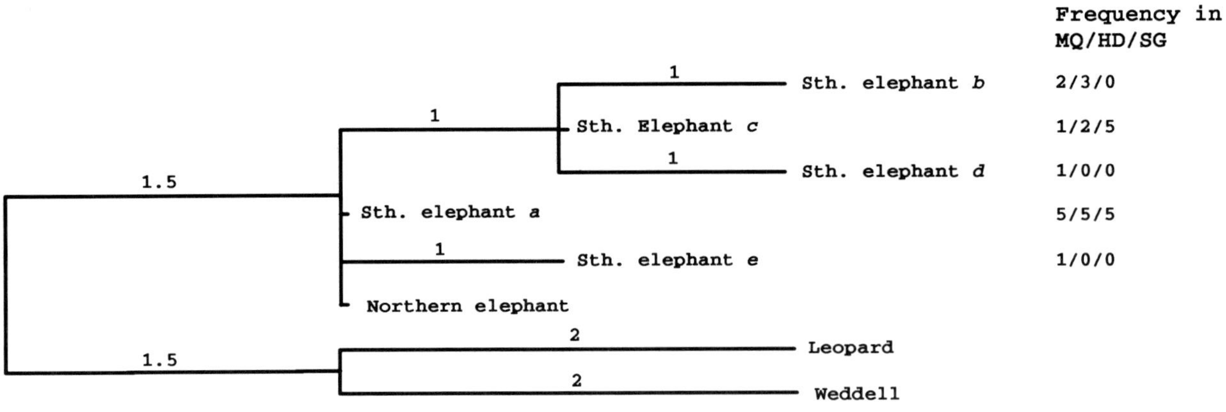

Fig. 3. Neighbour-joining tree (Saitou and Nei 1987) of *ALD-A* sequences showing the five alleles from the Southern Elephant Seal, the sequence from the Northern Elephant Seal with the outgroup sequences from the Leopard and Weddell Seals. The distribution of each Southern Elephant Seal allele among the three main populations of MI, HI and SG is shown. The length of each branch in number of substitutions is shown.

Table 4. Per cent diversity within and between populations for mtDNA control region I[a].

	PV (n = 21)	SG (n = 28)	HD (n = 6)	MQ (n = 5)
PV	*0.297*	3.365	3.071	5.224
SG	2.073	*2.289*	2.394	3.872
HD	1.908	0.235	*2.029*	3.801
MQ	4.273	1.926	1.985	*1.605*

[a]Diversity within populations in italics on diagonal and between populations above diagonal, and net divergence between populations below diagonal. Includes the data from Hoelzel *et al.* 1993 for SG and PV.

by one or two substitutions; of these four alleles two (*d, e*) were rare and unique to Macquarie Island while the remainder *(b, c)* were shared between two or more populations. The sharing of nDNA alleles between the Southern and Northern Elephant Seals was found for all the polymorphic nDNA sequences and reflects the low mutation rate of nDNA compared with mtDNA. It also indicates that there would be little, if any, possibility of defining populations within the southern elephant seal *via* a phylogeographic approach using nDNA.

Molecular variation within and between populations

This analysis was used on only the mtDNA variation. The levels of mtDNA nucleotide diversity and divergence within and among populations are shown in Table 4. These provide a quantitative picture of the qualitative impression given by the phylogenetic tree. Nucleotide diversity was highest within the SG and HI populations and very low by comparison in the PV population. The greatest divergence was between the PV and MI populations while the least is between the SG and HI populations. Although PV and SG are in the same oceanic region, there is almost an order of magnitude greater divergence between those two populations than between the separate oceanic populations of SG and HI.

The AMOVA procedure of Excoffier *et al.* (1992) was used to assess the amount of mtDNA variation distributed within and between the four populations and showed that 57% was distributed between populations while 43% was distributed within populations. This distribution of mtDNA variation is significantly different from random ($P < 0.001$), reflecting the presence of geographic structure. A hierarchical level above that of populations was constructed by placing populations into regions which allows then a comparison between regions as well as a comparison between populations within a region. A natural approach would be to compare oceanic regions, with SG and PV representing the south Atlantic Ocean, HI the south Indian Ocean and MI the south Pacific Ocean. This resulted in a decomposition of variation into; 5% among the three oceanic regions, 61% among SG and PV within the south Atlantic oceanic region and 44% within populations. The breakdown into oceanic regions thus explained none of the variation. This is due to heterogeneity within the south Atlantic Ocean between SG and PV and the close relationship between haplotypes from SG and HI representing different regions. For a three region comparison, the variation among regions is maximized by placing SD and HI in the same region (52%). In this case only 8% of the total variation is found between SD and HI,

Table 5. Analysis of marker frequencies between populations.

Locus	Marker	Comparison	X^2	p
mtDNA	Combined *Fok*I, *Rsa*I, *Bsl*I	Overall	224.76	< 0.001
		MQ/HD	59.31	<0.001
		MQ/SG	63.01	<0.001
		MQ/PV	53.21	<0.001
		HD/SG	32.51	<0.001
		HD/PV	31.98	<0.001
		SG/PV	38.23	<0.001
ALD-A	*Nsp*I site	Overall	2.00 (ns)	0.3360 ± 0.0149
Mhc-DQA	6bp indel	Overall	1.99 (ns)	0.4130 ± 0.0156
β-globin	microsatellites[a]	Overall	59.87	0.001 ± 0.001
		MQ/HD	23.42	0.0320 ±0.0056
		MQ/SG	42.82	<0.001
		HD/SG	20.75 (ns)	0.1000 ±0.0095

[a]There are two microsatellite loci, one in each of two *β-globin* genes (Slades, Unpubl. data). The analysis treated them as a single multiallelic locus in a tetraploid. This approach (Utter *et al.* 1992), or a similar one (Gharrett *et al.* 1987; Bartley and Gall 1990; Gall *et al.* 1992), has been used previously for duplicated loci in salmonid fish (see Waples 1988; and Waples and Aebersold 1990 for discussion).

although the distribution of that variation is still significantly different from random ($P = 0.027$).

Haplotype and allele frequency differences

Genetic markers that could be easily screened in a large sample were defined from the sequence data for mtDNA (three restriction enzyme sites — *Fok*I, *Rsa*I, *Bsl*I) and nDNA (a *Nsp*I restriction site in *ALD-A*, a six bp indel in *Mhc-DQA*, and two pentameric microsatellite loci in recently duplicated *β-globin* genes). These markers were screened in approximately 30 individuals from each of the three main populations. Additionally the mtDNA sequence data from 21 PV individuals were defined for the three restriction sites in control region I. The results of the χ^2 analyses that tested for heterogeneity in frequencies between populations are shown in Table 5.

There were significant haplotype frequency differences among all four populations and between each pair of populations for mtDNA. For the nDNA, there was no significant heterogeneity among the three main populations for alleles at the *Mhc-DQA* and *ALD-A* loci, whereas there was significant heterogeneity at the *β-globin* microsatellites (Table 5). For the microsatellites, pairwise comparisons showed significant differences in allele frequency between MI and the other two populations, but not between HI and SG (Table 5).

Summary

This is the first global survey of genetic variation in the Southern Elephant Seal. For mtDNA there was significant geographic structure among the four populations of the Southern Elephant Seal at three levels of analysis: distribution of lineages, distribution of genetic variation and frequencies of haplotypes. Most of the geographic structure was due to the divergent MI and PV populations, the haplotypes within which formed discrete monophyletic groups. The SG and HI populations shared haplotype lineages but there was, nonetheless, significant geographic structure for the distribution of those lineages as well as for the distribution of genetic variation and haplotype frequencies. The limited variation at the nDNA loci allowed only an analysis of allele frequency differences which, among the populations of SG, HI and MI, did not exhibit as strong a geographic structure as the mtDNA data. There were no significant differences in allele frequency for markers at the *ALD-A* and *Mhc-DQA* loci while for the *β-globin* microsatellites a significant difference was found between MI and the other two populations. The difference in the microsatellite allele frequencies between SG and HI was significant at the 10%, but not the 5%, level and suggests that further analysis of the distribution of nDNA variation between these two populations is warranted. The genetic discreteness of the HI and MI populations has also been shown in a previous study of allozyme variation among those two populations (Gales *et al.* 1989). It is clear, therefore, that there is strong geographic structure for the female (i.e., mtDNA) component among the four populations but for the biparental (i.e., nDNA) component this is clear only between MI and either HI or SG.

CAUSES OF THE OBSERVED DISTRIBUTION OF GENETIC VARIATION

The amount and distribution of genetic variation within a species is determined by mutation, selection, gene flow, genetic drift (determined by effective population size) and historical processes. What factors can best explain the observed pattern of geographic structure in the Southern Elephant Seal and particularly the difference between the mtDNA and nDNA results and the difference between populations in their level of genetic discreteness?

Selection

Does the geographic structure of genetic variation accurately reflect population structure (due to the influence of gene flow, drift and historical processes) or is there a strong influence of differential selection on genes within individuals? That is, are the observed genetic variants selectively neutral? Typically, neutrality is assumed and rarely tested (Wilson *et al.* 1985; Avise *et al.* 1987; Moritz *et al.* 1987). However, selection on mtDNA, either directly or *via* cytonuclear interactions, has been indicated in a few studies, such as *Drosophila*, mice and humans (e.g., MacRae and Anderson 1988; MacRae and Anderson 1990; Singh and Hale 1990; Hale and Hoffman 1990; Nigro 1994; Nachman *et al.* 1994a). In this study, two approaches were used to show that the observed genetic structure in the Southern Elephant Seal reflects demographic processes and not selection on genes.

Selection can be detected from patterns of DNA sequence variation using several recently developed methods (Kreitman and Aguadé 1986; Tajima 1989; Fu and Li 1993). For all methods, the distributions of the mtDNA and nDNA variants in the Southern Elephant Seal (with the exception of the *β-globin* microsatellites which were not tested) were compatible with neutrality (Slade, unpubl. data). It is also reasonable to assume that selection is not a factor if there are concordant geographic patterns among several unlinked loci. Although not as informative, the available data from nDNA loci are concordant with the mtDNA data. Three allozyme loci *(Ada, Pgam, Pgm)* showed significant differences in allele frequency between HI and MI (Gales *et al.* 1989) as did the *β-globin* microsatellites and there are non-significant indications from the latter of differences between HI and SG.

Comparison between mtDNA and nDNA

The vertebrate mtDNA molecule has several properties that make it a useful population genetic marker. These properties are its haploid nature, its inheritance from only the maternal parent and its rapid rate of evolution compared with nDNA (Moritz *et al.* 1987). Differences between mtDNA and nDNA in the level of observed geographic structure may reflect either differences in characteristics of the genes, such as mutation rate and/or rate of genetic drift, or differences in rates of gene flow between males and females.

Characteristics of the genes

Considering differences between genes, the greater sensitivity of mtDNA compared with nDNA is usually explained by two factors — its smaller effective population size and its higher mutation rate (Wilson *et al.* 1985). A smaller effective population size increases the rate of genetic drift and therefore the rate at which populations are expected to exhibit differences in allele frequency. MtDNA is haploid whereas nDNA genes exist in a diploid state (assuming all are autosomal) and so the comparison for rates of drift is N_f versus $2N_e$. From Table 3 it can be seen that for the Southern Elephant Seal N_f is roughly equal to N_e and so the rate of drift of mtDNA is expected to be roughly twice that of the nDNA, thereby contributing to the increased resolution of the mtDNA variation.

Perhaps a stronger influence on the increased sensitivity of mtDNA for detecting subdivision would come from the 50 times greater mutation rate compared with that in nDNA. A higher mutation rate would have two consequences, both of which would increase the chance of detecting geographic subdivision (Slade *et al.* 1994). First, it would result in a larger number of alleles at a lower frequency in the population compared with the fewer alleles found at nDNA loci. This would decrease the frequency of shared alleles when one population separates into two (Wu 1991). Second, it would increase the chance that isolated populations contain unique alleles that arose within each population after separation. That differences in mutation rate account for a significant part of the difference in detecting subdivision is supported by the results from the highly variable microsatellites, which alone among the nDNA loci showed evidence of geographic structure. Microsatellite loci are known to have mutation rates of the order of 10^{-2}–10^{-4} per locus per generation (Weber and Wong 1993). This is similar to the mutation rate per locus per generation for control region I which is 1.8×10^{-4} (i.e., $75 \times 10^{-9} \times 300$ bp $\times$ 8 years). Conversely, the per locus mutation rate of the other 250–450 bp nDNA gene fragments is 2.5–4.5×10^{-6}, some two orders of magnitude lower.

Female and male rates of gene flow

Under an equilibrium scenario of ongoing gene flow, we can estimate the amount of gene flow (in terms of either the proportion of migrants exchanged between populations per generation, m, or the number of migrants exchanged per generation, N_em) from K_{ST}, an analog of G_{ST} describing the distribution of variation among populations (Hudson *et al.* 1992). From the mtDNA variation we can obtain estimates of rates of female gene flow, N_fm_f, between pairs of populations and these are presented in Table 6. Considering the above discussion on the influence

Table 6. Rates of female gene flow between populations.

Comparison	N_f (x 10^4)[a]	K_{ST}[b]	With mutation rate incorporated: m_f^c (x 10^{-5})	With mutation rate incorporated: $N_f m_f$	Without mutation rate: $N_f m_f^d$
SG/PV	16.5	0.42	1.03	0.17	0.18
SG/HD	26	0.04	13.22	3.44	3.45
HD/PV	12.5	0.4	1.45	0.18	0.18
SG/MQ	24	0.31	1.13	0.27	0.28
HD/MQ	20	0.35	1.14	0.23	0.23
PV/MQ	10.5	0.71	0.47	0.05	0.05

[a] The arithmetic mean of the N_f's from the two populations.

[b] $K_{ST} = 1 - \frac{K_S}{K_T}$, and where K_S is the average nucleotide diversity within populations, and K_T is the total nucleotide diversity (Hudson *et al.* 1992). Nucleotide diversity was calculated from the Tamura-Nei adjusted mtDNA control region I data.

[c] The formula for m_f, with mutation rate incorporated, was derived from the finite one-dimensional stepping stone model of Maruyama (1970), where

$$m_f = \frac{K_{ST}(4N_f\mu - 2\mu + 1) + (2\mu - 1)}{4K_{ST}(2N_f - 1)(2\mu - 1) + (8\mu - 4)}$$ and where μ is the mutation rate of 6.04 substitutions per site per 10^7 generations (Slade, in prep.).

[d] The formula for $N_f m_f$ was modified from Crow and Aoki (1984) for haploid mtDNA and $n = 2$ populations where $N_f m_f = \frac{\frac{1}{K_{ST}} - 1}{16}$.

of mutation rate, it is important to note that incorporation of mutation rate does not affect the estimates of $N_f m_f$ (Table 6).

Estimates of total (i.e., male plus female) gene flow between HI and MI were obtained from allele frequency data of three polymorphic allozyme loci (*Acp-1, Ada-1, Pgm-1*) from Gales *et al.* (1989) and two polymorphic nDNA loci (*Mhc-DQA, ALD-A*) from this study. This resulted in an estimate of $N_e m$ between HI and MI of 1.83, some eight times greater than the level of maternal gene flow of 0.23 (i.e., $N_f m_f$ between HI and MI in Table 6). If males and females have an equal probability of migrating, then $N_e m$ is expected to be twice that of $N_f m_f$. The observed result thus suggests that the rate of male gene flow may be four times greater than female gene flow. The result must be viewed cautiously, principally because the information on levels of female gene flow were derived from one locus. There is a large variance among loci, reflecting primarily the stochasticity of the evolutionary process and migration rates, as well as the large variance within loci due to sampling of individuals and nucleotides (Nei 1987; Lynch and Crease 1990; Epperson 1994).

Nonetheless, male-biased gene flow is consistent with ecological studies on dispersal of male and female Southern Elephant Seals. There were some 20 000 individuals tagged in the south Indian and south Pacific Oceans between 1950 and 1980 (Carrick and Ingham 1962; Carrick *et al.* 1962a; Nicholls 1970; Condy and Bester 1975; Van Aarde and Pascal 1980; Burton 1985; Pascal 1986). Ther were roughly equal numbers of males and females tagged in all localities and nearly all were pups, except for the Vestfold Hills in which all were juvenile males. There were 48 resightings at localities outside the area in which tagging occurred. The direction and distance of all inter-island movements are summarized in Table 7. No movements were recorded between the three main breeding populations. There is a large bias towards males but this figure is heavily influenced by 20 males that moved from the Vestfold Hills, where only males were tagged, to Kerguelen Island. After removal of these resights and one of unknown sex, the male to female ratio is 18 to 9 and while this is still male-biased the difference is not significant ($\chi^2 = 3$, $0.05 < p < 0.1$; assuming equal numbers of males and females tagged). There is also sex-biased dispersal during the pelagic feeding phase of the annual cycle in which males generally travel further than females to reach the feeding grounds (Hindell *et al.* 1991b).

Comparisons between populations

What then were the relative contributions of mutation, gene flow, drift and historical processes to the observed geographic structure of genetic variation, in this case referring principally to the informative mtDNA variation. The primary question to address is whether the observed structure reflects primarily a balance between forces of contemporary gene flow and drift or whether it reflects mostly historical and not current, associations. Unfortunately, it is not possible to distinguish between these two situations from the genetic data alone. It may be useful, however, to consider the two scenarios separately and relate these to the known ecology of the Southern Elephant Seal.

Table 7. Inter-island movements of tagged Southern Elephant Seals (from Carrick *et al.* 1962a; Bester 1988).

Region	From	To	No. individuals, sex, age	Distance (kms)
South Pacific				
	Macquarie	Campbell	2,m,l; 1,f,l; 1,f,2	800
	Macquarie	Chatham	1,m,1	2 400
	Campbell	Macquarie	1,m,12;1,f,12	800
South Indian				
	Heard	Kerguelen	1,m,6	440
	Heard	Amsterdam	1,m,5	2 000
	Heard	Marion	1,m,2; 1,f,2	2 740
	Heard	Vestfold	1,m,5; 1,m,6	1 600
	Kerguelen	Heard	1,m,6; 1,m,3; 1,?,3	440
	Kerguelen	Marion	1,f,2	2 460
	Kerguelen	Vestfold	1,m,3; 1,m,8	2 000
	Vestfold	Kerguelen	20,m,juvenile	2 000
	Crozet	Marion	4,m,l; 1,m,?; 2,f,1; 1,f,2	1 140
	Marion	Crozet	1,f,4	1 140

The equilibrium scenario of ongoing gene flow

If we again assume an equilibrium situation, then female-mediated gene flow between SG and HI (roughly 3–4 individuals exchanged per generation) is over an order of magnitude greater than that between other populations (Table 6). Indeed, the amount of female gene flow between all other populations is so low that it is likely to be non-existent (this is also suggested by the monophyly of the MI and PV haplotypes). This is surprising considering that SG and HI are further apart than HI and MI. However, although the total distance between SG and HI is some 8 000 km, the greatest distance between known breeding islands dotted between SG and HI is 2 500 km between Marion/Prince Edward Islands and Bouvetøya (Fig. 1). This distance is well within the dispersal capacity of the Southern Elephant Seal. The longest inter-island movement recorded was some 3 000 km between HI and Marion Island (Table 7) and animals have made 4 000 km round trips from breeding to feeding areas (Hindell *et al.* 1991b). It is clear then that gene flow between SG and HI *via* intermediate breeding colonies is within the dispersal capacity of the Southern Elephant Seal.

Conversely, there are no known breeding colonies between HI and MI and the distance between them of 7 500 km is beyond the known dispersal capacity of the species. So it is perhaps not surprising that HI and MI are genetically further apart than are SG and HI. Nonetheless, there is potential for exchange *via* moulting colonies. Overlap of individuals from HI and MI has been recorded at the Windmill Islands near Casey station (unpublished results reported in Gales *et al.* 1989) and animals from HI and Kerguelen moult at the Vestfold Hills near Davis station (Burton 1985).

It is also surprising that SG and PV are not closer in terms of genetic distance considering that they are in the same oceanic region, that they are 2 400 km apart and that movements of animals between PV and the neighbouring Falklands have been recorded (Campagna and Lewis 1992). This isolation of the PV colony from subantarctic colonies in the south Atlantic Ocean is also consistent with several ecological, demographic and geographical features (Vergani 1985; Campagna *et al.* 1993 and references therein).

The historical association scenario

At the other extreme, we can consider a scenario in which the observed geographic structure in the Southern Elephant Seal reflects only historical connectedness and not ongoing gene flow subsequent to population separation. Under this scenario, levels of net sequence divergence between populations, combined with information on mutation rate, can provide estimates of time since population separation (Table 8). In this historical model, MI and PV last shared a common ancestral population some 600 000 years ago, SG and HI separated as recently as 20 000 years ago, while all other population divergences occurred some 200 000–300 000 years ago. Any gene flow occurring after population separation necessarily implies an older divergence time, but the extent of this increase is constrained by the estimated date of the divergence between the Northern and Southern Elephant Seal species of 800 000 years ago (Table 8).

Table 8. Time of population and species divergence[a].

Comparison	Time (years)
Between	
MQ & HD	270 000
MQ & SG	260 000
HD & SG	23 000
MQ & PV	600 000
HD & PV	215 000
SG & PV	270 000
Sth & Nth Elephant	800 000

[a]Calculated from the Tamura-Nei adjusted mtDNA control region I data.

A separation time between SG and HI of 20 000 years ago (assuming no gene flow), is similar to the estimated peak of the last ice-age

18 000 years ago (CLIMAP 1976). Would the distribution of Southern Elephant Seals during the last ice-age have been different to the current distribution and if so, would this provide any clues to the current geographic structure under the historical association scenario?

Ice cover on subantarctic islands during the last ice-age would likely have made them unsuitable as breeding habitat and would thus have significantly affected the distribution of the Southern Elephant. While some seals can be found on ice during the moulting season, breeding nearly always takes place on beaches (Laws 1960). The estimated extent of winter ice 18 000 years ago was at roughly 50°S (Hays *et al.* 1976) and would have extended north of South Georgia and Heard Island. Kerguelen Island at 52°S contains geological evidence of glaciation, as do other subantarctic islands (LeMasurier and Thomson 1990). Macquarie Island is below 50°S but there is some debate over whether it was glaciated at this time (see Adamson *et al.* 1988 and references therein). The estimated extent of summer ice at roughly 58°S (Hays *et al.* 1976) was below the latitude of most subantarctic islands, although the islands may well have remained frozen over during summer.

If South Georgia, Heard Island, Kerguelen Island and perhaps Macquarie Island did not contain major breeding populations of Southern Elephant Seals during the last ice-age then the main breeding areas were likely to have been further north, probably on the southern coasts of the continental land masses of South America, Africa and Australia. The colony at Península Valdés, the occasional breeding cow seen on the South African coast and the former breeding places on the coasts of South America and Australia (Laws 1960; Carrick and Ingham 1962), may be remnants of this former distribution. In the context of the current geographic structure, it is possible that SG and HI were derived from a common ancestral population that existed off the coast of southern Africa during the last ice-age, whereas the ancestral MI population was located off Australia. Given that SG and PV separated at least 200 000 years ago then it seems unlikely that both the PV and SG populations existed off the east coast of South America during the last ice-age, thus supporting a link between SG and South Africa.

Summary

For the Southern Elephant Seal, geographic structure of the mtDNA and nDNA variation assayed in this study is the result of a combination of mutation, gene flow, drift and historical processes. Selection has played no role in the distribution of these variants. There is strong geographic structure for mtDNA compared with nDNA and this reflects the higher mutation rate of the former, as well as male-biased gene flow and also a minor influence of differences in rate of genetic drift. The mtDNA data reveal that the Macquarie Island population has been isolated from the South Georgia and Heard Island populations for at least 200 000 years and the Península Valdés population has been isolated for at least the same period. The South Georgia and Heard Island populations have a more recent association, for which there are two possible explanations. First, there is ongoing gene flow between SG and HI at the rate of 3–4 females per generation, a rate which is sufficient to maintain similar haplotypes in each population but not to equilibrate haplotype frequencies (Allendorf and Phelps 1981). Second, the SG and HI populations diverged and became isolated as recently as the last ice-age some 20 000 years ago, during which time the distribution of the major populations was likely to have been different to the current distribution.

GENETIC VERSUS DEMOGRAPHIC STRUCTURE

Can we use information on the genetic relationship between populations to understand their current demographic relationships? Genes do not exist independent of the individuals carrying them so that, clearly, genetic structure must reflect to some extent the fate of individuals and thus the demographic structure of populations. However, the distribution and level of genetic variation is a very insensitive indicator of demographic processes. There are three principal reasons for this. The foremost is that the genetic structure of a population, influenced by effective population size (N_e) , mutation (m), migration (m), and selection (s), is always determined by composite parameters involving N_e (i.e., $N_e\,m$, N_em, N_es; see Hartl and Clark 1989). For example, genetic discreteness is a function of N_em, but demographic discreteness is a factor of m and there is not a linear relationship between them. A small amount of interchange from a demographic perspective (e.g., 3–4 individuals exchanged between SG and HI per generation) is a large amount of change from a genetic perspective, where generally if $N_em > 1$ then populations are genetically panmictic. Such populations may be genetically panmictic, but are essentially isolated from a demographic perspective. Conversely, populations of very small effective population size may appear genetically differentiated, while being demographically connected (see Allendorf and Phelps 1981 for further discussion). Second, genetic and demographic processes operate over vastly different time scales. Demographic properties can change within a single generation, whereas genetic properties take much longer. It takes on the order of $1/m$ generations for G_{ST} indices to achieve equilibrium, while overall genetic equilirium takes much longer, on the order of $1/\mu$ generations (Nei *et al.* 1975; Nagylaki 1976; Crow and Aoki 1984). Two populations may have recently become demographically isolated but not yet genetically differentiated. Third, the stochastic nature of the evolutionary process results in a large variance component to any population genetic parameter and this can only be overcome by further sampling of loci (Nei 1987).

For the Southern Elephant Seal, the main interest is in inferring the demographic connectivity between the two declining populations of HI and MI, compared with the stable population of SG. As discussed above, if two populations are shown to be genetically differentiated, then its reasonable to assume that they are demographically isolated, unless N_e is very small. Populations of the southern elephant seal have effective population sizes large enough that they can be considered demographically isolated from each other. This is so for female exchange, in which the proportion of female migrants exchanged per generation (m_f) is about 10^{-5} between most populations and 10^{-4} between SG and HI (Table 6). The nDNA data are also consistent with low levels of gene flow between at least MI and either HI or SG and there are indications of only limited gene flow between HI and SG. This highlights the value of obtaining information on N_e. It allows the parameter of demographic interest, m, to be quantified. Under the historical association scenario, the level of interchange between populations is likely to be even less than under the equilibrium, current gene flow scenario, so that demographic isolation of the populations is supported irrespective of the model of population structure.

Demographic discreteness of populations is also implied by ecological studies. Tagging studies have shown that Southern Elephant Seals tend to return to the natal island (Condy and Bester 1975; van Aarde and Pascal 1980). This appears to be so during all haul-outs but the tendency is most pronounced for the adults during the breeding season (Carrick *et al.* 1962a). On Macquarie Island it was estimated that 60% of adult males and 77% of adult females return to within four km of their birthsite during the breeding season (Nicholls 1970). It was also noted, however, that seals prove to be more mobile the longer their known history (Carrick *et al.* 1962a), but this appears to apply mainly to males. By contrast, dispersal did not change with age in the 4–11 year old females (Nicholls 1970), again supporting the inference of male-biased gene flow discussed above. The tagging studies are discussed further below. Also, other studies have shown that the feeding areas of the MI, HI and SG populations do not overlap (Hindell *et al.* 1991b; Slip and Burton 1993; McConnell *et al.* 1993).

CONTRIBUTION OF MOLECULAR GENETICS TO MANAGEMENT

The role of genetics in the conservation and management of biological resources is controversial (Lande 1988; Soulé and Mills 1992; Caughley 1994). The primary reason for the controversy is the historical focus of genetics on just one specific area of conservation biology — the consequences of small population size and the relative importance of genetic versus demographic factors in susceptibility to extinction. The central genetic factor in this debate is the standing level of genetic variation within a species and its relationship with current fitness and future evolutionary potential. In the early days of the emerging science of conservation biology, genetic factors were considered to be important (Soulé and Wilcox 1980) and, as genetic diversity is relatively easy to measure, gained widespread acceptance (e.g., O'Brien *et al.* 1985). The problem, though, is that both theoretical and empirical studies have since shown that the relationship between genetic diversity and fitness is not clear. For example, the size of the Northern Elephant Seal population increased rapidly following cessation of hunting (Bartholomew and Hubbs 1960; Hoelzel *et al.* 1993) despite having no detectable protein variation (Bonnell and Selander 1974). Not surprisingly then, it became considered by many that genetics has little relevance for the conservation and management of species. However, there is more to management and conservation than the consequences of small population size (Moritz 1994).

Two major directions in conservation biology were identified by Caughley (1994). The first is the small-population paradigm, which is described in the above paragraph. The second is the declining-population paradigm, which is concerned with identifying the agents of decline where such agents are not intrinsic properties of the population but rather external factors such as overkill, habitat change, impact of introduced species and chains of extinction. Caughley suggested that genetics did not have a role to play in addressing the declining-population paradigm because this paradigm is not concerned with the consequences of maintaining a population at a small size but rather with avoiding that situation altogether. Again, this is a misguided view of what genetics, especially molecular population genetics, can offer conservation biology and mirrors that which resulted in the controversy detailed above. For example, to identify the agents of decline, it would seem important to first identify the demographically independent units of the declining population and, for this, molecular population genetics is a valuable tool.

Since the 1950s, Southern Elephant Seal populations in the south Indian Ocean (Heard Island, Iles Kerguelen, Iles Crozet, Marion Island) and South Pacific Ocean (Macquarie Island) have declined by up to 50% while, in the South Atlantic Ocean, the South Georgia population has remained stable and the Península Valdés population has increased (Bester and Lenglart 1982; Condy 1984; Hindell and Burton 1987; McCann and Rothery 1988; Campagna and Lewis 1992). Hindell and Burton (1987) noted that the declines could be attributed to a myriad of agents that could be grouped into four, not mutually exclusive, categories: (i) intrinsic factors such as density-dependent population fluctuations; (ii) an increase in predation rates by the dominant predator, the Killer Whale *Orcinus orca*; (iii) depletion of resources, for example in numbers of the key food web species Krill *Euphausia* spp.; and (iv) environmental changes, for example by water temperature affecting levels of productivity. A key factor in determining the agent of decline is an understanding of the level to which the declining populations (represented here by MI and HI) are demographically discrete. If the declining populations are demographically discrete then the agent is operating on both populations. On the other hand, if the declining populations are behaving as a demographically panmictic unit then the agent may be operating on just one of the populations.

Tagging studies have implied that the MI and HI populations are demographically isolated, but have several shortcomings that caution against sole reliance on their results. First, the rate of tag return was very low. Precise figures are available for the study at Iles Kerguelen where the rate of tag return was 3.61% (van Aarde and Pascal 1980) and for MI the rate was less than 1% (M. Hindell, pers. comm.). Second, both tagging and search effort were extremely variable between years and between islands. Tagging on MI and HI was conducted mostly during the 1950s, where 6 661 and 1 486 individuals were tagged respectively (Carrick and Ingham 1962; Carrick *et al.* 1962a; Nicholls 1970). On MI, searches were concentrated in the isthmus area and only two systematic whole island counts and searches were made, in 1952 and 1959 (Carrick and Ingham 1962). On HI, just one whole island count was made in 1951, before the ANARE station closed in 1955 (Carrick and Ingham 1962). On Iles Kerguelen, the other dominant population in the south Indian Ocean, roughly 12 000 individuals were tagged, but this was during the 1970s (van Aarde and Pascal 1980; Pascal 1986). On SG, limited tagging was conducted during the 1920s and 1950s (Carrick *et al.* 1962a). Therefore, the ecological studies on dispersal between oceanic regions have to be viewed in light of the facts that the rate of tag return was low, that both tagging and searching were limited and that the studies in different oceanic regions were undertaken at different times.

This study has confirmed that the major populations of the Southern Elephant Seal are demographically independent. The conclusion, therefore, is that the agent of decline is operating in both the south Indian Ocean and south Pacific Ocean populations. The clue, then, to identifying the agent of decline is to identify which ecological factors are shared by populations in the south Pacific and south Indian Oceans, but not with populations in the south Atlantic Ocean. A decline in numbers of prey species is unlikely to be such a factor. The feeding grounds of the MI, HI and SG populations do not overlap (Hindell *et al.* 1991b; McConnell *et al.* 1993; Slip and Burton 1993) and, although little is known of the diet of the Southern Elephant Seal, the evidence suggests that HI and MI animals are feeding on different squid and fish species (Green and Burton 1993). A link with changes in numbers of krill is unlikely given that most high level consumers in the Antarctic are increasing (summarized in Hindell and Burton 1987). Other factors, such as environmental changes or an increase in predation rates cannot be addressed as easily, although they are, at the moment, difficult to reconcile with the observation of declines in the south Indian Ocean and south Pacific Ocean populations versus stability or increase in the south Atlantic Ocean populations.

However, one factor shared between the south Indian Ocean and south Pacific Ocean populations, but not with the south Atlantic Ocean populations is the degree of historical exploitation. Sealing activity was intense in the south Indian and south Pacific populations, but was controlled at SG and did not occur at PV (Laws 1960; van Aarde 1980; Vergani 1985; Hindell and Burton 1988). It was proposed by Hindell (1991) that the decline in the south Indian and south Pacific populations was the result of intrinsic population fluctuations in which the recovery after intense sealing activities resulted in an overshoot beyond the carrying capacity of the populations. The correction to pre-sealing numbers following the overshoot is the decline observed over the past 30 years. This intrinsic population fluctuation hypothesis is supported by the results of this study showing that all the major populations are demographically discrete.

ACKNOWLEDGEMENTS

This work was funded principally by grants to C. Moritz, R. W. Slade and H. I. McCallum from the Antarctic Science Advisory Committee of Australia. The MI and HI samples were provided by H. Burton (Australian Antarctic Division). The SG samples were provided by A. R. Hoelzel (National Cancer Institute, USA). The *A. forsteri* and *H. leptonyx* samples were provided by H. Burton (Australian Antarctic Division). The *M. angustirostris* sample was provided by H. Stanley

(Institute of Zoology, London). The *P. vitulina, L. weddelli* and *C. familiaris* samples were provided by Ú. Árnason (University of Lund, Sweden). S. Degnan provided a thorough review of the manuscript.

REFERENCES

Adamson, D. A., Selkirk, P. M. and Colhoun, E. A., 1988. Landforms of Aeolian, tectonic and marine origin in the Bauer Bay-Sandy Bay region of subantarctic Macquarie Island. *Pap. Proc. R. Soc. Tas.* **122:** 65–82.

Allendorf, F. W. and Phelps, S. R., 1981. Use of allelic frequencies to describe population structure. *Can. J. Fish. Aquat. Sci.* **38:** 1507–514.

Amos, W., Twiss, S., Pomeroy, P. P. and Anderson, S. S., 1993. Male mating success and paternity in the grey seal, *Halichoerus grypus*: a study using DNA fingerprinting. *Proc. R. Soc., Lond.* **B252:** 199–207.

Aquadro, C. F., 1992. Why is the genome variable? Insights from *Drosophila. Trends Genet.* **8:** 355–62.

Avise, J. C., Arnold, J., Ball, R. M., Bermingham, E., Lamb, T., Neigel, J. E., Reeb, C. A. and Saunders, N. C., 1987. Intraspecific phylogeography: The mitochondrial DNA bridge between population genetics and systematics. *Ann. Rev. Ecol. Syst.* **18:** 489–522.

Baker, C. S., Perry, A., Bannister, J. L., Weinrich, M. T., Abernethy, R. B., Calambokidis, J., Lien, J., Lambertsen, R. H., Urbán Ramírez, J., Vasquez, O., Clapham, P. J., Alling, A., O'Brien, S. J. and Palumbi, S. R., 1993. Abundant mitochondrial DNA variation and world-wide population structure in humpback whales. *Proc. Natl. Acad. Sci., USA.* **90:** 8239–243.

Bartholomew, G. A. and Hubbs, C. L., 1960. Population growth and seasonal movements of the northern elephant seal, *Mirounga angustirostris. Mammalia* **24:** 313–24.

Bartley, D. M. and Gall, G. A. E., 1990. Genetic structure and gene flow in chinook salmon populations of California. *Trans. Amer. Fish. Soc.* **119:** 55–71.

Begon, M., Harper, J. L. and Townsend, C. R., 1986. Ecology. Blackwell Scientific Publications: Oxford.

Bester, M. N., 1988. Marking and monitoring studies of the Kerguelen stock of southern elephant seals *Mirounga leonina* and their bearing on biological research in the Vestfold Hills. *Hydrobiologia* **165:** 269–77.

Bester, M. N. and Lenglart, P-Y., 1982. An analysis of the southern elephant seal *Mirounga leonina* breeding population at Kerguelen. *S. Afr. J. Antarct. Res.* **12:** 11–16.

Birky, C. W. Jr., Maruyama, T. and Fuerst, P., 1983. An approach to population and evolutionary genetic theory for genes in mitochondria and chloroplasts and some results. *Genetics* **103:** 513–27.

Bonnell, M. L. and Selander, R. K., 1974. Elephant seals: genetic variation and near extinction. *Science* **184:** 908–09.

Bowers, N., Stauffer, J. R., Kocher, T. D., 1994. Intra- and inter-specific mitochondrial DNA sequence variation within two species of rock-dwelling cichlids (Teleostei: Cichlidae) from Lake Malawi, Africa. *Mol. Phyl. Evol.* **3:** 75–82.

Brown, W. M., Prager, E. M., Wang, A. and Wilson, A. C., 1982. Mitochondrial DNA sequences in primates: tempo and mode of evolution. *J. Mol. Evol.* **18:** 225–39.

Brown, J. R., Beckenbach, A. T. and Smith, M. J., 1993. Intraspecific DNA sequence variation of the mitochondrial control region of white sturgeon *(Acipenser transmontanus). Mol. Biol. Evol.* **10:** 326–41.

Bryden, M. M., 1968. Control of growth in two populations of elephant seals. *Nature* **217:** 1106–108.

Bryden, M. M. and Lim, G. H. K., 1969. Blood parameters of the southern elephant seal (*Mirounga leonina,* Linn.) in relation to diving. *Comp. Biochem. Physiol.* **28:** 139–48.

Burton, H. R., 1985. Tagging studies of male southern elephant seals (*Mirounga leonina* L.) in the Vestfold Hills area, Antarctica and some aspects of their behaviour. Pp. 19–30 *in* Studies of Sea Mammals in South Latitudes, ed by J. K. Ling and M. M. Bryden. Proc. symp. 52nd. ANZAAS congress: South Australian Museum.

Campagna, C. and Lewis, M., 1992. Growth and distribution of a southern elephant seal colony. *Mar. Mamm. Sci.* **8:** 387–96.

Campagna, C., Lewis, M. and Baldi, R., 1993. Breeding biology of southern elephant seals in Patagonia. *Mar. Mamm. Sci.* **9:** 34–47.

Carrick, R. and Ingham, S. E., 1962. Studies on the southern elephant seal, *Mirounga leonina* (L.). I. Introduction to the series. *CSIRO Wildl. Res.* **7:** 89–101.

Carrick, R., Csordas, S. E., Ingham, S. E. and Keith, K., 1962a. Studies on the southern elephant seal, *Mirounga leonina* (L.). III. The annual cycle in relation to age and sex. *CSIRO Wildl. Res.* **7:** 119–60.

Carrick, R., Csordas, S. E. and Ingham, S. E., 1962b. Studies on the southern elephant seal, *Mirounga leonina* (L.). IV. Breeding and development. *CSIRO Wildl. Res.* **7:** 161–97.

Caughley, G., 1994. Directions in conservation biology. *J. Anim. Ecol.* **63:** 215–44.

CLIMAP Project Members, 1976. The surface of the ice-age earth. *Science* **191:** 1131–137

Condy, P. R., 1984. The population of the southern elephant seal, *Mirounga leonina,* at Marion Island, 1973–1983. *S. Afr. J. Sci.* **80:** 26–27.

Condy, P. R. and Bester, M. N., 1975. Notes on the tagging of seals on Marion and Gough Islands. *Sth. Afr. J. Antarct. Res.* **5:** 45–47.

Crow, J. F. and Aoki, K., 1984. Group selection for a polygenic behavioral trait: Estimating the degree of population subdivision. *Proc. Natl. Acad. Sci., USA.* **81:** 6073–077.

Easteal, S., 1990. The pattern of mammalian evolution and the relative rate of molecular evolution. *Genetics* **124:** 165–73.

Edwards, S. V., 1993. Long-distance gene flow in a co-operative breeder detected in genealogies of mitochondrial DNA sequences. *Proc. R. Soc., Lond.* **B252:** 177–85.

Epperson, B. K., 1994. Spatial and space-time correlations in systems of subpopulations with stochastic migration. *Theor. Pop. Biol.* **46:** 160–97.

Excoffier, L., Smouse, P. E. and Quattro, J. M., 1992. Analysis of molecular variance inferred from metric distances among DNA haplotypes: Application to human mitochondrial DNA restriction data. *Genetics* **131:** 479–91.

Frankham, R., 1994. Conservation of genetic diversity for animal improvement. *Proc. 5th World Cong. Genet. Applied to Lives. Prod.* **21:** 385–92.

Fu, Y-X., 1994. A phylogenetic estimator of effective population size or mutation rate. *Genetics* **136:** 685–92.

Fu, Y-X. and Li, W-H., 1993. Statistical tests of neutrality of mutations. *Genetics* **133:** 693–709.

Gales, N. J., Adams, M. and Burton, H. R., 1989. Genetic relatedness of two populations of the southern elephant seal, *Mirounga leonina. Mar. Mamm. Sci.* **5:** 57–67.

Gall, G. A. E., Bartley, D., Bentley, B., Brodziak, J., Gomulkiewicz, R. and Mangel, M., 1992. Geographic variation in population genetic structure of chinook salmon from California and Oregon. *Fish. Bull., US.* **90:** 77–100.

Gharrett, A. J., Shirley, S. M. and Tromble, G. R., 1987. Genetic relationships among populations of Alaskan chinook salmon *(Oncorhynchus tshawytscha)*. *Can. J. Fish. Aquat. Sci.* **44:** 765–74.

Green, K. and Burton, H. R., 1993. Comparison of the stomach contents of southern elephant seals, *Mirounga leonina*, at Macquarie and Heard Islands. *Mar. Mamm. Sci.* **9:** 10–22.

Hale, L. R. and Hoffmann, A. A., 1990. Mitochondrial DNA polymorphism and cytoplasmic incompatibility in natural populations of *Drosophila simulans*. *Evolution* **44:** 1383–386.

Harris, A. S., Young, J. S. F. and Wright, J. M., 1991. DNA fingerprinting of harbour seals *(Phoca vitulina concolor)*: male mating behaviour may not be a reliable indicator of reproductive success. *Can. J. Zool.* **69:** 1862–866.

Hartl, D. L. and Clark, A. G., 1989. Principles of population genetics. Sinauer Associates: Massachusetts.

Hays, J. D., Lozano, J. A., Shackleton, N. and Irving, G., 1976. Reconstruction of the Atlantic and western Indian ocean sectors of the 18 000 B.P. Antarctic ocean. *Geol. Soc. Amer. Mem.* **145:** 337–72.

Hindell, M. A., 1991. Some life-history parameters of a declining population of southern elephant seals, *Mirounga leonina*. *J. Anim. Ecol.* **60:** 119–34.

Hindell, M. A. and Burton, H. R., 1987. Past and present status of the southern elephant seal *(Mirounga leonina)* at Macquarie Island. *J. Zool., Lond.* **213:** 365–80.

Hindell, M. A. and Burton, H. R., 1988. The history of the elephant seal industry at Macquarie Island and an estimate of the pre-sealing numbers. *Pap. Proc. R. Soc. Tas.* **122:** 159–76.

Hindell, M. A., Slip, D. J. and Burton, H. R., 1991a. The diving behaviour of adult male and female southern elephant seals, *Mirounga leonina* (Pinnipedia: Phocidae). *Aust. J. Zool.* **39:** 595–619.

Hindell, M. A., Burton, H. R. and Slip, D. J., 1991b. Foraging areas of southern elephant seals, *Mirounga leonina*, as inferred from water temperature data. *Aust. J. Mar. Freshw. Res.* **42:** 115–28.

Hoelzel, A. R., Halley, J., O'Brien, S. J., Campagna, C., Arnbom, T., Le Boeuf, B., Ralls, K. and Dover, G. A., 1993. Elephant seal genetic variation and the use of simulation models to investigate historical population bottlenecks. *J. Hered.* **84:** 443–49.

Horai, S., Hayasaka, K., Kondo, R., Tsugane, K. and Takahata, N., 1995. Recent African origin of modern humans revealed by complete sequences of hominoid mitochondrial DNAs. *Proc. Natl. Acad. Sci., USA.* **92:** 532–36.

Hudson, R. R., Slatkin, M. and Maddison, W. P., 1992. Estimation of levels of gene flow from DNA sequence data. *Genetics* **132:** 583–89.

Ishida, N., Oyunsuren, T., Mashima, S., Mukoyama, H. and Saitou, N., 1995. Mitochondrial DNA sequences of various species of the genus *Equus* with special reference to the phylogenetic relationship between Przewalskii's wild horse and domestic horse. *J. Mol. Evol.* **41:** 180–88.

Jukes, T. H. and Cantor, C. R., 1969. Evolution of protein molecules. Pp. 21–132 *in* Mammalian Protein Metabolism, vol. III, ed by H. N. Munro. Academic Press: New York.

Kimura, M. and Crow, J. F., 1963. The measurement of effective population number. *Evolution* **17:** 279–88.

Kreitman, M. E. and Aguadé, M., 1986. Excess polymorphism at the *Adh* locus in *Drosophila melanogaster*. *Genetics* **114:** 93–110.

Lambert, D. M., Millar, C. D., Jack, K. Anderson, S. and Craig, J. L., 1994. Single- and multilocus DNA fingerprinting of communally breeding pukeko: Do copulations or dominance ensure reproductive success? *Proc. Natl. Acad. Sci., USA.* **91:** 9641–645.

Lande, R., 1988. Genetics and demography in biological conservation. *Science* **241:** 1455–460.

Laws, R. M., 1960. The southern elephant seal (*Mirounga leonina* Linn) at South Georgia. *Norsk Hvalf. Tid.* **10:** 466–76; **11:** 520–42.

Le Boeuf, B. J., Naito, Y., Huntley, A. C. and Asaga, T., 1989. Prolonged, continuous, deep diving by northern elephant seals. *Can. J. Zool.* **67:** 2514–519.

LeMasurier, W. E. and Thomson, J. W., eds., 1990. Volcanoes of the antarctic plate and southern oceans. American Geophysical Union: Washington, DC.

Li, W-H., Tanimura, M. and Sharp, P. M., 1987. An evaluation of the molecular clock hypothesis using mammalian DNA sequences. *J. Mol. Evol.* **25:** 330–42.

Li, W-H. and Sadler, L. A., 1991. Low nucleotide diversity in man. *Genetics* **129:** 513–23.

Ling, J. K. and Bryden, M. M., 1992. *Mirounga leonina*. No. 391. *In* Mammalian Species, ed by C. S. Hood. The American Society of Mammalogists.

Lydekker, R., 1909. On the skull-characters in the southern sea-elephant. *Proc. Zool. Soc., Lond.* **1909:** 600–06.

Lynch, M. and Crease, T. J., 1990. The analysis of population survey data on DNA sequence variation. *Mol. Biol. Evol.* **7:** 377–94.

MacRae, A. F. and Anderson, W. W., 1988. Evidence for non-neutrality of mitochondrial DNA haplotypes in *Drosophila pseudoobscura*. *Genetics* **120:** 485–94.

MacRae, A. F. and Anderson, W. W., 1990. Can mating preferences explain changes in mtDNA haplotype frequencies? *Genetics* **124:** 999–1001.

Maddison, W. P. and Maddison, D. R., 1992. MacClade: Analysis of phylogeny and character evolution. Sinauer Associates: Massachusetts.

Maddison, W. P. and Slatkin, M., 1991. Null models for the number of evolutionary steps in a character on a phylogenetic tree. *Evolution* **45:** 1184–197.

Maldonado, J. E., Orta Davila, F., Stewart, B. S., Geffen, E. and Wayne, R. K., 1995. Intraspecific genetic differentiation in California sea lions *(Zalophus californianus)* from southern California and the Gulf of California. *Mar. Mamm. Sci.* **11:** 46–58.

Maruyama, T., 1970. Analysis of population structure. I. One-dimensional stepping-stone models of finite length. *Ann. Hum. Genet., Lond.* **34:** 201–19

McCann, T. S., 1985. Size, status and demography of southern elephant seals *(Mirounga leonina)* populations. Pp. 1–17 *in* Studies of Sea Mammals in South Latitudes, ed by J. K. Ling and M. M. Bryden. South Australian Museum: Adelaide.

McCann, T. S. and Rothery, P., 1988. Population size and status of the southern elephant seal *(Mirounga leonina)* at South Georgia, 1951–1985. *Pol. Biol.* **8:** 305–09.

McConnell, B. J., Fedak, M. and Hunter, C., 1993. Foraging strategy of southern elephant seals. Abstract, Tenth Biennial Conference on Marine Mammals, Galveston, Texas.

McDermid, E. M., Ananthakrishnan, R. and Agar, N. S., 1972. Electrophoretic investigation of plasma and red cell proteins and enzymes of Macquarie Island elephant seals. *Anim. Blood Grps. biochem. Genet.* **3:** 85–94.

Moritz, C., 1994. Applications of mitochondrial DNA analysis in conservation: A critical review. *Mol. Ecol.* **3:** 401–11.

Moritz, C., Dowling, T. E. and Brown, W. M., 1987. Evolution of animal mitochondrial DNA: Relevance for population biology and systematics. *Ann. Rev. Ecol. Syst.* **18:** 269–92.

Nachman, M. W., Boyer, S. N. and Aquadro, C. F., 1994a. Nonneutral evolution at the mitochondrial NADH dehydrogenase subunit 3 gene in mice. *Proc. Natl. Acad. Sci., USA.* **91:** 6364–368.

Nachman, M. W., Boyer, S. N., Searle, J. B. and Aquadro, C. F., 1994b. Mitochondrial DNA variation and the evolution of robertsonian chromosomal races of house mice, *Mus domesticus. Genetics* **136:** 1105–120.

Nagylaki, T., 1976. The decay of genetic variability in geographically structured populations. II. *Theor. Pop. Biol.* **10:** 70–82.

Nei, M., 1987. Molecular Evolutionary Genetics. Columbia University Press: New York.

Nei, M., Maruyama, T. and Chakraborty, R., 1975. The bottleneck effect and genetic variability in populations. *Evolution* **29:** 1–10.

Nevo, E., Beiles, A. and Ben-Shlomo, R., 1984. The evolutionary significance of genetic diversity: Ecological, demographic and life history correlates. Pp. 13–213 *in* Evolutionary Dynamics of Genetic Diversity, ed by G. S. Mani. Lecture Notes in Biomathematics, 53.

Nicholls, D. G., 1970. Dispersal and dispersion in relation to the birthsite of the southern elephant seal, *Mirounga leonina* (L.), of Macquarie Island. *Mammalia* **34:** 598–616.

Nigro, L., 1994. Nuclear background affects frequency dynamics of mitochondrial DNA variants in *Drosophila simulans. Heredity* **72:** 582–86

O'Brien, S. J., Roelke, M. E., Marker, I., Newman, A., Winkler, C. A., Meltzer, D., Colly, I., Evermann, J. F., Bush, M. and Wildt, D. E., 1985. Genetic basis for species vulnerability in the cheetah. *Science* **227:** 1428–434.

Pascal, M., 1986. Numerical changes in the population of elephant seals (*Mirounga leonina* L) in the Kerguelen Archipelago during the past 30 years. Pp. 170–86 *in* Marine Mammal Fishery Interactions, ed by J. R. Beddington, R. J. H. Beverton and D. M. Lavigne. George Allen and Unwin: London.

Pemberton, J. M., Albon, S. D., Guiness, F. E., Clutton-Brock, T. H. and Dover, G. A., 1992. Behavioural estimates of male mating success tested by DNA fingerprinting in a polygynous mammal. *Behav. Ecol.* **3:** 66–75.

Ray, C. E., 1976. Geography of phocid evolution. *Syst. Zool.* **25:** 391–406.

Roff, D. A. and Bentzen, P., 1989. The statistical analysis of mitochondrial DNA polymorphisms: c^2 and the problem of small samples. *Mol. Biol. Evol.* **6:** 539–45.

Rzhetsky, A. and Nei, M., 1992. A sample method for estimating and testing minimum-evolution trees. *Mol. Biol. Evol.* **9:** 945–67.

Sarich, V. M., 1985. Rodent macromolecular systematics. Pp. 423–52 *in* Evolutionary Relationships Among Rodents: A Multidisciplinary Analysis, ed by W. P. Luckett and J-L. Hartenberger. NATO ASI Series, Plenum Press: New York.

Seal, U. S., Erickson, A. W., Siniff, D. B. and Hoffman, R. J., 1971. Biochemical, population genetic, phylogenetic and cytological studies of Antarctic seal species. Pp. 77–95 *in* Symposium on Antarctic Ice and Water Masses, ed by G. Deacon. Science Committee for Antarctic Research: Cambridge.

Shaughnessy, P. D., 1974. An electrophoretic study of blood and milk proteins of the southern elephant seal, *Mirounga leonina. J. Mammal.* **55:** 796–808.

Singh, R. S. and Hale, L. R., 1990. Are mitochondrial DNA variants selectively non-neutral. *Genetics* **124:** 995–97.

Slade, R. W., Moritz, C., Heideman. A. and Hale, P. T., 1993. Rapid assessment of single copy nuclear DNA variation in diverse species. *Mol. Ecol.* **2:** 359–73.

Slade, R. W., Moritz, C. and Heideman, A., 1994. Multiple nuclear-gene phylogenies: Application to pinnipeds and comparison with a mitochondrial DNA gene phylogeny. *Mol. Biol. Evol.* **11:** 341–56.

Slip, D. J. and Burton, H. R., 1993. Movements and diving behaviour of southern elephant seals from Heard Island. Abstract, Tenth Biennial Conference on Marine Mammals, Galveston, Texas.

Soulé, M. E. and Wilcox, B. A., eds, 1980. Conservation Biology. Sinauer Associates: Sunderland, MA.

Soulé, M. E. and Mills, L. S., 1992. Conservation genetics and conservation biology: A troubled marriage. Pp. 55–69 *in* Conservation of Biodiversity for Sustainable Development, ed by O. T. Sandlund, K. Hindar and A. H. D. Brown. Scandinavian University Press: Oslo.

Stoneking, M., Sherry, S. T., Redd, A. J. and Vigilant, L., 1992. New approaches to dating suggest a recent age for the human mtDNA ancestor. *Philos. Trans. R. Soc. Lond.* **B337:** 167–75.

Tajima, F., 1983. Evolutionary relationship of DNA sequences in finite populations. *Genetics* **105:** 437–60.

Tajima, F., 1989. Statistical method for testing the neutral mutation hypothesis by DNA polymorphism. *Genetics* **123:** 585–95.

Tajima, F., 1993. Statistical analysis of DNA polymorphism. *Jpn. J. Genet.* **68:** 567–95.

Takahata, N., 1993. Allelic genealogy and human evolution. *Mol. Biol. Evol.* **10:** 2–22.

Tamura, K. and Nei, M., 1993. Estimation of the number of nucleotide substitutions in the control region of mitochondrial DNA in humans and chimpanzees. *Mol. Biol. Evol.* **10:** 512–26.

Thomas, W. K., Pääbo, S., Villablanca, F. X. and Wilson, A. C., 1990. Spatial and temporal continuity of kangaroo rat populations shown by sequencing mitochondrial DNA from museum specimens. *J. Mol. Evol.* **31:** 101–12.

Utter, F. M., Waples, R. S. and Teel, D. J., 1992. Genetic isolation of previously indistinguishable chinook salmon populations of the Snake and Klamath rivers: Limitations of negative data. *Fish. Bull., US.* **90:** 770–77.

van Aarde, R. J., 1980. Fluctuations in the population of southern elephant seals *Mirounga leonina* at Kerguelen Island. *S. Afr. J. Zool.* **15:** 99–106.

van Aarde, R. J. and Pascal, M., 1980. Marking southern elephant seals on Iles Kerguelen. *Pol. Rec.* **20:** 62–65.

Vergani, D. F., 1985. Comparative study of populations in Antarctica and Patagonia of the southern elephant seal *Mirounga leonina* (Linné, 1758) and its methodology. *Direccion Nacional Del Antartico Instituto Antartico Argentino,* Publication **15.**

Vigilant, L., Stoneking, M., Harpending, H., Hawkes, K. and Wilson, A. C., 1991. African populations and the evolution of human mitochondrial DNA. *Science* **253:** 1503–507.

Waples, R. S., 1988. Estimation of allele frequencies at isoloci. *Genetics* **118:** 371–84.

Waples, R. S., Aebersold, P. B., 1990. Treatment of data for duplicated gene loci in mixed-stock fishery analysis. *Can. J. Fish. Aquat. Sci.* **47:** 2092–098.

Ward, R. H., Frazier, B. L., Dew-Jager, K. and Pääbo, S., 1991. Extensive mitochondrial diversity within a single Amerindian tribe. *Proc. Natl. Acad. Sci., USA.* **88:** 8720–724.

Weber, J. L. and Wong, C., 1993. Mutation of human short tandem repeats. *Hum. Mol. Genet.* **2:** 1123–128.

Wenink, P. W., Baker, A. J. and Tilanus, M. G. J., 1993. Hypervariable-control-region sequences reveal global population structuring in a long-distance migrant shorebird, the Dunlin *(Calidris alpina). Proc. Natl. Acad. Sci., USA.* **90:** 94–98.

Wilson, A. C., Cann, R. L., Carr, S. M., George, M., Gyllensten, U. B., Helm-Bychowski, K. M., Higuchi, R. G., Palumbi, S. R., Prager, E. M., Sage, R. D. and Stoneking, M., 1985. Mitochondrial DNA and two perspectives on evolutionary genetics. *Biol. J. Linn. Soc.* **26:** 375–400.

Wilson, A. C., Ochman, H. and Prager, E. M., 1987. Molecular time scale for evolution. *Trends Genet.* **3:** 241–47.

Wright, S., 1931. Evolution in mendelian populations. *Genetics* **16:** 97–159.

Wu, C-I., 1991. Inferences of species phylogeny in relation to segregation of ancient polymorphisms. *Genetics* **127:** 429–35.

A yacht-based cetacean research cruise to Antarctic waters south of Australia

P. C. GILL[1]

During the 1992–93 austral summer cetacean research was conducted from a yacht while traversing Antarctic coastal waters from 162°E to 90°E. Research focused on the opportunistic use of non-lethal techniques such as behavioural observation, acoustic monitoring and photo-identification. There was no attempt to survey cetacean abundance. A total of 55 Baleen and 143 Toothed Whale (including dolphins) sightings were made. South of 60°S, 84% of cetaceans of identified species were Killer or Minke Whales. Only 17 large whale sightings were made. Possible reasons for the observed paucity of large cetaceans include the limitations of using such a small vessel. Yachts can be used to complement existing research in the Southern Ocean Whale Sanctuary by studying social, acoustic and feeding behaviour, interspecific interactions, habitat requirements and individual movements. Conventional vessels may be unsuitable or too expensive for such research.

Key words: Yacht survey, Antarctic, Southern Ocean Whale Sanctuary, Cetacean research.

INTRODUCTION

WATERS south of the Antarctic Convergence are the summer feeding ground of seven species of mysticete cetaceans, and eight species of odontocetes, including small cetaceans (Brown and Lockyer 1984). When the sea ice edge retreats in early summer, Baleen Whales move south to feed on euphausiids, copepods, and other invertebrates (Mackintosh 1970; Kawamura 1980), and Sperm and Beaked Whales to feed on squid and fish (Matthews 1938). The over-exploitation of this seasonal abundance of cetaceans during this century is well-documented (e.g., Brown 1963; Tonnessen and Johnsen 1982). Antarctic whaling generated volumes of useful data such as the distribution of whales in the Antarctic (Omura 1973; Mizroch *et al.* 1985); an understanding of migration routes between the Antarctic and other areas, from *Discovery* marking (Chittleborough 1959; Dawbin 1966); and a mass of anatomical and biological data (e.g., Mackintosh and Wheeler 1929; Matthews 1938; Mackintosh 1965; Kawamura 1980).

However, those aspects of the biology of living cetaceans was little studied during the commercial whaling era. Since the whaling "moratorium" was implemented in 1986, Antarctic cetacean research has consisted largely of lethal research, or sightings surveys, many of the latter under the aegis of the IWC/IDCR Minke Whale Assessment Cruises (Kasamatsu *et al.* 1988). Sightings surveys focus on distribution and abundance of cetaceans, in order to monitor the recovery of stocks from whaling, and to estimate potential for future whaling. Non-lethal research techniques, such as photo-identification and tissue biopsy, have been included in sightings surveys in recent years. Vessels typically used are whale catcher type or larger. Base resupply vessels have sometimes been used for cetacean sightings programmes (Parker 1978), as well as for marine science programmes. Antarctic shipping, while extensively used for population studies of whales, has very rarely been used for behavioural or acoustic studies (e.g., Thomas *et al.* 1981). The remoteness of Antarctic waters has been a barrier to more varied studies of cetacean biology.

Yachts have been increasingly used for cetacean research in other parts of the world during the past decade. This paper describes a cruise through Antarctic coastal waters to the south of Australia during the summer of 1992–93. Its aim was to use a small sailing vessel for opportunistic Antarctic cetacean research, employing such techniques as photo-identification, acoustic monitoring, and behavioural observation. Humpback Whales *Megaptera novaeangliae* were of special interest but other species were to be studied opportunistically.

VOYAGE DESCRIPTION AND METHODS

The 15-metre steel sloop *Iniquity*, with six crew, was equipped with GPS navigation, radar, weatherfax, depth sounder (100 m), and seawater thermometer. The vessel's speed was usually between 5 and 8 knots (9.3–14.8 km h^{-1}) under sail or power. It left Hobart, Tasmania, on 31 December 1992 and sailed via Macquarie Island for the Balleny Islands (which extend south-east from 66°15′S, 162°10′E). Due to pack ice the vessel was unable to approach the Balleny Islands, where past heavy catching of Humpbacks and other species took place (Mizroch *et al.* 1985). From the region of the Ballenys it sailed west at about latitude 64–65°S, to a western limit of 63°50′S, 90°E, reached on 18 February 1993. The vessel then returned to Adelaide, South Australia on 9 March

[1]Oceanic Research Foundation, c/- 148 Copeland Road East, Beecroft, New South Wales, Australia 2119.

Pages 30–39 *in* MARINE MAMMAL RESEARCH IN THE SOUTHERN HEMISPHERE VOLUME 1: STATUS, ECOLOGY AND MEDICINE ed by M. Hindell and C. Kemper. Surrey Beatty & Sons, Chipping Norton. 1997.

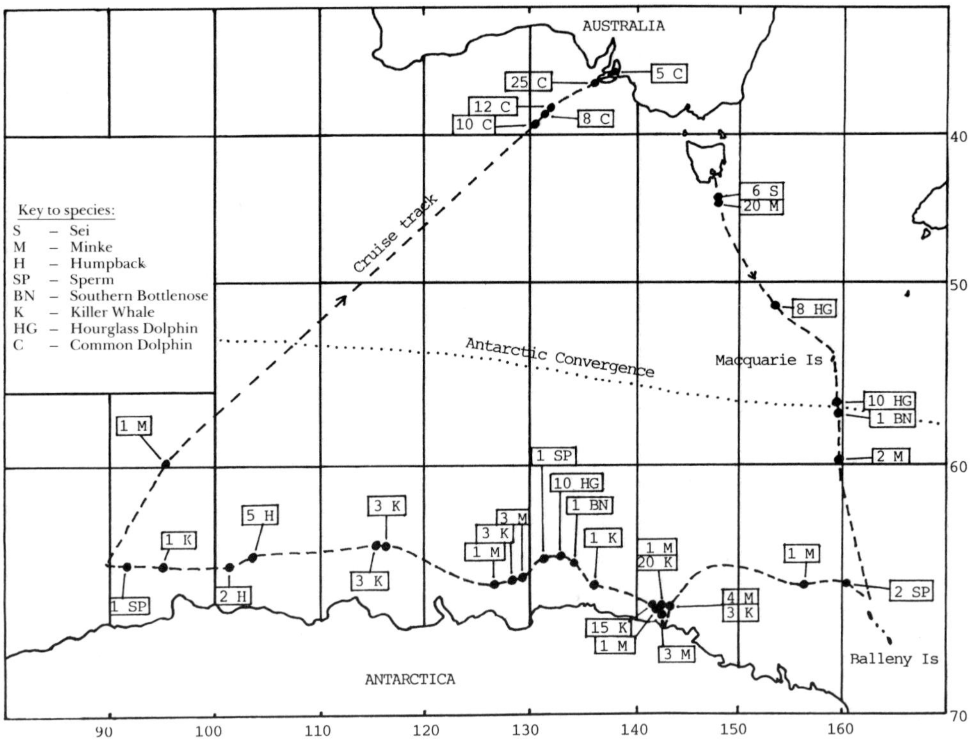

Fig. 1. Cruise track of the yacht *Iniquity*, and locations of cetaceans encountered.

1993. The voyage covered 12 765 km, of which about 5 550 km were south of 60°S, between longitudes 162° and 90°E (see Fig. 1). This covered significant portions of Antarctic Whaling Management Areas IV (70–130°E) and V (130°E–170°W). Bad weather frequently made cetacean sightings impossible during the return voyage to Australia.

A watch was kept for cetaceans, usually from deck level (3 m a.s.l.) and occasionally from a point 12 m up the mast. The two people on watch also attended to sailing the vessel but prudent sailing in Antarctic waters requires an alert watch to be kept at all times as a matter of course. The cruise was conducted in "closing mode" (Joyce *et al.* 1988), i.e., whales were approached when sighted, for species identification, estimation of group size and composition, observation of behaviour, photo-identification and sound recording. The voyage track attempted to cover areas of known past abundance of cetaceans, but was often determined by ice and weather conditions. For safety reasons, the ice edge was usually avoided. The vessel was stopped during rough conditions, and in fog, low light or darkness to avoid collision with floating ice, and to avoid missing whales. Whenever possible, attempts were made to approach the 1 000 m isobath, a zone where krill has often been reported (S. Nicol, pers. comm.).

A Nikon FM2 with motordrive and 300 mm and 70–210 mm lenses was used with Kodachrome 200 transparency film for photo-identification. Observations of behaviour were recorded in real time on a handheld cassette recorder, and vidoetaped whenever possible. Sound monitoring was conducted on an irregular basis (when conditions were calm enough) both in the visual prsence of cetaceans and in order to detect them at times when none were visible. There was no regular schedule of monitoring. The recording apparatus had a system reponse of 30 Hz–6 kHz.

SUMMARY OF SIGHTINGS

Fifty-five Baleen Whale sightings of three species (Sei, Minke, Humpback), 143 Toothed Whale sightings of five species (Sperm, Southern Bottlenose, Killer, Hourglass Dolphin and Common Dolphin), and six unidentified whales were made during the voyage. The majority were south of 60°S. Details of sightings are shown in Figure 1 and Table 1. Sightings occurred in all conditions up to Beaufort sea state 7 (see Fig. 2), with a marked decrease above sea state 4 (windspeed 30 km h^{-1}). Observations are described by species, below.

Sei Whales *Balaenoptera borealis*

A single group of at least six Sei Whales was located at 44°07′S, 148°09′E on 1 January 1993. One animal appeared markedly smaller than the others. They were in deep water (4 000 m) and appeared to be swimming purposefully, with apparently shallow dives and frequent changes of direction, in a localized area in which a flock (100+) of Prions *(Pachyptila* sp.) was feeding.

Table 1. Summary of cetacean sightings made on cruise of *Iniquity*. Time is Eastern Australian Summer Time. Depth obtained from nautical charts.

Date/time	Position	Species	No. animals	Tw (°C)	Depth (m)	Beaufort sea state
1/1/93 1430	44°07′S 148°09′E	Sei	6	13.6	4 000	1
1/1/93 1535	44°08′S 148°09′E	Minke	20	13.6	4 000	1
4/1/93 1145	51°11′S 153°21′E	Hourglass	8	9.1	4 200	4
8/1/93 0315	56°46′S 159°34′E	Hourglass	10	5.4	4 000	0
8/1/93 1105	57°08′S 159°42′E	Southern Bottlenose	1	5.4	3 900	0
9/1/93 1400	59°41′S 159°58′E	Minke	2	4.0	3 600	4
14/1/93 1220	65°03′S 160°32′E	Sperm	2	0.7	2 200	2
17/1/93 0600	65°12′S 156°15′E	Minke	1	0.5	2 200	5
20/1/93 0635	64°52′S 151°24′E	Unidentified whale	1	0.5	3 200	3
23/1/93 0710	65°52′S 143°14′E	Killer	3	—0.3	1 500	4
23/1/93 0710	65°52′S 143°14′E	Minke	4	—0.3	1 500	4
26/1/93 0915	65°53′S 142°45′E	Killer	20	—0.4	1 000	3
26/1/93 0915	65°53′S 142°45′E	Unidentified	2	—0.4	1 000	3
26/1/93 1100	65°57′S 142°42′E	Minke	1	—0.4	250	1
26/1/93 1800	66°24′S 142°32′E	Minke	3	—0.4	200	0
1/2/93 1700	66°11′S 142°01′E	Minke	1	—1.0	200	1
1/2/93 1800	66°09′S 141°57′E	Killer	15	1.0	200	2
3/2/93 0840	65°02′S 136°05°E	Killer	1	—	400	7
3/2/93 2050	63°55′S 134°05′E	Southern Bottlenose	1	0.2	3 000	6
4/2/93 0630	63°50′S 133°00′E	Hourglass	10	0.9	3 600	5
4/2/93 1935	63°55′S 131°05′E	Sperm	1	1.1	3 000	3
5/2/93	64°39′S 129°40′E	Minke	3	—0.1	1 700	0
5/2/93 2115	64°50′S 128°33′E	Killer	3	—0.1	1 800	4
5/2/93 2115	64°50′S 128°33′E	Minke	5	—0.1	1 800	4
6/2/93 0845	65°04′S 126°47′E	Minke	1	—	1 000	7
9/2/93 1700	63°22′S 116°14′E	Unidentified whales	2	1.6	1 500	1
9/2/93 1840	63°23′S 116°05′E	Killer	3	—	3 000	2

Table 1—continued

Date/time	Position	Species	No. animals	Tw (°C)	Depth (m)	Beaufort sea state
9/2/93 2200	63°21′S 115°49′E	Killer	3	1.6	3 000	0
13/2/93 2200	63°55′S 103°38′E	Humpback	2	0.7	3 000	2
13/2/93 2320	63°53′S 103°36′E	Humpback	3	0.7	3 000	3
14/2/93 2200	64°21′S 101°12′E	Humpback	2	0.2	1 000	3
16/2/93 1845	64°21′S 95°06′E	Killer	1	0.3	2 000	3
18/2/93 0100	64°01′S 91°23′E	Sperm	1	0.7	3 000	3
21/2/93 1340	60°10′S 95°18′E	Minke	1	1.7	4 300	2
6/3/93 0500	39°06′S 128°57′E	Unidentified dolphins	3	15.3	5 000	4
6/3/93 1550	38°54′S 130°13′E	Common	10	15.5	—	5
7/3/93 0100	38°17′S 131°11′E	Common	8	16.2	—	2
7/3/93 0311	38°11′S 131°21′E	Common	12	16.2	—	4
8/3/93 2010	35°43′S 136°30′E	Common	25	—	—	—
9/3/93 1140	34°56′S 138°15′E	Common	5	17.4	200	0

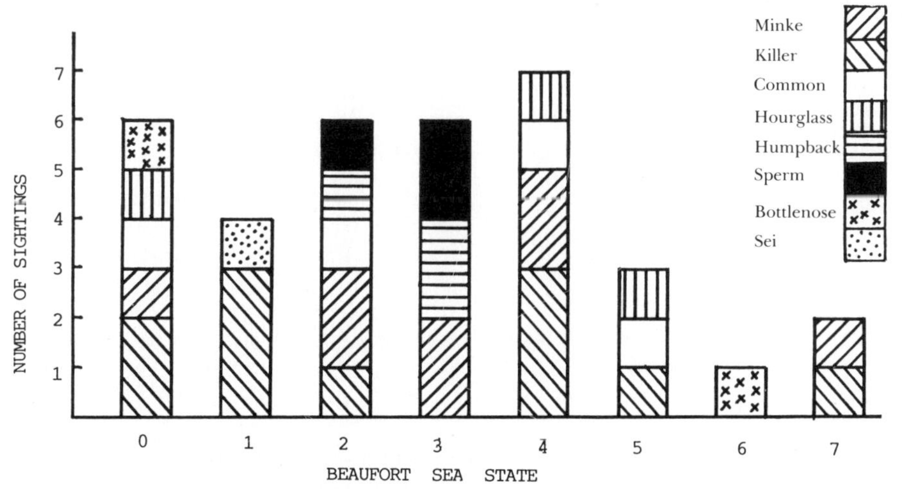

Fig. 2. Number of cetacean sightings (by species) according to Beaufort sea state.

Minke Whales were also seen in this vicinity (see below). This sighting occurred near the Subtropical Convergence, in an area where the Copepod *Calanus tonsus* (a preferred prey) is abundant, and where heavy catching of Sei Whales occurred in the past (Kawamura 1974). No plankton swarms were visible at the water surface. One whale appeared to investigate the vessel, passing within 30 m.

Minke Whales *Balaenoptera acutorostrata*

Forty-two Minkes were sighted in 10 groups, (average = 4.2). However, 20 of these were in one group, and were believed to be the diminutive form (Arnold *et al.* 1987). These were noticeably smaller than Minkes sighted nearer the Antarctic continent. They were sighted on 1 January 1993 at 44°08′S, 148°09′E, about 2 km south of the pod of Sei Whales (above). Some young calves were present.

The remainder of sightings occurred either south of 60°S or just north of it, between 159°58′E and 95°18′E, and the animals appeared to be the dark shoulder form (Arnold *et al.* 1987). Minkes were seen both in open water and (were the only species observed) in open drift ice. Twenty-two were sighted in 10 groups (average = 2.2); the most common observed group size was one (five times), then three (two times), two (one), four (one), and five (one).

Behaviour: Minkes Whales frequently approached to investigate the vessel. On 23 January four animals accompanied it for 20 min while it sailed through open drift ice, occasionally cutting close across the bow. On 5 February three Minkes were observed at 64°39′S, 129°40′E, milling slowly and diving for 3–4 min to resurface nearby. Sooty Shearwaters *Puffinus griseus* were active in this area. In an observation later on 5 February at 64°50′S, 128°33′E, shortly after three to four Killer Whales *Orcinus orca* were seen, a group of five to six Minkes was observed splitting into two groups which swam rapidly away from this vicinity. The Killer Whales were not resighted. Minke Whales may constitute an important prey item for Killer Whales in Antarctic waters (Mikhalev *et al.* 1981; Jefferson *et al.* 1991).

Humpback Whales *Megaptera novaeangliae*

The low density of Humpbacks in oceanic waters around East Antarctica results from dispersal into a vast area between the Antarctic Convergence and the ice edge (Omura 1973). This contrasts with relatively high concentrations occurring in the fjords of the Antarctic Peninsula (Stone and Hamner 1988). After sighting no Humpbacks in Area V, groups of two, three and two were encountered in Area IV (between 103°38′E and 101°12′E). One animal appeared immature. The remainder were presumed adults (*c.* 12 m in length). All three groups were located late in the evening, so they could not be followed for extended periods. All animals appeared in good condition.

Behaviour: During encounters with the first two groups on 13 February, the most obvious feature of their behaviour was apparent curiosity towards the stationary vessel, mixed with initial caution. They all approached to within 5 m, swimming under and around it for periods of 1 hour, and 20 min, respectively. The first group approached immediately, while the second approached after energetically circling the vessel in tight formation. One of the animals in the first group was subadult size (*c.* 10 m).

Observed behaviour included threat postures such as cocking the caudal peduncle, "growl" blowing, and bubble streams, as well as frequent spyhopping. One animal lay quietly at the stern of the vessel for several minutes, apparently looking at the swinging rudder and mimicking its movements with its body. Flukes were rarely shown when diving. No underwater sounds were heard while these whales were near the vessel. Both groups gave the impression that they were resting when approached and left the vessel slowly when they moved away.

Feeding behaviour: After being approached on 14 February; the third group of two animals carried out a series of synchronous dives of 4, 4, 5 and 4 min each, fluking high when diving, and resurfacing within about 200 m of the point of diving. In feeding Alaskan Humpbacks, these dive durations indicate feeding at depths between 30 and 40 m (Dolphin 1987). This location was close to both the ice edge and the 1 000 m isobath. No plankton was observed at the surface but an unusual concentration of seabirds was active (some of them apparently feeding) in this area, including Snow Petrels *Pagodroma nivea*, Antarctic Petrels *Thalassoica antarctica,* Antarctic Fulmars *Fulmarus glacioides,* Wilson's Storm Petrels *Oceanites oceanicus*, Cape Pigeons *Daption capensis*, Sooty Shearwaters *Puffinus griseus*, Blue Petrels *Halobaena caerulea,* White Chinned Petrels *Procellaria aequinoctialis*, and Lightmantled Sooty Albatrosses *Phoebetria palpebrata*. These birds are known to feed on Antarctic Krill *Euphausia superba* and other euphausiids (Carrick and Ingham 1967; Obst 1985).

After 33 min of diving behaviour, these two whales closely approached the vessel for 10 min, swimming quietly with none of the aggressive/ curious body displays exhibited by the two earlier groups. Two "chirps" only were heard via the hydrophone while the whales were near the vessel; these were the only vocalizations detected in the presence of Humpbacks. One of these animals exhibited fresh scarring on its dorsal surface, probably from surfacing under loose sea ice; this has been neglected as a source of scarring in Humpbacks (Chu and Nieukirk 1988).

Photo-identification: Humpback Whale migratory links were first established between Area V and Eastern Australia, and Area IV and Western Australia, by the use of "Discovery" marks (Chittleborough 1959; Dawbin 1966). More recently, photo-identification has confirmed or established migratory connections between Antarctic Area V and East Australia (Kaufman *et al.* 1990), and between the Antarctic Peninsula and Colombia (Stone *et al.* 1990), but until now, not between Area IV and Western Australia. Five Humpbacks were identified photographically: three by ventral fluke pigmentation patterns (Katona *et al.* 1979), and two by lateral body markings (Kaufman *et al.* 1987). A whale identified by lateral body markings on 14 February 1993 had previously been photographed off Perth, Western Australia in 1989 (Gill and Burton 1995).

Sperm whales *Physeter macrocephalus*

Sperm Whales were detected on three occasions, all south of 63°55′S and in depths of

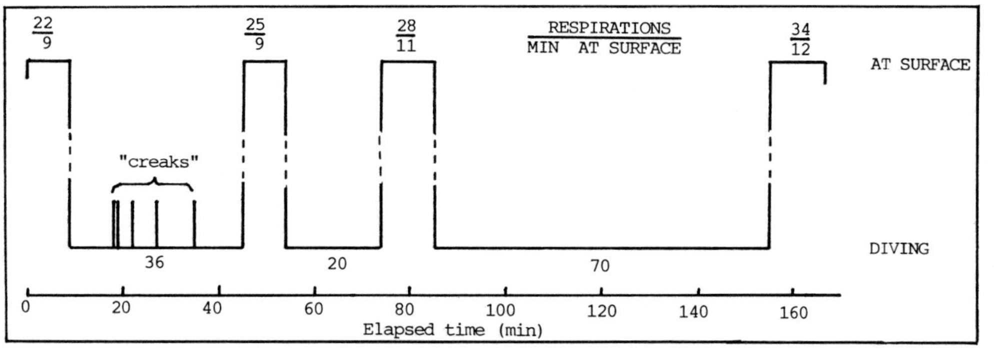

Fig. 3. Dive durations, respiration rates and "creak" vocalizations of a mature male Sperm Whale, 14 January 1993. Vocalizations were recorded during one dive.

2 220 m or more. On 14 January a solitary animal was detected visually, and this animal and another unseen animal were detected acoustically. On 4 February a solitary animal was detected visually and acoustically. On 18 February a single whale was detected by sound alone.

Feeding behaviour and vocalizations: The first Sperm Whale sighted, on 14 January was to the north of the Balleny Seamount in an approximate depth of 2 250 m. It was estimated at 18 m in length. The whale was observed during three dive sequences, over a period of 167 min (see Fig. 3). During four surfacing periods (average 10.25 min), the whale respired an average 2.66 observed blows min^{-1}; this rate is within the range normally observed in Sperm Whales (Gaskin 1964; Mullins *et al.* 1988). It is possible that some blows may have been missed immediately after the whale surfaced, although spotting conditions were excellent. The three intervening dives were of 36, 20 and 70 min (average 42 min). The animal only fluked-up when sounding and photographs of its fluke shape (Arnbom 1987) confirmed that the same animal was observed in all dives. While there was marked variation in dive duration, all durations were within the reported range for Sperm Whales (Watkins *et al.* 1985). On resurfacing the whale invariably swam away from the vessel.

During the 36 min. dive the whale was continually monitored acoustically. Its vocalizations consisted of "usual clicks" (interclick interval, or ICI less than 3 sec) both at the surface and while diving, "slow clicks" (ICI greater than 3 sec) and silences (Mullins *et al.* 1988; Weilgart and Whitehead 1988). In addition, there were five "creaks" (trains of very rapid clicks), after 9, 10, 12.5, 18 and 26 min (see Fig. 3). Creaks suggest close echolocation contact with objects, possibly prey items (Mullins *et al.* 1988). A second, more distant Sperm Whale was heard during this episode but was not seen.

The second Sperm Whale sighted, on 4 February was also large but was not followed for behavioural observation. It was seen swimming at the surface, and then sounded. Brief acoustic monitoring detected "usual clicks". The whale detected by sound alone on 18 February also emitted "usual clicks".

Southern Bottlenose Whale *Hyperoodon planifrons*

The two sightings of this species were very brief, just long enough for species identification. Both occurred when water depth was 3 000 m or more. The sighting on 8 January occurred in the presumed vicinity of the Antarctic Convergence (Fig. 1). This animal was about 7–8 m in length, and was observed to surface twice about 30 m behind the vessel, swimming slowly. The size of the second animal sighted was not estimated, as it only surfaced once close beside the vessel in rough conditions. The distinctive melon and beak of *Hyperoodon* were easily distinguished in both sightings.

Killer Whale *Orcinus orca*

This was the most commonly sighted species throughout the cruise. All were sighted south of 60°S. A total of 49 animals were sighted in eight groups (average = 6.13). Most frequent group size was three (four times), then one (two), 15 (one) and 20 (one). All animals had a yellow diatom film covering their bodies.

Behaviour: Killer Whales were usually seen at a distance, moving rapidly relative to the vessel and were difficult to approach due to their speed. On 26 January about 20 animals were observed swimming rapidly in three groups, spread over about 2 km. Several mature males were present. These three groups coalesced into a single pod, which moved south in the direction of distant blows of at least two large whales. When the vessel headed in that direction no whales could be found. Killer Whales have been observed to attack or harass Blue, Sei and Minke Whales in Antarctic waters, and all other great whale species elsewhere (Jefferson *et al.* 1991).

On 1 February a group of 15 Killer Whales was sighted within 1.85 km of a field of drift ice, among which Emperor *Aptenodytes forsteri* and Adélie *Pygoscelis adeliae* Penguins, as well as a Minke Whale, were observed. This was the only slowly moving group observed during the cruise. There were two mature males present, and at least two calves, one of which appeared quite small (3–4 m). After the vessel approached a group of eight animals, including a calf, passed about 15 m behind its stern. Mature males came no closer than about 400 m. No sound monitoring was conducted while the animals were near the vessel, but faint Killer Whale whistles were detected after they had moved away.

On 3 February a single large Killer Whale (not a mature male) swam close alongside for about 10 min while the vessel was running in a rough sea. This animal appeared to mimic the rolling of the vessel and made eye contact with the two observers.

On 5 February three to four Killer Whales were seen briefly, just before three Minke Whales were seen rapidly leaving the area (see above).

Hourglass Dolphins *Lagenorhynchus cruciger*

Twenty-eight were seen in three groups. Average group size was 9.33, compared to an average of 7.01 from 112 groups observed during IWC/IDCR cruises (Kasamatsu *et al.* 1990). Two groups of eight and 10 were sighted north of 60°S, one of them near the Antarctic Convergence, and a group of 10 at 63°50′S. No obvious young were noted in the three groups.

Behaviour: All groups encountered accompanied the vessel for periods up to 1 hour. On 8 January, 10 animals bowrode briefly in calm, foggy conditions until the vessel's engine was stopped. The dolphins remained within 50 m of the stationary vessel for about 20 min, often approaching to within 5 m, and swimming quietly. Faint whistles and click trains were recorded during this encounter but have not been analysed.

Common Dolphins *Delphinus delphis*

These were the only cetaceans sighted to the north of the Antarctic Convergence during the return voyage to Australia (except a group of dolphins, unidentifiable in the dark, not far from the first *Delphinus* sighting). All *Delphinus* were sighted north of 38°54′S, in water of 15.5°C and warmer. Sixty-seven were seen in six groups (average = 1.17). No calves were noted. This species bowrode for up to 20 min in all observations.

DISCUSSION

Yachts *vs.* ships

An obvious advantage of using small sailing vessels for cetacean research is that of cost. This study was conducted for a small fraction of the cost of conventional Antarctic shipping, such as IWC/IDCR survey vessels, or resupply vessels. For example, this entire expedition was undertaken for less than the cost of a day's charter of a vessel currently used for marine science and resupply. However, yachts and ships are suited to different types of studies. Resupply vessels are committed to their resupply role, as well as to diverse programmes of marine research, but are also capable of cetacean line transect surveys en route (Creet *et al.* 1993). IWC/IDCR vessels are dedicated to surveys. Yachts are not suitable for Antarctic surveys, for reasons discussed below. However, it is proposed that they can complement existing Antarctic research by enabling greater diversity of cetacean study. Small sailing vessels have successfully been used elsewhere for a range of cetacean studies (e.g., Whitehead and Moore 1982; Whitehead and Gordon 1986; Gordon 1987; Mattila and Clapham 1989; Clapham and Mattila 1990; Whitehead and Waters 1990; Dawbin and Gill 1991), including studies on high latitude feeding grounds in the northern hemisphere (Whitehead *et al.* 1982; Mullins *et al.* 1988; Whitehead *et al.* 1992). Increasing numbers of yachts operate safely each year in Antarctic waters.

While low cost is an advantage *per se,* there are other advantages on yachts, such as considerably more flexibility of time, which can be spent with selected groups of whales, and of movement. In addition, small, quiet vessels have much less impact on cetacean behaviour and acoustic communication than large vessels.

Small vessels are not without disadvantages. Visibility is reduced due to the often low viewing height for the observer and such vessels are vulnerable to sea ice. These are discussed in the following section. Most yachts travel at lower speeds than ships: while the normal range of speeds for the study vessel was between 5 and 8 knots (9.3–14.8 km h^{-1}), IWC/IDCR vessels typically travel at 12 knots (22.2 km h^{-1}). However, experiments conducted during IWC/IDCR cruises compared sightings results at 6 and 12 knots (11.1 and 22.2 km h^{-1}), with inconclusive results (Joyce *et al.* 1988).

Possible reasons for low numbers of cetaceans detected

The total number of sightings was disappointing (Fig. 1, Table 1); only 17 large whales were detected. One reason is undoubtedly the present relative paucity of large whales resulting from the era of large-scale Antarctic whaling. The most commonly sighted species were Killer Whales (49) and Minkes (42), both commercially unimportant during most of this era.

It was occasionally possible to search for whales from a point about 12 m a.s.l. up the mast of *Iniquity* (distance to horizon 13.3 km), but this position was tenable only for short periods, due to cramping or excessive motion. This shortcoming would be largely overcome by the addition of a secure crow's nest. From the usual observation positions on deck the height of eye was about 3 m a.s.l. (distance to horizon 6.7 km), compared to observer heights of 17–20 m a.s.l. (distance to horizon 15.9–17.4 km) on IWC/IDCR cruises (Joyce *et al.* 1988). It is likely that the relatively low height of the observers on the yacht partially contributed to the low number of sightings. However, the advantage of height was frequently negated by poor visibility which limited the field of view.

Continental Antarctica is fringed by sea ice. The ice edge is believed to be an important feeding zone for baleen whales (Mackintosh 1970), some species of which penetrate well into the sea ice (pers. obs.). The greatest density of cetaceans (Killer and Minke Whales) observed in this study was in the vicinity of sea ice while entering Commonwealth Bay (about 142°E), probably partly because the vessel was in this area for several days. The Humpback Whales apparently feeding on 14 February were near the ice edge, as indicated by the presence of Snow Petrels, a pagophilic species, and ice blink (light reflected off ice into the sky) to the south. No Baleen Whales were seen near the ice edge to the north of the Balleny Islands

Our vessel, which was not ice-strengthened, occasionally encountered the ice edge in conditions of rough seas and poor visibility, when sailing conditions became so potentially hazardous that immediate detours were made to the north to clear it. To avoid these situations, an attempt was made to keep a safe distance from the ice edge. While small scale (large area) ice charts were available by weatherfax, these gave insufficient information on local ice conditions, and ice was occasionally encountered where none was shown on faxes. The vessel's distance from the ice edge was usually impossible to estimate. On one occasion large numbers of icebergs and their floating debris also necessitated a detour to the north. It is likely that encounters with whales would have been more frequent had the ice edge been followed, but prudent navigation ruled this out. The use of an ice-strengthened yacht (of which there are few) would enable safer examination of the ice edge zone.

It is also possible that the cruise design itself contributed to the low number of sightings. The basis of the route was determined by the decision to examine two areas where large numbers of Humpback and other whales have been previously reported. These are around the Balleny Islands near 163°E in Area V, and around 90°E in Area IV (Mizroch *et al.* 1985; Kasamatsu *et al.* 1988). The region in between is also of interest with regard to Humpback Whales, as the extent of geographical overlap of feeding Area IV and V stocks is unknown. The great distance (3 500 km) between these two end points necessitated constant travel throughout much of the cruise, and the brief Antarctic summer was a deterrent to spending more time in areas where whales were found. It is possible that more information may have been gained by a more prolonged examination of selected areas, at the expense of a wider coverage.

The future

As a pilot project, this study highlighted some problems of small vessel cetacean research in Antarctic waters. Operating a yacht in such a hostile environment is mentally and physically taxing, and coupled with the elusive nature of cetaceans, there are many difficulties. However, with a suitable vessel and a determined crew, economical access to this remote ocean was possible, and although limited data were obtained, the feasibility of yachts as Antarctic research platforms was demonstrated. Examples of this were the feeding observations of Humpback and Sperm Whales, including acoustic observations, and the photo-identification resight of a Humpback Whale.

After the Indian Ocean Whale Sanctuary was adopted in 1979, the most significant subsequent programmes of whale research in that ocean were conducted from a yacht (Whitehead and Gordon 1986; Gordon 1987). The adoption of a Southern Ocean Whale Sanctuary by the International Whaling Commission in May 1994 provides an opportunity for the implementation of diverse programmes of research and monitoring, based on non-lethal methods (IWC 1992). While sightings surveys are to continue in the Sanctuary, and can provide population estimates useful for management purposes, there is also a role for small vessels, operating on relatively low budgets, and employing a variety of non-lethal techniques such as telemetry, photo-identification, tissue biopsy, behavioural observation and acoustic monitoring to examine a wide range of ecological questions.

Such studies can be undertaken not only in Antarctic waters, but also in the more northerly parts of the Sanctuary area, around the coasts of the southern continents and New Zealand. In conjunction with conventional studies such as sightings surveys, work of this type could contribute information valuable for the management of cetaceans as a component of a marine ecosystem vulnerable to increasing anthropogenic threats.

ACKNOWLEDGEMENTS

Thanks to *Iniquity's* crew — Tim Edkins, Peter Drew, Chook Wall-Smith, Peter Manthorpe and Phil Kelly — for their dedicated assistance and support in the field. Special thanks to Di Davidson. Thanks also to Bill Dawbin and Don Richards for assistance prior to the cruise. The

The study was funded by a grant from the W. V. Scott Estate, and was conducted under *Whale Protection Act* Permit 1992/029 issued by Australian National Parks and Wildlife Service, Canberra.

REFERENCES

Anon., 1988. Report of the planning meeting for IWC/ IDCR southern hemisphere minke whale assessment cruise 1988/89. Institute of Cetacean Research, Tokyo (unpublished).

Arnbom, T., 1987. Individual identification of sperm whales. *Rep. Int. Whal. Commn. (Special Issue 37)* 201–04.

Arnold, P., Marsh, H. and Heinsohn, G., 1987. The occurrence of two forms of minke whales in east Australian waters with a description of external characters and skeleton of the diminutive or dwarf form. *Sci. Rep. Whales Res. Inst.* **38:** 1–46.

Brown, S. G., 1963. A review of Antarctic whaling. *Polar Rec.* **11:** 555–66.

Brown, S. G. and Lockyer, C. H., 1984. Whales. Pp. 717–81 *in* Antarctic Ecology, Volume Two, ed by R. M. Laws. Academic Press: London.

Carrick, R. and Ingham, S. E., 1967. Antarctic seabirds as subjects for ecological research. *JARE Scientific Reports (Special Issue)* **1:** 151–84.

Chittleborough, R. G., 1959. Australian marking of humpback whales. *Norsk Hvalf.-Tid.* **48:** 47–55.

Clapham, P. J. and Mattila, D. K., 1990. Humpback whale songs as indicators of migration routes. *Mar. Mam. Sci.* **6**(2): 155–60.

Creet, S. M., Green, K. and Burton, H. R., 1993. Whale observations from ANARE ships. Pp. 131–36 *in* Encounters with Whales '93, ed by D. Postle and M. Simmons. Workshop Series No 20, Great Barrier Reef Marine Park Authority: Townsville.

Chu, K. and Nieukirk, S., 1988. Dorsal fin scars as indicators of age, sex, and social status in humpback whales *Megaptera novaeangliae*. *Can. J. Zool.* **66:** 416–20.

Dawbin, W. H., 1966. The seasonal migratory cycle of humpback whales. Pp. 145–70 *in* Whales, Dolphins and Porpoises, ed by K. S. Norris. University of California Press: Berkeley.

Dawbin, W. H. and Gill, P. C., 1991. Humpback whale survey along the west coast of Australia: a comparison of visual and acoustic observations. *Mem. Qld Mus.* **30**(2): 255–57.

Dolphin, W. F., 1987. Dive behavior and energy expenditure of foraging humpback whales in south-east Alaska. *Can. J. Zool.* **65:** 354–62.

Gaskin, D. E., 1964. Recent observations in New Zealand waters on some aspects of the behaviour of the sperm whale (*Physeter macrocephalus*). *Tuatara* **12:** 106–14.

Gill, P. C. and Burton, C. L. K., 1995. Photographic resight of a humpback whale between Western Australia and Antartic Area IV. *Mar. Mam. Sci.* **11**(1): 96–100.

Gordon, J. C. D., 1987. Sperm whale groups and social behaviour observed off Sri Lanka. *Rep. Int. Whal. Commn. (Special Issue 37)* 205–17.

Iinternational Whaling Commission, 1992. Proposal for a Whale Sanctuary in the Southern Hemispere. IWC/44 Agenda Item 12.2, Glasgow, May 1992.

Jefferson, T. A., Stacey, P. J. and Baird, R. W., 1991. A review of killer whale interactions with other marine mammals. *Mammal Rev.* **21**(4): 151–80.

Joyce, G. G., Kasamatsu, F., Rowlett, R. and Tsunoda, L., 1988. The IWC/IDCR Southern Hemisphere Minke Whale Assessment Cruizes: the first ten years. Paper no. SC/40/O16 submitted to the Scientific Committee of the International Whaling Commission (unpublished).

Katona, S., Baxter, B., Brazier, O., Kraus, S., Perkins, J. and Whitehead, H., 1979. Identification of humpback whales by fluke photographs. Pp. 33–44 *in* Behavior of Marine Mammals Volume 3 Cetaceans, ed by H. E. Winn and B. L. Olla. Plenum Press: New York.

Kasamatsu, F., Hembree, D., Joyce, G., Tsunoda, L., Rowlett, R. and Nakano, T., 1988. Distribution of cetacean sightings in the Antarctic: results obtained from the IWC/ IDCR Minke Whale Assessment Cruizes, 1978/79 to 1983/84. *Rep. Int. Whal. Commn. (Special Issue 38)* 449–87.

Kasamatsu, F., Joyce, G. G., Ensor, P. and Mermoz, J., 1990. Current occurrence of cetacea in the southern hemisphere: results form the IWC/IDCR southern hemisphere Minke Whale Assessment Cruizes, 1978/79–1987/88. (Unpublished report SC/42/O 15 to the Scientific Committee of the International Whaling Commission).

Kaufman, G. D., Osmond, M. G., Ward, A. J. and Forestell, P. H., 1990. Photographic documentation of the migratory movement of a humpback whale *(Megaptera novaeangliae)* between east Australia and Antarctic Area V. *Sci. Rep. Int. Whal. Commn. (Special Issue 12)* 265–67.

Kawamura, A., 1974. Food and feeding ecology in the southern sei whale. *Sci. Rep. Whales Res. Inst.* **26:** 25–144.

Kawamura, A., 1980. A review of food of balaenopterid whales. *Sci. Rep. Whales Res. Inst.* **32:** 155–97.

Mackintosh, N. A., 1965. The stocks of whales. Fishing News (Books) Ltd: London.

Mackintosh, N. A., 1970. Whales and krill in the twentieth century. Pp. 195–212 *in* Antarctic Ecology Vol. 1, ed by M. W. Holdgate. Academic Press: London.

Mackintosh, N. A. and Wheeler, J. F. G., 1929. Southern blue and fin whales. *Discovery Rep.* **I:** 257–540.

Matthews, L. H., 1938. The sperm whale, *Physeter catodon*. *Discovery Rep.* **XVII:** 93–168.

Mattila, D. K. and Clapham, P. J., 1989. Humpback whales, *Megaptera novaeangliae,* and other cetaceans on Virgin Bank and in the northern Leeward Islands, 1985 and 1986. *Can. J. Zool.* **67:** 2201–211.

Mizroch, S. A. *et al.*, 1985. Preliminary atlas of balaenopterid whale distribution in the Southern Ocean based on pelagic catch data. Selected Paper SC-CAMLR-IV/BG/ 21 presented to the Scientific Committee of CCAMLR (unpublished).

Mullins, J., Whitehead, H. and Weilgart, L. S., 1988. Behaviour and vocalizations of two single sperm whales, *Physeter macrocephalus*, off Nova Scotia. *Can. J. Fish. Aquat. Sci.* **45:** 1736–743.

Obst, B. S., 1985. Densities of Antarctic seabirds at sea and the presence of the krill *Euphausia superba*. *Auk* **102:** 540–49.

Omura, H., 1973. A review of pelagic whaling operations in the Antarctic based on the effort and catch data in 10° squares of latitude and longitude. *Sci. Rep. Whales Res. Inst.* **25:** 105–203.

Parker, D. A. A., 1978. Observations of whales on ANARE voyages between Australia and Antarctica. *Aust. Wildl. Res.* **5:** 25–39.

Stone, G. S. and Hamner, W. M., 1988. Humpback whales *Megaptera novaeangliae* and southern right whales *Eubalaena australis* in Gerlache Strait, Antarctica. *Polar Rec.* **24:** 15–20.

Stone, G. S., Florez-Gonzalez, L. and Katona, S., 1990. Whale migration record. *Nature* **346:** 705.

Thomas, J. A., Leatherwood, S., Evans, W. E. and Jehl, J. R., 1981. Ross Sea killer whale distribution, behavior, color pattern, and vocalizations. *Antarctic J. U.S.* **16:** 157–58.

Tonnessen, J. N. and Johnsen, A. O., 1982. The history of modern whaling. University of California Press: Berkeley.

Watkins, W. A., Moore, K. E. and Tyack, P., 1985. Sperm whale acoustic behaviours in the south-east Caribbean. *Cetology* **49:** 115.

Weilgart, L. S. and Whitehead, H., 1988. Distinctive vocalizations from mature male sperm whales *(Physeter macrocephalus). Can. J. Zool.* **66:** 1931–937.

Whitehead, H., Brennan, S. and Grover, D., 1992. Distribution and behaviour of male sperm whales on the Scotian Shelf, Canada. *Can. J. Zool.* **70:** 912–18.

Whitehead, H. and Gordon, J. C. D., 1986. Methods of obtaining data for assessing and modelling sperm whale populations which do not depend on catches. *Rep. Int. Whal. Commn. (Special Issue 8)* 149–65.

Whitehead, H. and Moore, M. J., 1982. Distribution and movements of West Indian humpback whales in winter. *Can. J. Zool.* **60:** 2203–211.

Whitehead, H., Silver, R. and Harcourt, P., 1982. The migration of humpback whales along the northeast coast of Newfoundland. *Can. J. Zool.* **60:** 2173–179.

Whitehead, H. and Waters, S., 1990. Social organization and population structure of sperm whales off the Galapagos Islands, Ecuador (1985 and 1987). *Rep. Int. Whal. Commn. (Special Issue 12)* 249–57.

Above: Killer Whale *Orcinus orca*, Sceale Bay, South Australia, 25th December 1992. Photo W. Guiddera.

Top right: Southern Right Whale *Eubaleana australis*, Bicheno, Tasmania, 1st June 1993. Adult female. Photo: L. Geeves.

Bottom right: Humpback Whale *Megaptera novaeangliae*, interacting with yacht "Iniquity", Antarctic Waters (63°55′S, 103°38′E), 13th February 1993. Photo: P. Gill.

Southern Right Whales in southeastern Australia — aerial surveys during 1991–93 and incidental information from 1904

C. M. KEMPER,[1] J. MOLE[1], R. M. WARNEKE,[2] J. K. LING,[1] D. J. NEEDHAM[3] and J. E. WAPSTRA[4]

The distribution, abundance and movements of *Eubalaena australis* were studied during 1991–93 by conducting aerial surveys and opportunistic flights along the coast of eastern South Australia, Victoria, southern New South Wales and Tasmania, and by collating incidental reports from the beginning of the century. Whales were individually identified from photographs of callosity patterns. Public reports of Southern Right Whales were analysed for trends and their frequency rose sharply in the early 1980s. In Tasmania, sightings have been recorded in all months except February and along the mainland between April and December. For all states studied, peak numbers occurred in winter and spring. Aerial surveys showed that the most frequented regions were western Victoria and along the Coorong of South Australia. Both are south to south-east-facing, experience high wave energy and have moderate water depth up to 2 km offshore. Regions rarely frequented were northern Kangaroo Island/Adelaide coast and central Bass Strait. The highest number seen on an aerial survey along the mainland coast was 16, including calves. A total of 54 adults were individually identified, of which 41% were seen with calves. Within southeastern Australia no confirmed repeat sightings were made between years. Many movements within years were recorded and some females with calves were seen over extended periods, up to 16 weeks, in the Warrnambool area. Other females with calves were resighted at distances of up to 300 km. Some adults without calves, and a yearling, moved over long distances (east/west) in short periods of time, including two observed at Head of Bight after they were seen in southeastern Australia. No movements were recorded between Tasmania and the mainland. It was estimated that a minimum of 10% of the Australian population of Southern Right Whales frequented southeastern Australia during 1991–93.

Key words: Southern Right Whale, *Eubalaena australis,* Australia, Numbers, Movements, Survey, Distribution.

INTRODUCTION

THE Southern Right Whale *Eubalaena australis* is an endangered species whose numbers were greatly reduced by whaling during the 19th century. At least 26 000 were taken from southeastern Australia and New Zealand (Dawbin 1986). Unlike many of the "great whales" which were hunted and studied intensively during the 20th century, Southern Right Whales received little scientific attention at the time they were being hunted. The little direct evidence available (see Chittleborough 1956) suggests that numbers remained very low until the 1960s or 1970s. In the last 10–20 years significant increases in numbers have been noted off southern Africa (Best 1990a), Argentina (Payne *et al.* 1990) and Western Australia (Bannister 1990; 1994). The extent of interchange, if any, between populations off the southern continents is not known.

Studies of Southern Right Whales off Australia have been made at the head of the Great Australian Bight in South Australia (Ling and Needham 1985–91; Bannister 1994; S. Burnell, pers. comm.) and southwestern Western Australia (Bannister 1990; 1994). From these studies there is emerging a body of information which clearly demonstrates the importance of some areas in Western Australia and at Head of Bight as major nursery sites. A large number of whales seen in these regions has been individually identified from photographs of head callosity patterns and other body markings. There is recent evidence to show that some individuals travel between Head of Bight and Western Australia but the full extent of this interchange is not known (J. Bannister, pers. comm; S. Burnell, pers. comm.). In New South Wales, Victoria, Tasmania and South Australia east of Eyre Peninsula, anecdotal evidence suggests that in recent times there have been fewer Southern Right Whales visiting these waters than the western part of Australia.

Before 1991 there was no co-ordinated research on Southern Right Whales in southeastern Australia. Several sets of data concerning public reports of Southern Right Whales existed and Bannister (1986a) published a summary of most of those records up to 1982. Dawbin (1986)

[1]South Australian Museum, North Terrace, Adelaide, South Australia, Australia 5000.
[2]R.S.D. 273, Mount Hicks Road, Yolla, Tasmania, Australia 7325.
[3]294 Greenhill Road, Glenside, South Australia, Australia 5065.
[4]Parks and Wildlife Service, Hobart, Tasmania, Australia 7000.

Pp. 40–55 *in* MARINE MAMMAL RESEARCH IN THE SOUTHERN HEMISPHERE VOLUME 1: STATUS, ECOLOGY AND MEDICINE ed by M. Hindell and C. Kemper. Surrey Beatty & Sons, Chipping Norton. 1997.

detailed the over 13 000 whaling catches in southeastern Australia and New Zealand. Historical information suggests that initial numbers of Southern Right Whales were high but that these were rapidly depleted by whaling, particularly during the 1830s and 1840s.

During the period 1983 to 1991 the Australian Nature Conservation Agency (ANCA) funded seven aerial surveys which covered differing portions of the mainland coast of southern Australia, several of which originated in Western Australia or South Australia and traversed the Victorian coast to the New South Wales border or beyond. The maximum number of whales seen along the southeastern coast on any of these survey flights was two adults and one calf. The results for Victoria were indicative of the generally low and widely fluctuating levels of right whale activity in southeastern Australia. The survey which covered the whole of the south coast of Australia, from Cape Leeuwin to Sydney, in 1984 clearly demonstrated the relative importance of Western Australia and western South Australia (42, including calves) compared with southeastern South Australia (3 whales), Victoria (2) and New South Wales (0). In addition to the ANCA flights, the Victorian Department of Conservation and Natural Resources (CNR) flew six limited aerial searches in relation to local whale activity, one in each year from 1982 to 1986 and one in 1989. These more specifically-targeted searches encountered as many as seven adults and four calves in the vicinity of Warrnambool (on 21 August 1985).

During 1991–93 BHP Petroleum funded the South Australian Museum to conduct aerial surveys and solicit public reports of Southern Right Whale sightings in South Australia, Victoria, Tasmania and to a lesser extent southern New South Wales. The aims of the study were: (1) to assess the numbers and locations of Southern Right Whales in southeastern Australia and the importance of certain locations for breeding, (2) to identify individual animals from photographs of head callosities and other recognizable features and to track the movements of individuals within and between years, (3) to compare identified animals with those from outside the study area, (4) to bring together all past records of Southern Right Whales in southeastern Australia and (5) to conduct a preliminary investigation into the relationship between coastal and oceanographic conditions and the distribution of Southern Right Whales in South Australia and Victoria. This paper draws together the results of those investigations.

METHODS

Aerial surveys 1991–93

Six aerial surveys were conducted between early to mid June and early October in each of the three years of study. They were spaced 3–5 weeks apart, in order to span the whales' inshore season relatively evenly. The surveys were 1½–4 days duration. Two surveys (one in both 1991 and 1993) were aborted before completion because of deteriorating weather. Survey destinations and hours searching varied within and between years (Figs 1, 5) and were standardized at the beginning of 1993 to a coastal route between Adelaide and Cape Howe (New South Wales/Victorian border), including Kangaroo Island. Observations were made on both outward and return flights of each survey. During 1991 and 1992 Kangaroo Island was surveyed only once, usually on the outward route but in 1993 it was also surveyed on the return journey. The total survey hours for each year were: 1991 — 87.2, 1992 — 89.1, 1993 — 116.6.

Two types of aircraft were used on surveys. In 1991, a single-engine Cessna 177B was flown at 185 km/h (100 knots) and in 1992 and 1993, a twin-engine Cessna 337 was flown at 222 km/h (120 knots). Both were high-winged and permitted a clear view of the survey transect. Survey height was 460 m (1 500 feet) except for times when cloud cover was below this altitude or when air turbulence was not conducive to observations at 460 m. The latter was often the case around Wilsons Promontory and along south-facing coasts of high relief when the wind was offshore, particularly for example near Cape Otway.

An attempt was made to conduct the surveys in optimum viewing conditions, e.g., Beaufort scale sea state of 0–2, cloud cover 0–4/8 and wind speeds of less than 15 knots. However, weather conditions were difficult to predict for the full length of the coast on both outgoing and return journeys because the survey area spanned three very different weather regions, South Australia, western Bass Strait and eastern Bass Strait (Fig. 1). The best way to achieve optimum conditions was to start the survey at the beginning of a high pressure system and move with this system as it travelled east. Even so, there were many times when the weather conditions deteriorated for parts of the surveys. Few weather observations were made in 1991 but during 1992 and 1993 more regular observations were made. These observations showed that mean daily Beaufort scale records were less than 3.0 in 1992 (except 3 of 15 survey days) and 2.6 or less during 1993, and cloud cover daily means were less than 6/8 (except on 6 of 15 days during 1992 and 4 of 18 days in 1993); wind speed was generally less than 20 knots except on several occasions in both years when either storms arose quickly or there was a strong northerly blowing offshore. Strong northerlies affected observer comfort and photography more than whale viewing conditions because the sea was calm inshore in

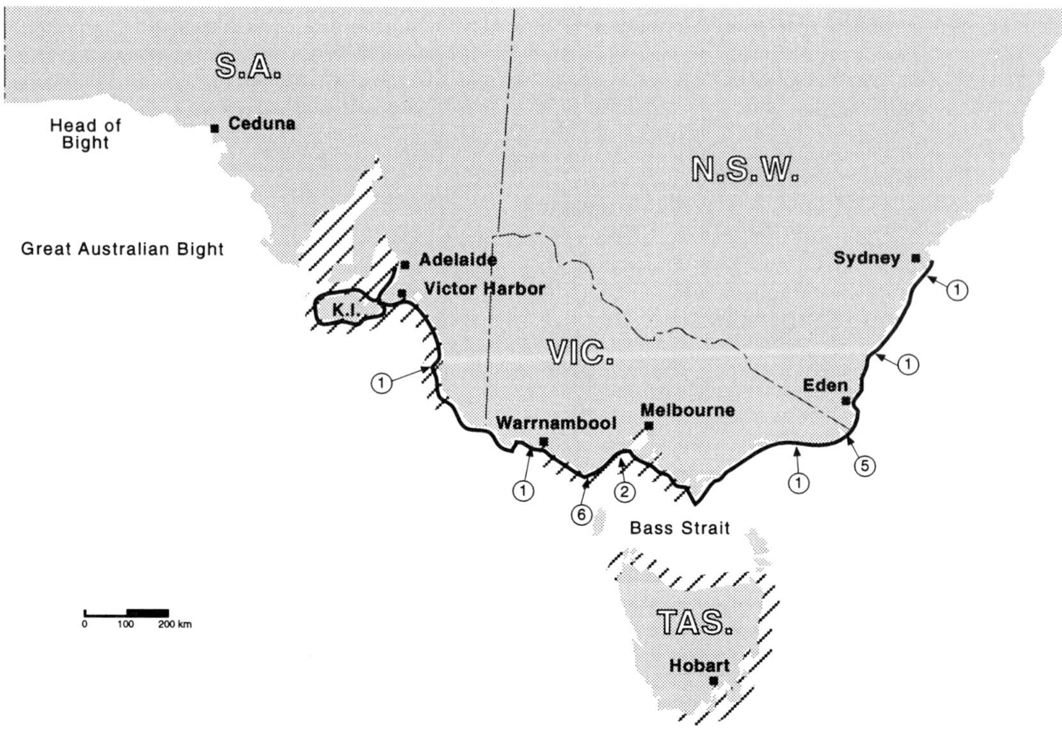

Fig. 1. Area covered on aerial surveys and opportunistic flights for Southern Right Whales during 1991–93. Area covered is variable within and between years. Numbers in circles indicate number of surveys to particular destinations. ———— route for aerial surveys. ////// opportunistic flights in response to public reports of sightings of Southern Right Whales.

such conditions. Weather conditions were on the whole poorest during the following surveys: August 1991; late September 1992; mid-September 1993.

Surveys were generally carried out between 0800 and 1630 hours and the best time for photography was between 1000 and 1500 hours. One or two rest stops on the ground were taken each day in order to maintain concentration on viewing.

During the 1991 surveys, the four-seater aircraft was manned by two observers/photographers and the pilot. Both observers sat on the seaward side on the outgoing journey whereas on the return journey the pilot became an observer and another sat behind him. During 1992 and 1993, in a six-seater aircraft, there were usually three observers and the pilot. While surveying, the aircraft maintained a position directly above the coast so that the observers had a clear view directly below to the water's edge. Observers concentrated their search effort on an area 2 km seaward as suggested by Bannister (1990). Occasionally our flight path would be 1–2 km out to sea, looking coastward, in order to avoid sun glare on the water.

Opportunistic flights

Opportunistic flights began in 1992 and were conducted in response to public reports of whales within a range of about two hours flying time of Adelaide, Melbourne, Warrnambool, Devonport or Hobart (Fig. 1). The aim was to locate and photograph whales to identify them individually in order to increase the number of individuals recorded in the study area and to get a better idea of movements within the region. Several different aircraft, pilots and observers were used. The following opportunistic flights were carried out: in 1992; South Australia – 13 flights, Victoria – 1, Tasmania – 5: in 1993; South Australia – 7 flights, Victoria – 3, Tasmania – 5. A total of 67 hours was flown on opportunistic flights. The success rate for locating reported whales was 82%. Flights were conducted if the weather was good for viewing and photography (see aerial surveys), and the behaviour of the whales suggested that they were likely to remain in the area. Generally these were socializing adults or adults with calves.

Public reports

Public reports of whale sightings were available from several sources: (1) The South Australian Museum had about 600 reports (sightings, not number of whales) dating back to 1904, the majority made after 1982; (2) R. Warneke and the Victorian Wildlife Atlas contributed about 1 000 records for Victoria dating back to 1942; (3) Parks and Wildlife Service, Tasmania, contributed about 190 records from 1977 onwards; (4) New South Wales National Parks and Wildlife Service listed 20 records from 1986 onwards; (5) Bannister (1986a) recorded incidental sightings in Australia between 1909 and 1982. The type and accuracy of information recorded for each public report varied. In most cases a reliability of identification (1, 2 or 3) was assigned to each record based on the information provided and expertise of the observer. Only confirmed/certain (1) and acceptable/ probable (2) records were used in the analyses. Reports of whales at the same place on the same day were treated as the same animal(s).

Photography

To photograph whales, three types of circuit were flown. The first was at a wide angle of approximately 45°, while the second was closer to the whale, at approximately 60°. A third type of circuit (80° and more vertical) was used in situations where whales were close to the coastline and high headlands made circling in the usual manner difficult. The altitude for photographing whales increased during the study period because of a concern that flying low could disturb the animals. During 1991 and early 1992 the average height for photography was 150 m (500 feet) and it was noted that whales, especially cows with calves, often submerged when the aircraft circled at this height, particularly if many circuits were flown. For the remainder of 1992 and 1993 the majority of photographs was taken at 210–40 m (700–800 feet) and the number of circuits was restricted to less than five in most cases.

Planned surveys and opportunistic flights provided most of the photographs used to identify and match individuals but some were also obtained from other sources, e.g., public observers, media sources. Videos taken by television crews were the most useful and from these, prints were taken and converted to slides.

Several types of equipment and film were used during the three years of the study. Cameras included a Bronica ETRSi (56 × 42 mm) with a 250 mm lens and a 2× teleconverter, Nikon 301 (35 mm) with 300 mm lens and 2× teleconverter, Nikon 801S (35 mm) with 300 mm lens. Slide films were: Ektachrome 200 ASA EPD 120 and ASA EPL, Fujichrome RH 400 ASA. Some film used in 1991 and early 1992 was "pushed" during processing to double the recommended ASA and allow more contrast. This method was abandoned during 1992 as it produced "grainy" slides which were poor quality for matching.

Matching and identification

Slides used to identify and match whales were chosen according to the photograph angle (as close to vertical as possible, or looking slightly forward or back onto the whale's head) and clarity of the callosities. These slides were projected and enlarged onto a daylight screen and the callosity pattern traced. Tracings were then attached to filing cards with the whales' details.

Matches and identifications were made by comparing all parts of the callosity pattern. Priority was given, in order, to presence/absence and size of lip patches, to number, size and position of rostral islands, and lastly to bonnet, coaming and post-blowhole callosities. If a match looked likely, any other visible callosities were checked, including, when possible, number and position of mandibular callosities. The reliability of a match was classed as 1 (with certainty), 2 (very likely) or 3 (possible). The only time the latter was used was in Figure 9. All identified and matched animals were rechecked by at least two people several times in an attempt to eliminate mismatches.

Coastal regions

The coast along the aerial survey route (i.e., Adelaide to Cape Howe) area was divided into seven regions (Fig. 2), based on marine regionalization studies by the governments of Victoria (Land Conservation Council and Department of Conservation and Natural Resources 1994) and South Australia (Edyvane and Baker 1994). Physical, biological, oceanographic and coastal geomorphological information was used to describe these regions. As the length of the coast was not available for each region in the publications, it was determined by us using a "map measure" and 1:250 000 map sheets.

RESULTS

Reporting frequency

The number of public reports, from the sources available to us, of Southern Right Whales in South Australia (east of 135°30′E), Victoria, Tasmania and New South Wales was very low until the early 1980s (Fig. 3). The earliest reports (not associated with whaling) date back to the first decade of the 20th century (two in New South Wales and one in South Australia). Between 1904 and 1969, only 18 reports were documented. In 1982 and 1983 the number of reports rose sharply in all four states except New South Wales. Overall, the number of reports was highest in South Australia and Victoria and lowest in New South Wales, possibly because until recently there has been no co-ordinated effort to record sightings in the latter state. There was great variability between years in the number of sightings in each state. In Victoria, where recording effort was the most consistent, there were low numbers of reports in

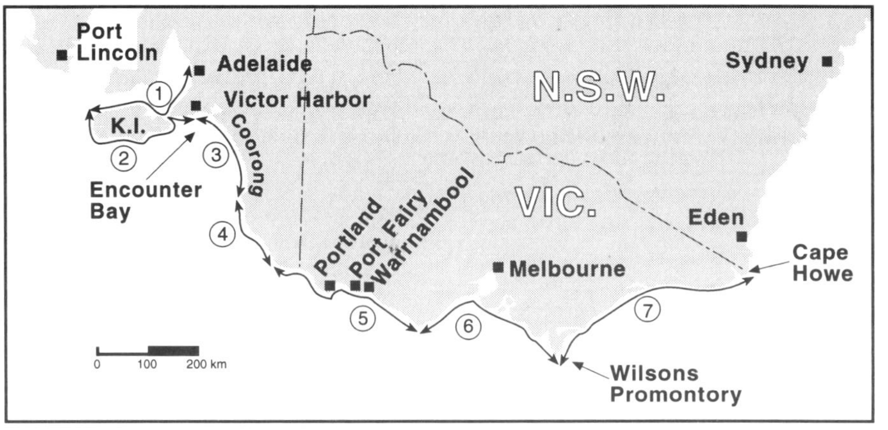

Fig. 2. Seven coastal regions adopted for assessment of possible relationship between oceanographic and coastal features and Southern Right Whale sightings on aerial surveys.

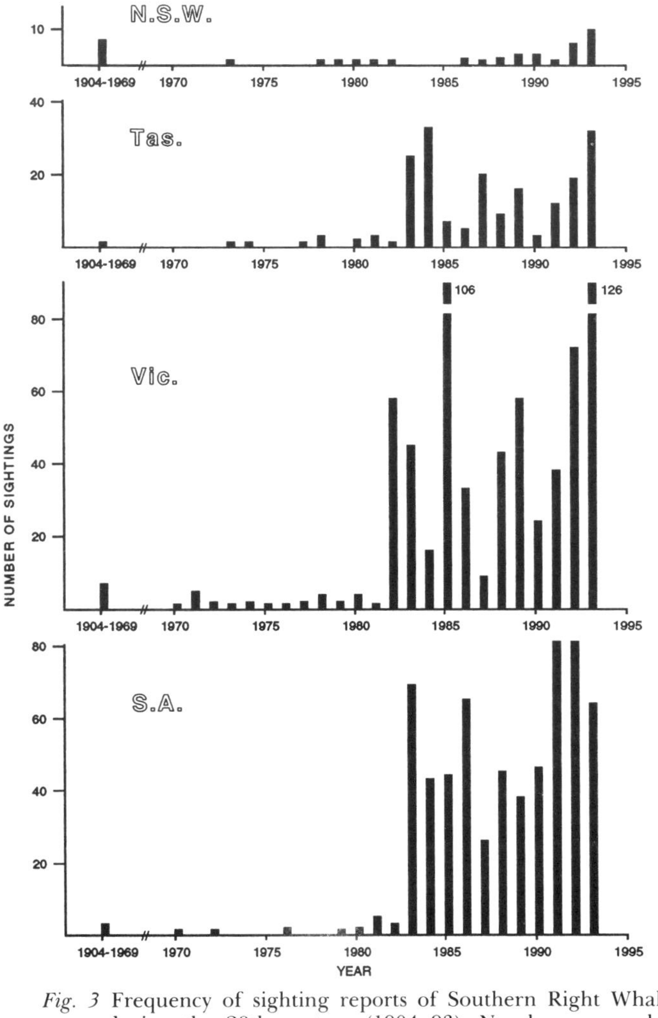

Fig. 3 Frequency of sighting reports of Southern Right Whales during the 20th century (1904–93). Numbers are public reports of sightings, not individuals. South Australian reports are east of 135°30′E.

1984, 1987 and 1990, but not in 1993, as might be expected from the 3-year inshore cycle of calving Southern Right Whales.

Inshore season

According to the public reports, Southern Right Whales were seen in southeastern Australia during all months of the year except February (Fig. 4). Their appearance in inshore waters of South Australia (April–December), Victoria (May–November) and New South Wales (June–November) was less protracted than in Tasmania (all months except February). Outside the period of study, but relevant to this discussion, is the record of a distinctively coloured adult and large calf which traversed part of Victoria's west coast early in January 1994. In South Australia and Victoria the peak season, here defined as those months with >10% of total sightings, was June to September. In Tasmania the peak season was June to August with 7% of sightings being made in November, at a time when the other states had few sightings. These may have been of animals migrating southward for the austral summer.

A more accurate assessment of the peak months for inshore numbers can be gained from the results of our aerial surveys (Fig. 5) and from counts of individually identified whales (Fig. 9). These are largely from the coast of South Australia and Victoria. On aerial surveys, whales were seen between July and September 1991, July and October 1992, and June and October 1993. Adjusted aerial survey counts of whales (Fig. 5) showed peaks in August 1991, September 1992 and August 1993. Except for 1991, most were females with calves. No animals were seen during June surveys in 1991 and 1992, although incidental sightings were made by CNR staff and the public. Counts of individually identified whales, combined by month for all three years (Fig. 9), showed peak numbers were recorded during July (21) and August (25), with lower numbers in June (14), September (18) and October (6).

Distribution

On aerial surveys, Southern Right Whales were more frequently (71% of

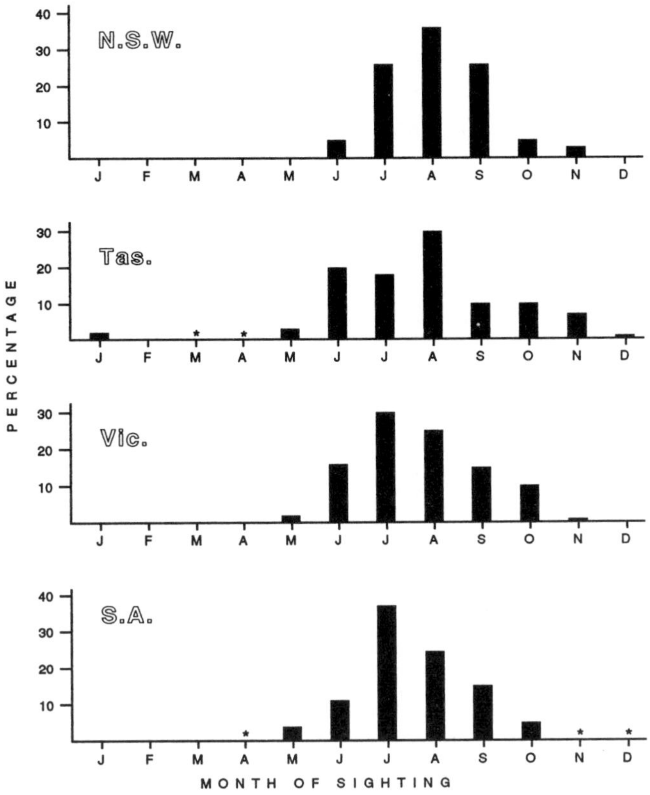

Fig. 4. Time of year Southern Right Whales sighted in southeastern Australia (1909–93). Data are public reports pooled by month within states. Sample sizes: New South Wales, 39; Tasmania, 190; Victoria, 673; South Australia, 612. * less than 1%.

adults) seen within 500 m of the shore than 500 m to 2 km offshore. This is in agreement with the results of Best (1990b), who found that 90% of Southern Right Whales within about 11 km of the southern African coast were within 1.85 km of the shore. It is possible to see beyond the 2 km transect during good viewing conditions, as evidenced by the sighting of a Humpback Whale *Megaptera novaeangliae* at about 3 km offshore in far western Victoria. No right whales were seen beyond the 2 km zone.

Southern Right Whales have been recorded, by aerial surveys and public reports, along most of the coast of southeastern Australia (Fig. 6). Few or no sightings along some remote or sparsely inhabited coasts (e.g., western Tasmania) may be attributable to low observer effort. Aerial surveys during 1991–93 along the mainland coast recorded most whales in the western half of the study area, between Kangaroo Island and Cape Otway (Fig. 7). On these surveys frequent sightings were made at Warrnambool (especially in 1992 and 1993) and south-east of Victor Harbor (especially in 1991).

Survey results suggest that certain areas of the coast are being visited by whales more frequently than others. To investigate this, raw aerial survey counts of animals (including repeat sightings) were tallied for each of the seven coastal regions adopted. To illustrate these data graphically, it was necessary to adjust the counts to survey effort in each region because not all regions were surveyed each time, or were partially surveyed, and because the regions were of different length. This was accomplished by dividing the animal counts by the number of surveyed kilometres, per region, for all three years (Fig. 8). Raw counts were used to test the differences between observed and expected sighting frequencies, relative to total survey effort, and these were highly significant for all animals ($\chi^2 = 57.34$, $P < 0.005$), for adults with calves ($\chi^2 = 30.40$, $P < 0.005$) and for adults without calves ($\chi^2 = 63.88$, $P < 0.005$). Southern Right Whales were most frequently seen in Region 5, which includes Warrnambool, and

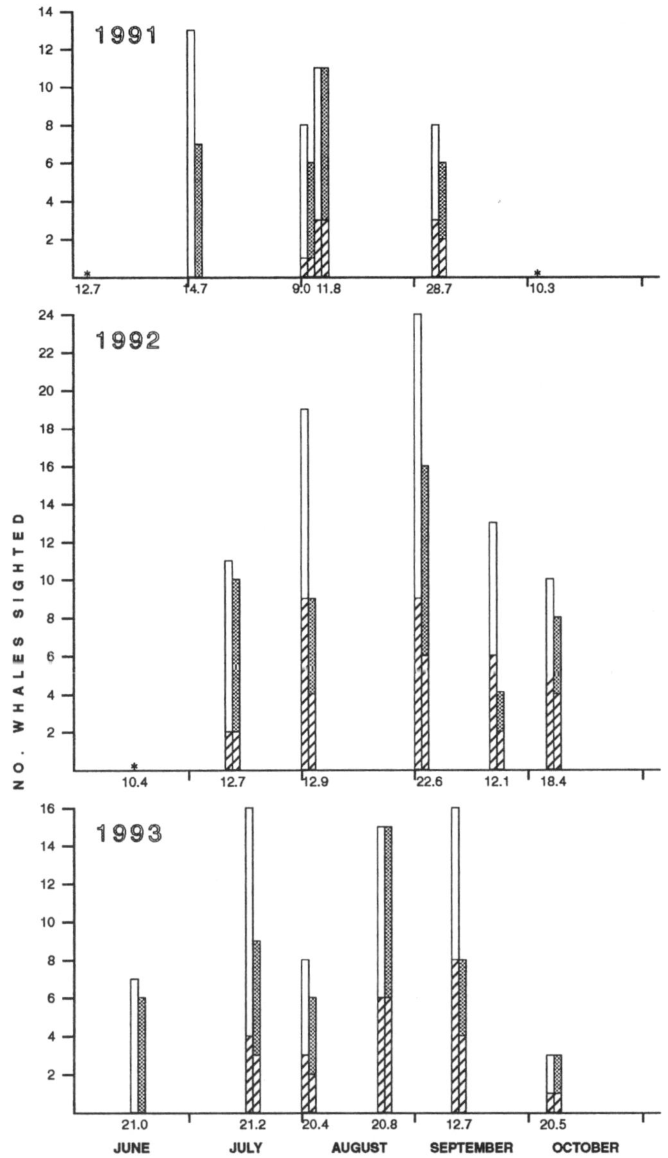

Fig. 5. Aerial survey results for Southern Right Whales, 1991–93. Number of whales seen is illustrated in unadjusted (left bar) and adjusted (right bar) form. * no whales seen. Numbers below bars are flying hours. Open bar, unadjusted adults; dotted, adjusted adults; hatched, calves.

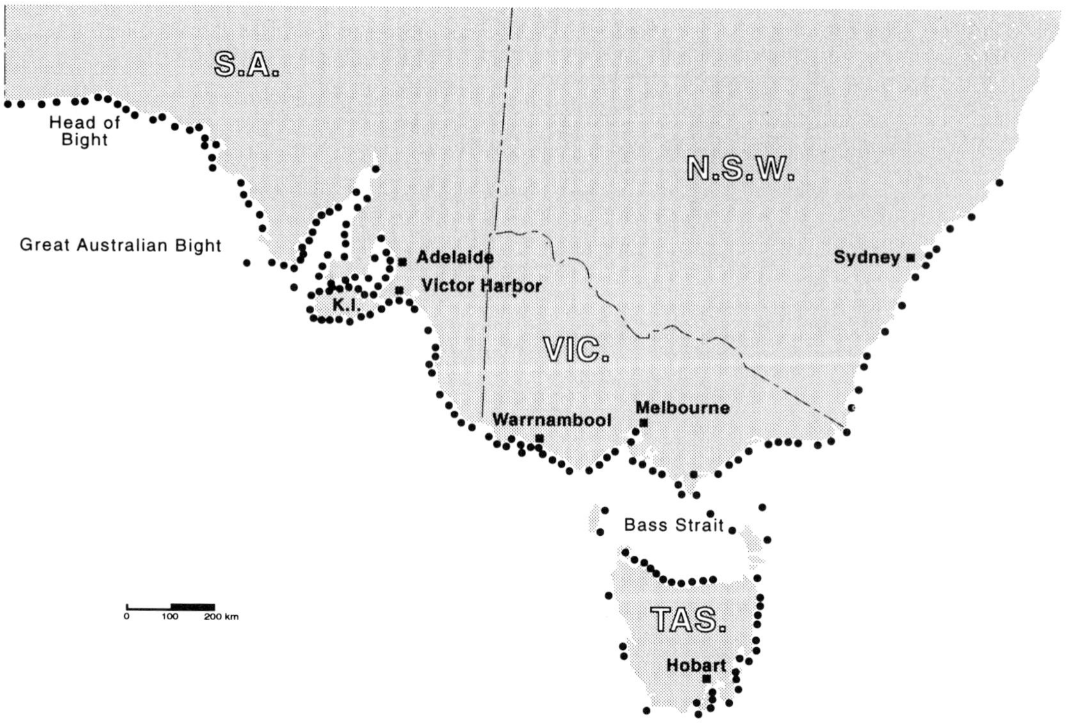

Fig. 6. Distribution of sightings of Southern Right Whales, 1904–93. Data from all sources. Each closed circle represents a locality, not the number of sightings.

Region 3, the Coorong coast (Fig. 8). The striking difference between the two, however, was that in Region 5 the number of adults with calves was higher than adults without calves whereas in Region 3 it was the converse. Whales, usually adults without calves, were also seen quite frequently in Region 2, the south coast of Kangaroo Island and Encounter Bay. Animals were rarely seen in Region 1 (north coast of Kangaroo Island) and Region 6 (between Cape Otway and Wilsons Promontory). There is an increasing trend, from both east and west, of sightings of adults with calves as Warrnambool is approached.

Table 1. Southern Right Whales individually identified using images from all sources during 1991–93. "Adults without calves" includes possible subadults, since accurate size was not determined. Effort increased during 1992–93 because of inclusion of photographs from other sources and opportunistic photography in Victoria, South Australia, Tasmania and New South Wales. No aerial surveys in Tasmania.

Year	State	Adults without calves	Females with calves
1991	SA and Vic.	6	1
1992	SA and Vic.	4	9
	Tas.	4	0
1993	SA and Vic.	14	9
	Tas.	4	1
	NSW	0	2
Total		32	22

Population size

Fifty-four adults (i.e., non-calves) were individually identified during 1991–93 (Table 1). Most were photographed on aerial surveys and opportunistic flights. States such as New South Wales and Tasmania, where very little or no aerial survey work was carried out, had very few identified individuals. Calves accompanied 22 of the identified adults, thus making the minimum number of animals visiting south-eastern Australia during 1991–93, 76 Southern Right Whales. It was not possible to determine sex ratios except to the extent that the minimum proportion of mature females was 41% of the non-calf population. The actual figure is likely to be higher because animals seen without calves at the beginning of the season but not resighted may have calved without this being detected.

The proportion of identified adults with and without calves differed between Tasmania (11%) and the mainland (44%) (Table 2), however sample sizes were small.

Table 2. Southern Right Whales individually identified during aerial surveys (not opportunistic flights) off the mainland coast, 1991–93. "Adults without calves" includes possible subadults, since accurate size was not determined for these animals. * high proportion of photographs unsuitable for identification of animals.

Year	Adults without calves	Females with calves
1991*	6	1
1992	3	7
1993	11	6

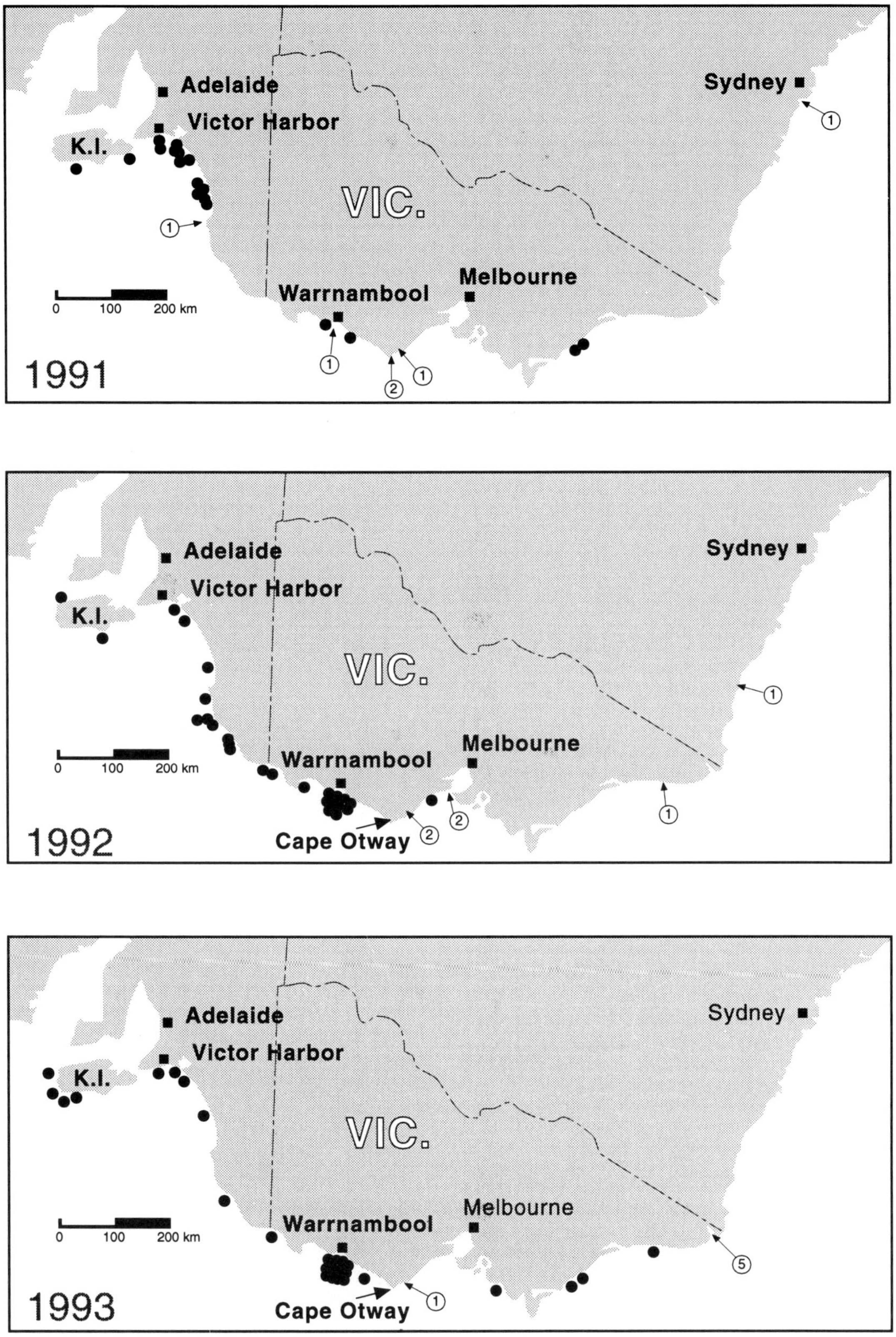

Fig. 7. Sightings (closed circles) of Southern Right Whales made on aerial surveys during 1991, 1992 and 1993. Repeat sightings (same individual at same place) within survey excluded. Destination and number of surveys to that destination included in open circles.

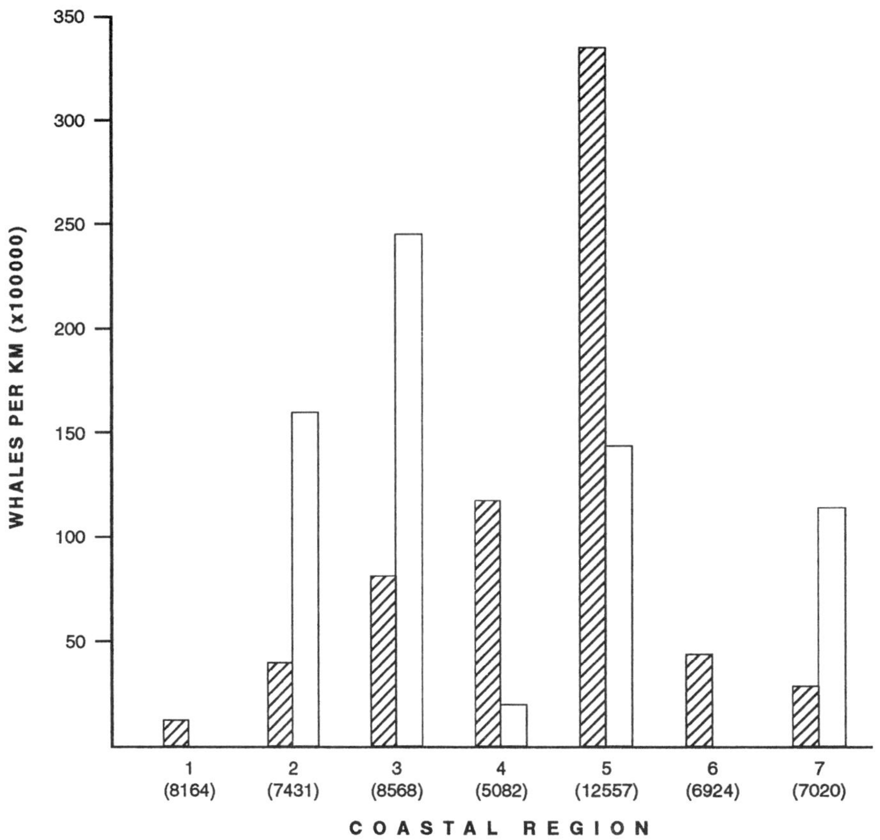

Fig. 8. Southern Right Whale sightings made in coastal regions 1–7 (see Fig. 2) during aerial surveys, 1991–93. Counts are raw sightings, not adjusted for repeats within surveys and are whales per survey kilometre of coast. Total survey kilometres per region appears at base of graph. Open bar, adults with calves; hatched bar, "adults without calves".

Possible differences between years in the proportions of adults with and without calves were evident in the results obtained during aerial surveys (Table 2). Females with calves represented 14% (1991), 70% (1992) and 35% (1993) of the identified animals on planned surveys. A possible reason for such inter-year variation is differential strength of year-classes. Best (1981) showed that the proportion of calving females varied between 26% and 56% over a 10-year period off southern Africa.

Reproduction

It was possible to determine the month of birth of five calves. Three were born in July, one in August and one in September. Those born in July and August were timed by observing identified adults first without, and later with, calves (Fig. 9). The September birth was at Bermagui, New South Wales, and was observed and filmed by local residents. The earliest reported calf was in early May but this is probably too early in the season for calves to be present and, being a public report, may have been a mistaken yearling. The proportion of identified whales which were females with calves increased from 7% in June to 83% in October.

Movements

Positive resightings of identified animals between years within the study area were not made, however one adult (93/28) was possibly seen on two successive years (Fig. 9). Several good photographic matches were recorded between the study area and Head of Bight, both within and between years. One individual (adult male 92/10, S. Burnell, pers. comm.) was seen in southeastern Australia during 1992 then at Head of Bight in 1993 (Fig. 9). This male has also been seen at Head of Bight in 1989 (J. Ling and D. Needham, unpubl. data). A female seen at Warrnambool with a calf in 1993 was seen at Head of Bight with a calf in 1984, 1987 and 1990. No matches were recorded between southeastern Australia and Western Australia (J. Bannister, pers. comm.). Within years, two individuals were first recorded along the central South Australian coast and later at Head of Bight, one each in 1992 and 1993. Both had moved in an east to west direction. One was a full grown male (S. Burnell, pers. comm.), the other was a yearling, first seen as a calf at Head of Bight in 1991 (J. Ling and D. Needham, unpubl. data).

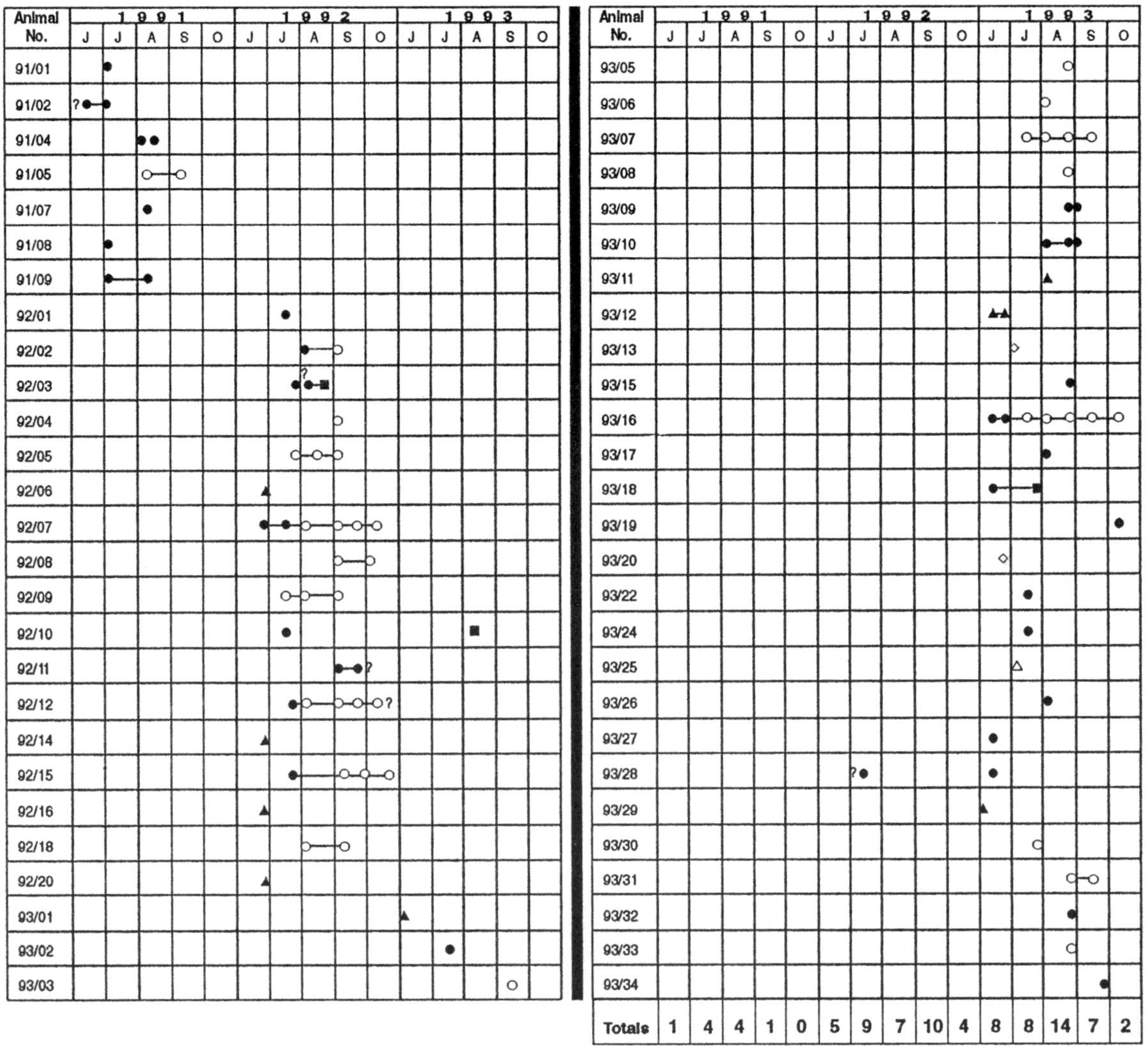

Fig. 9. Presence of individually identified Southern Right Whales in southeastern Australia during 1991–93. Solid symbols are "adults without calves", open symbols are adults with calves. Each symbol represents a different identified sighting. Monthly totals do not include Head of Bight animals. ●, ○ South Australia and Victoria, ▲ Tasmania, ■ Head of Bight, ◇ New South Wales. ? reliability of match = 3.

In the study area, many movements of known individuals were recorded within years (Figs 10–12). None was recorded between Tasmania and the mainland but since only nine of the 54 identified individuals were recorded from Tasmania, that sample may be too small to detect such movements. The number of repeat sightings within the study area was variable between years (Table 3). In 1991, when only aerial surveys were carried out, no animal was recorded more than twice. In 1992, with additional information from opportunistic flights, four known animals, all females with calves, were seen four or more times in the study area. Two of these were long-term within-year residents of the Warrnambool area and two were seen primarily in the South Australian gulfs. In 1993 two animals at Warrnambool, both females with calves, were sighted more than three times, one over a period of nine weeks , the other over 16 weeks.

Table 3. Number of times individual Southern Right Whales identified in southeastern Australia (including New South Wales) during each year of study. Aerial surveys and opportunistic flights included. Some were repeat sightings during same aerial survey. Opportunistic flights introduced in 1992 and 1993.

No. times identified	1991	1992	1993
1	3	7	24
2	4	5	2
3	0	1	2
4	0	1	0
5	0	1	0
6	0	0	1
7	0	1	0
8	0	1	1
Total	7	17	30

Long-range movements were recorded for four individuals without calves (one of which was a yearling) and all were in an east/west direction. The longest was for animal number 93/10, first recorded off northeastern Victoria

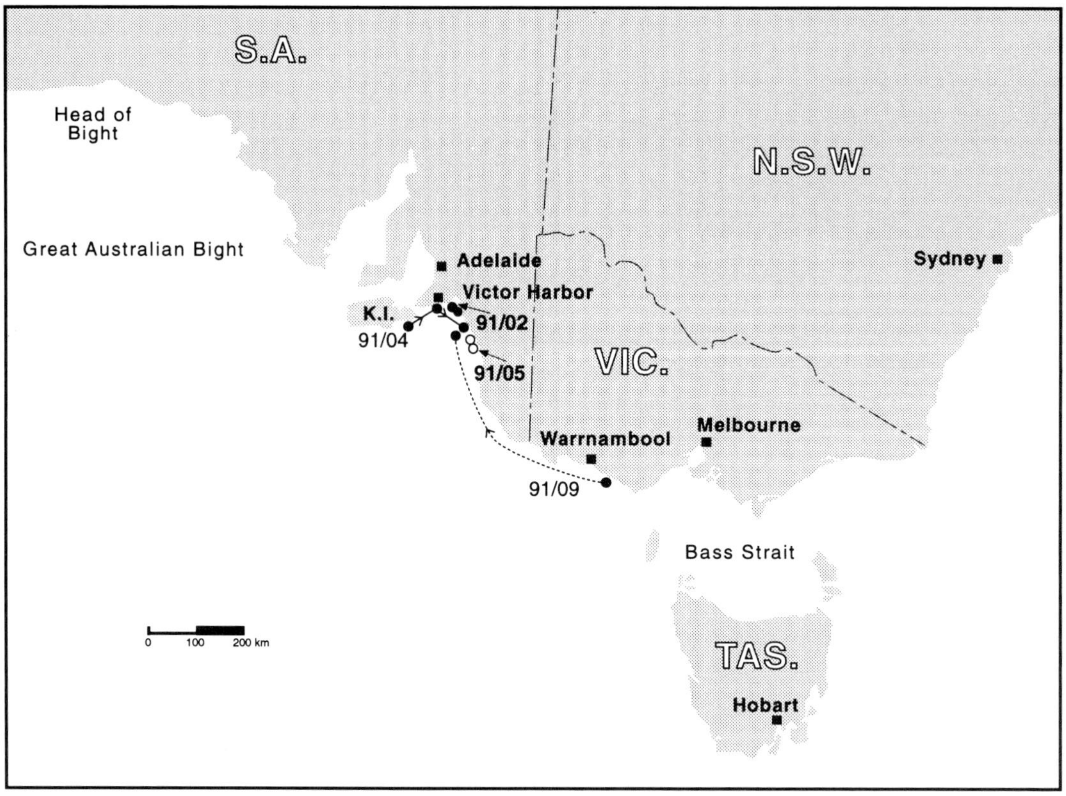

Fig. 10. Movements of individually identified Southern Right Whales during 1991. Animal number cross-referenced to Figure 9, showing month sighted. Only animals with >1 sighting mapped. Broken and solid lines which join consecutive sightings help to distinguish individuals and do not imply routes of travel. ● "adults without calves", ○ adults with calves.

on 4 August 1993 then off the northern part of the Coorong, South Australia on 25 August 1993, a distance of about 1 000 km (Fig. 12). The only known adult female (92/15) which made a long-range movement was first seen at Port Fairy, just west of Warrnambool, without a calf on 17 July 1992, and then on 8 September 1992 with a calf east of Victor Harbor, having travelled at least 500 km (Fig. 11). All other repeat sightings of females calving during the season were either in a very restricted area, e.g., Warrnambool, or within a range of less than 300 km.

As a result of the large and unprecedented influx of seven females with calves to Warrnambool in 1982, observer interest in Victoria has remained high and sightings by the public and CNR staff have been conscientiously logged. As indicated in Table 4, the number of whales present has fluctuated widely from year to year, as well as the overall period of attendance. Peak numbers of female:calf pairs as well as years of zero attendance appear to be cyclical at intervals of three or four years. There is a close correlation between length of tenure and number of female:calf pairs. During this sequence of 13 seasons, except in those years when no female:calf pairs occupied the Warrnambool nursery area, fewer sightings of Southern Right Whales were reported elsewhere on the Victorian coast and their recorded tenure at any specific locality was usually very short, from one to about 10 days.

DISCUSSION

Population size

Hammond (1986) reviewed the methods of estimating population size for whales where natural markings, such as the callosities of Southern Right Whales, allow individuals to be identified. However, the assumptions which must be met in applying the models for populations size estimation are quite restrictive and there is disagreement (Hammond 1986; Whitehead *et al.* 1986; Best and Underhill 1990) as to which methods and how many samples should be used.

Best and Underhill (1990) developed a purpose-built model based on the Bayesian method for a closed population in their study of Southern Right Whales off southern Africa. This model took into account the unequal "recapture" probabilities and increasing population size of the southern African population. Their method requires three years' data of identified females with calves before the first estimate can be made. In the fourth year all females then would have equal probability of being present and reidentified. Since the present

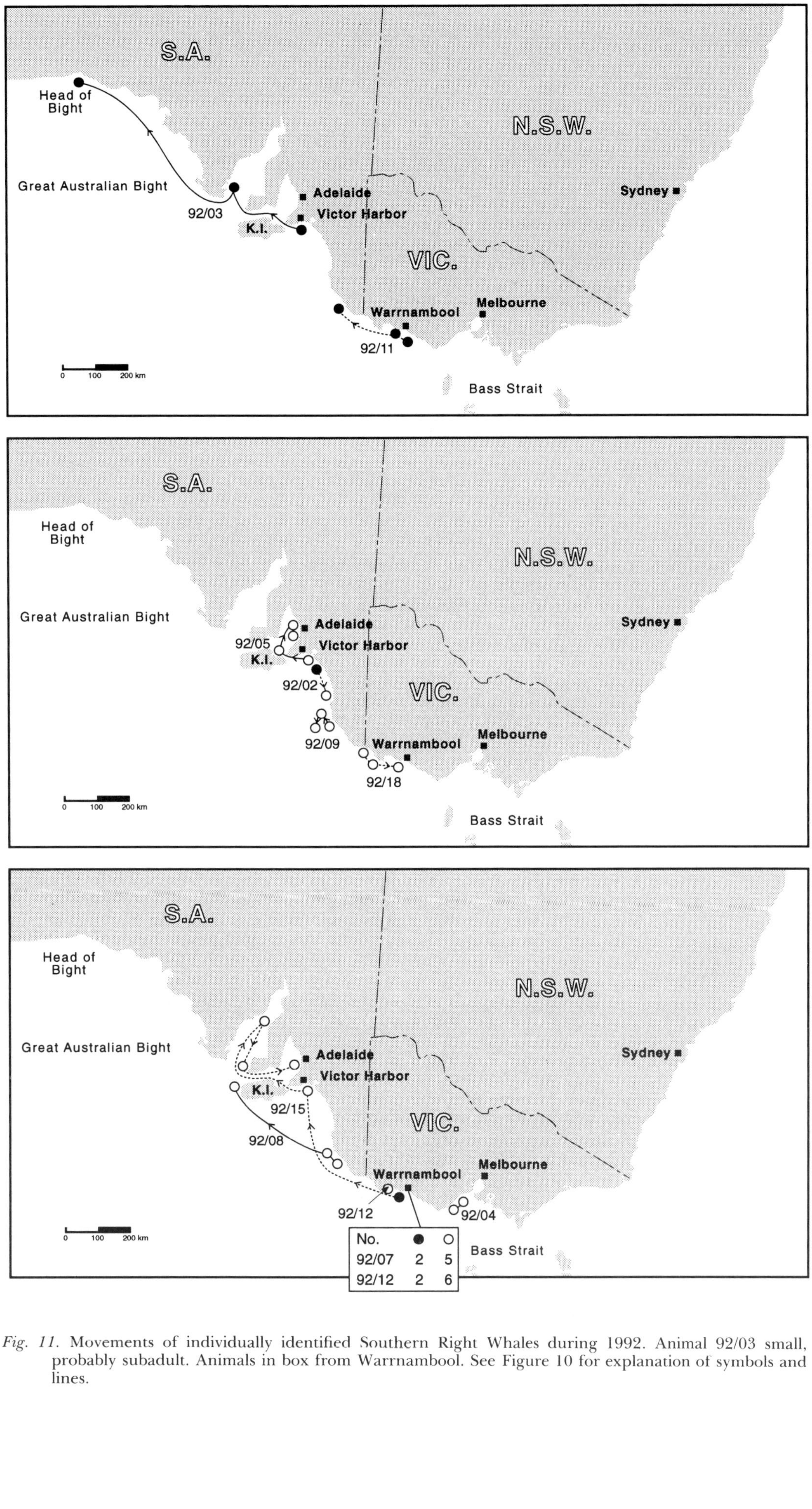

Fig. 11. Movements of individually identified Southern Right Whales during 1992. Animal 92/03 small, probably subadult. Animals in box from Warrnambool. See Figure 10 for explanation of symbols and lines.

Table 4. Attendance of Southern Right Whales in the Warrnambool area, 1981–93. Animals not photo-identified during 1981–90. Period = number of days from first to last sighting for season. Maximum pairs = greatest number of pairs seen at one time, except during 1991–93, when animals were photo-identified.

Year	First sighting	Last sighting	Period (days)	Maximum female:calf pairs
1981	21 July	30 July	10	1
1982	27 May	11 October	138	7
1983	1 July	28 August	59	1
1984	—	—	—	0
1985	23 June	22 October	122	5
1986	14 June	11 September	90	1
1987	—	—	—	0
1988	10 July	25 September	78	1
1989	2 June	17 October	138	4
1990	8 June	15 August	69	1
1991	1 June	31 July	30	0
1992	26 June	20 October	107	3
1993	15 June	8 October	116	4

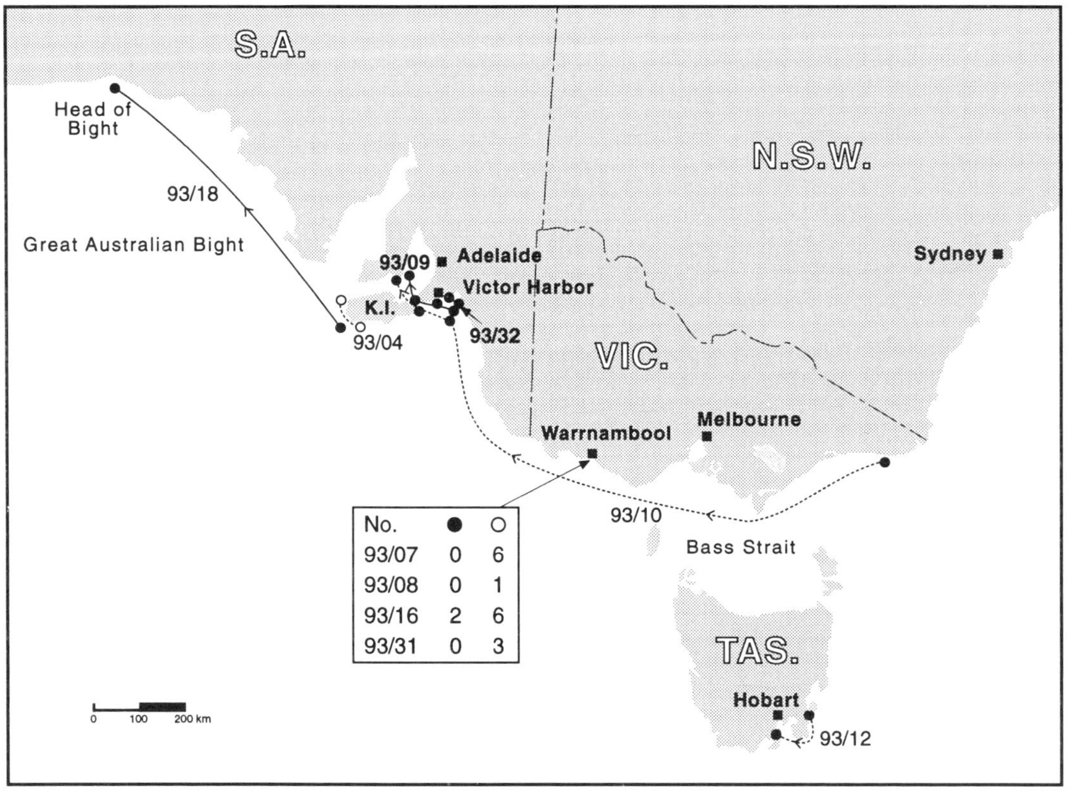

Fig. 12. Movements of individually identified Southern Right Whales during 1993. See Figures 10 and 11 for explanation of symbols and lines.

study involved only three years of sampling and very low numbers of animals, we have not attempted to estimate absolute numbers of Southern Right Whales in southeastern Australia using that methodology.

The 54 adult (i.e., non-calf) Southern Right Whales identified in the study area during 1991–93 represented the minimum number of adults visiting the region. At least 22 were females with calves. Other studies (Payne *et al.* 1990; Best 1990b; Bannister 1990) have shown that females return to the coast to give birth, on average, once every three years. For this reason and because females with calves are relatively sedentary and easily identifiable as belonging to such a group, this part of the population is often used when comparing population size within and between areas. Bannister (1994) recorded 63 adults with calves between Cape Leeuwin, Western Australia, and Ceduna, South Australia (about 2 000 km of coast) during the peak of the season in 1993 and estimated that over a three-year period there would be a minimum of 189 females with calves (assuming no difference in the size of the calving female cohort each year) visiting that part of the Australian coast. In southeastern Australia, females with calves (identified in our overall study area of roughly 3 500 km of coast) therefore made up only 10% (22/211) of the minimum estimate of Australian females with calves. It is likely that the actual number of females with calves in the south-eastern states was higher because our study put very little effort into survey flights off New South Wales and only incidental flights were conducted in Tasmania. Bannister (1994) estimated the total "population" of Southern Right Whales in the western half of Australia to be about 800 animals.

Female:calf numbers off Western Australia have been increasing at a rate of about 8–10% per annum (Bannister 1994). This rate, which may be compounded by immigration, e.g., from South Australia, is somewhat higher than estimates of increase (6.8%) for southern Africa (Best 1990a) but similar to rates of increase for South America (Payne *et al.* 1990). Trends in population abundance of Southern Right Whales can only be determined by medium to long-term monitoring, taking in several three-year cycles of this long-lived and species with a low reproductive rate. In southeastern Australia, where numbers are very low, a statistically significant increase may still be difficult to demonstrate for some time but it is critical that aerial surveys be carried out in this early stage of a presumed recovery.

Although public reports of whale sightings can be misleading because they are affected by observer effort, the rarity of Southern Right Whale reports before 1970 seems likely to be a reflection of their status at that time. Recently-released Russian whaling catch statistics have shown that substantial numbers (at least 3 212) of Southern Right Whales were taken during the 1950s and 60s (Brownell, pers. comm.). The sudden increase in reported sightings off south-eastern Australia during the early 1980s could be explained by both the increased public and media attention given to whales at that time and the real increase in whale numbers that would have occurred as a result of cessation of whaling about 20 years previously.

Distribution

Southern Right Whales have been seen along almost all parts of the coast of southeastern Australia. The results presented in this study, however, point to some areas being more frequently used than others and more years of study of this possibly-expanding population may change the patterns observed during 1991–93. Even within the short period of study there was variation from year to year in which areas were most frequently used, e.g., the reduced importance of Warrnambool during 1991 compared to 1992 and 1993. Interestingly, there were again almost no Southern Right Whales, and no females with calves, at Warrnambool during the 1994 season. In an expanding population off southern Africa, Best (1990a) noted shifts in the locations of peak densities of adults with calves. Major 19th century whaling stations in south-eastern Australia were located at Port Lincoln (2 stations), Victor Harbor (2), Portland (1), Port Fairy (1), Eden (1) and southeastern Tasmania (8) (Dawbin 1986).

Since 1982 the Warrnambool region appears to have been the most frequently used calving area in southeastern Australia. Like Head of Bight, an important calving region in the western part of the continent, Warrnambool is on a high-energy coast, facing the Southern Ocean and experiencing more powerful swells than anywhere along the Victorian coast (Land Conservation Council and Department of Conservation and Natural Resources 1994). The substrate is rock overlaid with sand and the beach profile slopes moderately. The females and calves observed at Warrnambool and many other places in southeastern Australia were often seen loafing just outside the breakers where the water depth is 5–6 m. Best (1990a) noted that the peak density of right whales occurred at same distance offshore as the average distance that the breakers occurred (0.17 km) in southern African and that water depth at 0.93 km offshore, the outer edge of the adult/calf distribution, was usually less than 15 m and often less than 10 m. Payne (1986) carefully documented the preference of adults with calves for the 5 m water depth contour in Argentina. Bannister (1990) concluded that several areas along the Western Australian coast were intensively used by calving Southern Right Whales, especially the Albany to Hopetown region. This coast predominately faces the south-east and also experiences heavy swells. Big surf with breakers at a distance of up to 500 m from the water's edge (Best 1990a) and a beach, rather than cliff environment, may be important components of the habitat of females with calves. If this is so, coastal regions such as eastern and central Bass Strait, the Adelaide area and northern Kangaroo Island would not be favoured because they have lower wave energy. A possible advantage of a high energy coast to calving Southern Right Whales is that such a habitat has high levels of sound, which could make it difficult for acoustically-sensitive predators such as Killer Whales to detect the presence of vulnerable whales. Warrnambool and Head of Bight have another, and possibly important, feature in common. They both have cliffs and deeper water adjacent to the main nursery site which appear to be used as "exercising" areas for young whales (S. Burnell, pers. comm.; Ling and Needham 1985–91).

Movements

In southeastern Australia a yearling and adults without calves moved over much longer distances and were less likely to be seen regularly along the coast than females calving during a season (Figs 9–12). They are pobably transient animals. In contrast, females with calves were more restricted in their seasonal movements. Those at Warrnambool were sighted over long periods (up to almost 20 weeks) in the area. Other females with calves moved within an area of up to 300 km. Bannister (1990) found that Western Australian females with calves stayed up to 13 weeks within his study area and adults without calves did not stay longer than six weeks. He also noted that 68% of females with calves did not move more than 100 km from where they were first seen in a particular season.

Long distance movements have been recorded between Western Australia and Head of Bight, and between Head of Bight and southeastern South Australia both within and between years, including one female which was seen with a calf in 1984, 1987 and 1990 at the Head of Bight (Ling and Needham, unpubl. data) and in 1993 at Warrnambool. This indicates that fidelity to a nursery site is not absolute. Payne (1986) showed that most mature females returned on average every three years to the Península Valdés area of Argentina. Bannister (1990) demonstrated that 81% of adults with calves in Western Australia returned to places within 300 km of where they were first sighted. No individuals from southeastern Australia have been seen in Western Australia. Best *et al.* (1993) reported on an instance of widely-spaced nursery sites (2 000 km apart) for a female in South America. These authors also noted very long range movements made by other animals, some spanning over 50 latitudinal degrees in the south Atlantic, thus showing the potential for mingling between "populations".

The migratory path(s) of Southern Right Whales between presumed summer feeding grounds in the sub-Antarctic and winter breeding grounds along southern Australia is not known. As summarized by Bannister (1986b) and Dawbin (1986), nineteenth-century whalers believed that the whales appeared off Tasmania in early April then moved north to either western Victoria and on to the Great Australian Bight or to New South Wales. There was apparently also an arrival of whales directly from the south to the coast of the continent. The return route was believed to have been well offshore (several hundred miles) and to the south-east, again taking in Tasmania. Townsend's (1935) charts agree with these hypotheses. The results of the present study were too few to support or refute the above suggestions of migration routes. The few long range movements that were recorded were adults without calves travelling in an east–west direction and some went as far as the Head of Bight (S. Burnell, pers. comm.). These were coastal movements and may not be comparable to the whaling ship catch data, which would have been offshore. Calving females may have different routes and destinations from other migrating parts of the population. No movements were recorded between Tasmania and the mainland but many more animals need to be identified from that state before a conclusion can be reached as to whether coastal Tasmania is part of a migration path for Southern Right Whales.

ACKNOWLEDGEMENTS

We gratefully acknowledge the funding support of BHP Petroleum and the help of their staff, Rick Nash, Jane Cutler and Robert Bell. We thank the many people who contributed public reports. Special thanks in this regard go to Robert and Helen Ulmann of Allansford who contributed extensive series of daily observation, to the Friends of the Southern Right Whale, to staff of the Victorian Department of Conservation and Natural Resources (especially Philip Du Guesclin, Peter Goldstraw, Graeme Hanel, Ian Westhorpe and Brad Dale) and of the South Australian National Parks and Wildlife Service (especially Chris Halstead) and of the South Australian Department of Primary Industry, Fisheries (Trevor Whibley). Les and Pauline McDiarmid of the Whale Information Network, Victor Harbor, and Ian Milne of Victor Harbor Whale Watch Centre passed on many useful reports.

Our pilots, especially John Rogers, did an admirable job. Special thanks to Rossair and Master Flight and The Aviation Authority. We thank the many people who generously contributed photographic material, including Irynej Skira, Channel 7, 9 and 10 and Prime TV (Newcastle). Many volunteers helped with various aspects of the project, including Martin Verco, Walter Hammon and Bruce Wilkinson. We thank John Bannister, Steve Burnell and Stuart Anstee for assistance with matching photographs of identified individuals and allowed us to use information from their studies of Southern Right Whales. South Australian and Victorian wildlife authorities gave permission to photograph whales below the guideline limits of 300 m. Debbie Churches and Debbie Van Weenen typed the manuscript and Pen Roberts prepared the figures.

REFERENCES

Bannister, J. L., 1986a. Southern Right Whales: status off Australia from twentieth-century "incidental" sightings and aerial survey. *Rep. Int. Whal. Commn. (Special Issue 10)* 153–58.

Bannister, J. L., 1986b. Notes on nineteenth century catches of Southern Right Whales *(Eubalaena australis)* off the southern coasts of Western Australia. *Rep. Int. Whal. Commn. (Special Issue 10)* 255–59.

Bannister, J. L., 1990. Southern Right Whales off Western Australia. *Rep. Int. Whal. Commn. (Special Issue 12)* 279–88.

Bannister, J. L., 1994. Report on aerial survey for Southern Right Whales off southern Australia, 1993. Report to Australian Nature Conservation Agency, Canberra.

Best, P. B., 1981. The status of right whales *(Eubalaena glacialis)* off South Africa, 1969–1979. *Invest. Rept Sea Fish. Inst.* **123:** 1–44.

Best, P. B. and Underhill, L. G., 1990. Estimating population size in Southern Right Whales *(Eubalaena australis)* using naturally marked animals. *Rep. Int. Whal. Commn. (Special Issue 12)* 183–89.

Best, P. B., 1990a. Trends in the inshore right whale population off South Africa, 1969–1987. *Mar. Mammal Sci.* **6:** 93–108.

Best, P. B., 1990b. Natural markings and their use in determining calving intervals in right whales off South Africa. *S. Afr. J. Zool.* **25:** 114–23.

Best, P. B., Payne, R., Rowntree, V., Palazo, J. T. and Both, M. D. C., 1993. Long-range movements of South Atlantic right whales *Eubalaena australis. Mar. Mammal Sci.* **9:** 227–34.

Chittleborough, R. G., 1956. Southern Right Whale in Australian waters. *J. Mammal.* **37:** 456–57.

Dawbin, W. H., 1986. Right whales caught in waters around south eastern Australia and New Zealand during the nineteenth and early twentieth centuries. *Rep. Int. Whal. Commn. (Special Issue 10)* 261–67.

Edyvane, K. and Baker, J., 1994. The South Australian Marine Regionalization Project, ANZECC's Marine Biogeographic Regionalization Workshop, 23–25 March 1994. South Australian Research and Development Institute, Adelaide.

Hammond, P. S., 1986. Estimating the size of naturally marked whale populations using capture-recapture techniques. *Rep. Int. Whal. Commn. (Special Issue 8)* 253–82.

Land Conservation Council and Department of Conservation and Natural Resources, 1994. Environmental inventory of Victoria's marine ecosystems — stage one biophysical classification final report. Melbourne.

Ling, J. K. and Needham, D. J., 1985–1991. Annual reports on consultancy agreements with Australian National Parks and Wildlife Service, Canberra.

Payne, R., 1986. Long term behavioural studies of the southern right whale *(Eubalaena australis). Rep. Int. Whal. Commn. (Special Issue 10)* 161–67.

Payne, R., Rowntree, V., Perkins, J. S., Cooke, J. G. and Lankester, K., 1990. Population size, trends and reproductive parameters of right whales *(Eubalaena australis)* off Península Valdés, Argentina. *Rep. Int. Whal. Commn. (Special Issue 12)* 161–67.

Townsend, C. H., 1935. The distribution of certain whales as shown by logbook records of American whaleships. *Zoologica, N.Y.* **19:** 1–50.

Whitehead, H., Payne, R. and Payne, M., 1986. Population estimate for the right whales off Península Valdés, Argentina, 1971–1976. *Rep. Int. Whal. Commn. (Special Issue 10)* 169–71.

A census of fur seals at Montagu Island, New South Wales

A. IRVINE,[1] M. M. BRYDEN,[1] P. J. CORKERON[1] and R. M. WARNEKE[2]

The colony of Australian Fur Seals *Arctocephalus pusillus doriferus* at Montagu Island was censused every fourth week from May 1993 to May 1994. There was a maximum of 305 animals in September 1993 and a minimum of nine in December 1993. Prior to this study *A. pusillus doriferus* was thought to be the only species of fur seal on the island, but in 1993 the New Zealand Fur Seal *Arctocephalus forsteri*, was identified there as well, both in the main colony and on its own. One to three New Zealand Fur Seals were seen each month from December to June, 13 to 18 each month from July to November. An Australian Fur Seal pup born in October 1993 demonstrated that the colony is not comprised exclusively of bachelors, but the frequency of occurrence and prevalence of females on the island have yet to be determined.

INTRODUCTION

BREEDING colonies of the Australian Fur Seal *Arctocephalus pusillus doriferus* are restricted to nine islands within Bass Strait (Warneke and Shaughnessy 1985; Pemberton and Kirkwood 1994) (Fig. 1). Haulout sites extend the range much further, from as far west as the North Casuarina Islets at Kangaroo Island (Shaughnessy *et al.* 1994), south to Maatsuyker Island off southern Tasmania (Brothers and Pemberton 1990), and north to Montagu Island in New South Wales (Warneke 1982) (Fig. 1).

The New Zealand Fur Seal *Arctocephalus forsteri* breeds in New Zealand, on the southern coasts of Western and South Australia (King 1969; Warneke 1982), and at Maatsuyker Island off the south western coast of Tasmania (Brothers and Pemberton 1990). *Arctocephalus forsteri* is increasing in numbers throughout southern Australian waters (Shaughnessy *et al.* 1994), but hitherto there was no record of this species on Montagu Island.

Montagu Island (36°15′S, 150°14′E) is now recognized as the major haulout site of *A. pusillus doriferus* in New South Wales. Seal Rocks (32°28′S, 152°33′E), once a breeding site, diminished from 500 individuals in 1925 to 12 in 1974 (Warneke 1982), and has not been occupied for at least 10 years (Warneke, unpubl. data).

Arctocephalus pusillus doriferus prefer bare rock or pebble beach substrates at their breeding sites (Brothers and Pemberton 1990). *Arctocephalus forsteri* are most often seen on irregular rocky habitat such as jumbled boulders, often avoiding open rock platforms and pebble beach habitats (Goldsworthy *et al.* 1997), Montagu Island provides all these habitats with direct access to the ocean. The island is 6 km offshore, it is roughly rectangular, 1 400 m by 525 m and 49 ha in area (Fig. 2). It is a well vegetated island practically divided into two by a gulch, creating a smaller northern end of basalt composition and a larger southern end of granite. The dominant vegetation is matrush *Lomandra longifolia* often with creepers *Eustrephus* sp. and blady grass *Imperata cylindrica* (Fullagar 1973). At the northern end there are considerable areas of exposed rock shelf while the southern end is characterized by many large granite outcrops. All parts of the island are readily accessible on foot.

No seasonal census of fur seals at Montagu Island has been conducted prior to this study, although P. Fullagar recorded numbers present in March for several years between 1965 and 1994 (Fig. 3). The aim of this paper was to record the seasonal variation in total numbers, and the sex and age classes of Australian Fur Seals on Montagu Island, by carrying out a regular census of the population over a year (May 1993 to May 94). As a consequence of observing New Zealand Fur Seals for the first time on the island early in the study, the seasonal variation in numbers of that species was recorded also.

METHODS

The area where seals haulout on Montagu Island is confined to the northern end of the island. The area was identified simply by occupation and/or evidence of occupation (that is, faeces and white staining on the rocks), and to facilitate censusing, divided into five sections (see Fig. 2). Observation points were chosen to provide the best view of the colony while disguising the observer to minimize the disturbance. As Section 1 was relatively large

[1]Department of Veterinary Anatomy, The University of Sydney, New South Wales, Australia 2006.
[2]"Blackwood Lodge", R.S.D. 273 Mount Hicks Road, Yolla, Tasmania, Australia 7325.

Pp. 56–62 *in* MARINE MAMMAL RESEARCH IN THE SOUTHERN HEMISPHERE VOLUME 1: STATUS, ECOLOGY AND MEDICINE ed by M. Hindell and C. Kemper. Surrey Beatty & Sons, Chipping Norton. 1997.

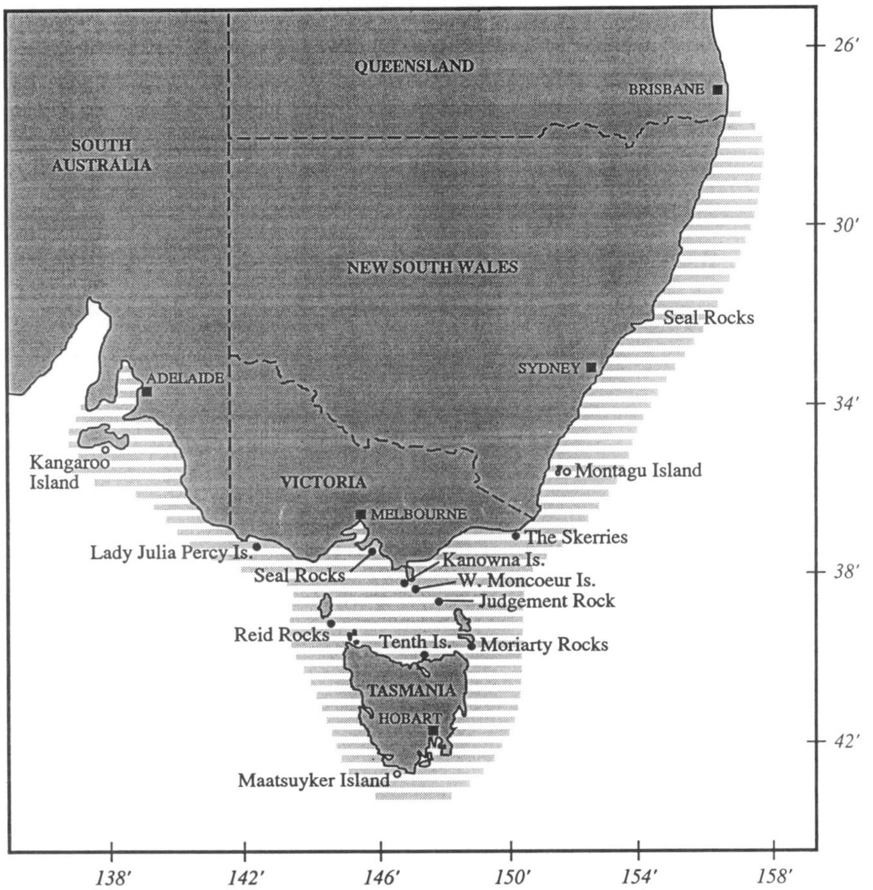

Fig. 1. The distribution of Australian Fur Seals *Arctocephalus pusillus doriferus,* closed circles are breeding sites, open circles haulout sites mentioned in the text.

and complex two fixed observation points were established. The other four sections had one fixed observation point each.

Each section was censused for four days every fourth week from 2 May 1993 to 5 May 1994. Each four-day census will be referred to as an observation period for the remainder of this paper. The number and age class of animals in each section for Sections 1 to 4 was recorded every two hours between 0800 h and 1600 h. Because Section 5 was remote, and a full census involved a considerable length of time, it was surveyed only twice each day, at 0730 h and 1630 h. The number and location of animals in the water and the number of boats were also recorded during a single scan from west to east. This scan was made from the same fixed observation point immediately after sections 1 to 4 were censused.

The age classes and sex of *A. pusillus doriferus* were defined as follows:

Pups of Both Sexes: Pups were readily identified by a black coat up to four months of age, before moulting to a female-like pelage with black vibrissae. At birth pups measure 62–80 cm and weigh up to 12.5 kg.

Juveniles of Both Sexes: Males and female pre-puberal juveniles, one to four years of age are similar in form and colour to adult females. Sex may be determined by the presence or absence of a penile opening and males may be wet around this area. These animals measure up to 1.7 m nose to tail and weigh 30–110 kg.

Adult Females: Adult females have a pale throat and chest with brown ventral surfaces and grey/ brown dorsal surfaces, and a relatively small gracile head. Their vibrissae are pale fawn in colour with darker tips in young adults. These animals measure 1.2–1.7 m nose to tail, and weigh 36–110 kg (King 1983).

Sub-adult Males: Between four to eight years of age males develop a large head and shoulders, a course mane, and uniform grey pelage. Their vibrissae are pale fawn vibrissae with darker tips in younger individuals. They measure 1.5 m–2.0 m nose to tail and weigh 110–200 kg.

Adult Male: Adult males are between eight and 19 years of age. They have very heavy shoulders and an obvious coarse mane, with pale fawn vibrissae, The head, flippers, neck and shoulders

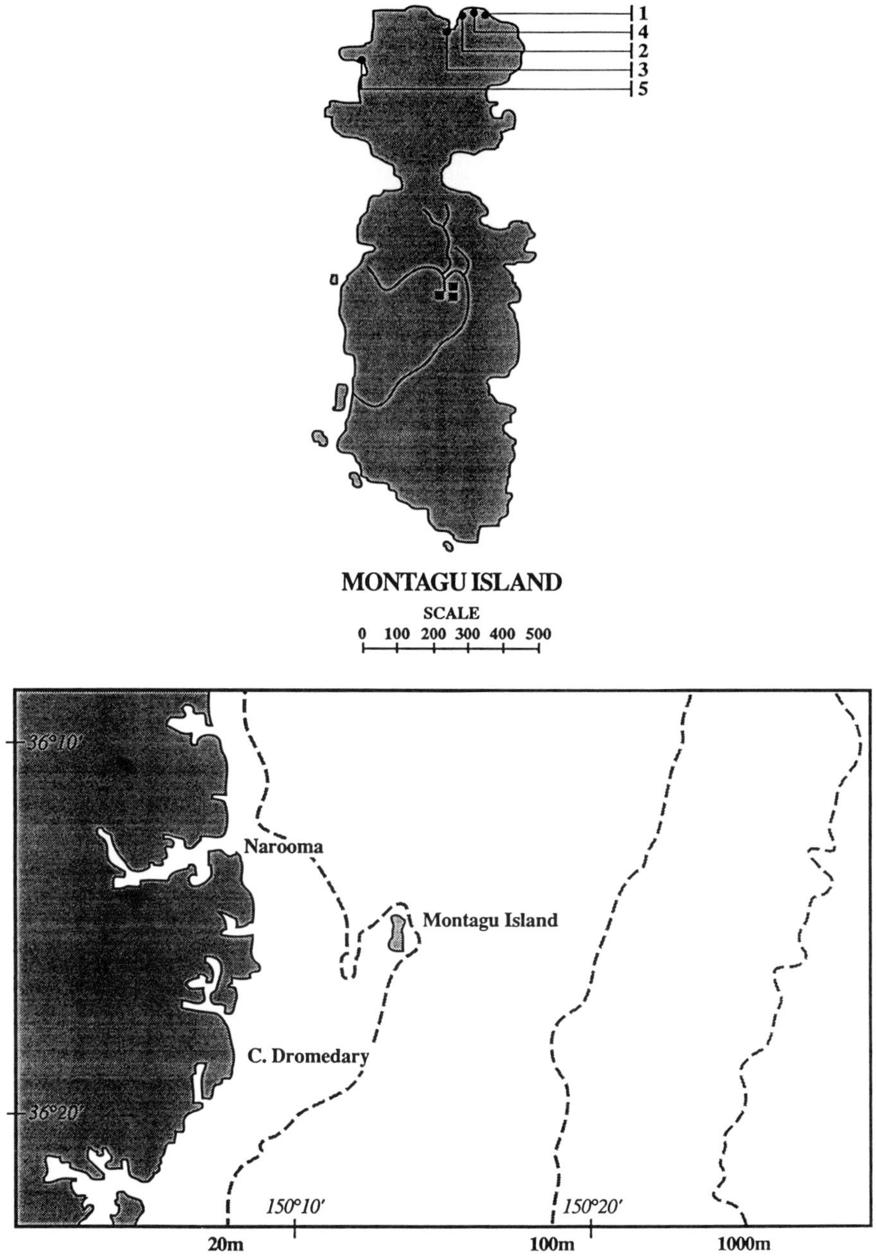

Fig. 2. Map of Montagu Island, showing the area where fur seals haulout. The area was divided into five sections (1 to 5).

are often heavily scarred. These animals measure 2–2.2 m nose to tail and weigh 218–360 kg (King 1983).

Arctocephalus pusillus doriferus and *A. forsteri* were distinguished using a number of morphological and behavioural characteristics, including their different vocalizations (Stirling and Warneke 1971) which was particularly useful when observing animals from a distance. When animals were in close proximity the different head shape helped distinguish the two species. Field identification of these species is addressed specifically in Goldsworthy *et al.* (1997).

RESULTS

The maximum number of fur seals (*A. pusillus doriferus* and *A. forsteri*) recorded for each observation period is presented in Figure 4. These are minimum counts, as there are areas of Sections two and four that cannot be seen from the observation points. Counts of animals in the water are only of animals seen at the surface during the single scan from east to west.

The number of seals of both species counted for May 1993 and May 1994 were 12 and 16 respectively. In the intervening period, numbers increased, reaching a maximum of 322 in September, then decreased to a minimum of nine in December. The numbers remained low (less than 40) from December 1993 to May 1994 (Fig. 4).

Counts of adult Australian and New Zealand Fur Seals (Fig. 5) follow a similar trend to the entire population (Fig. 4), increasing from low

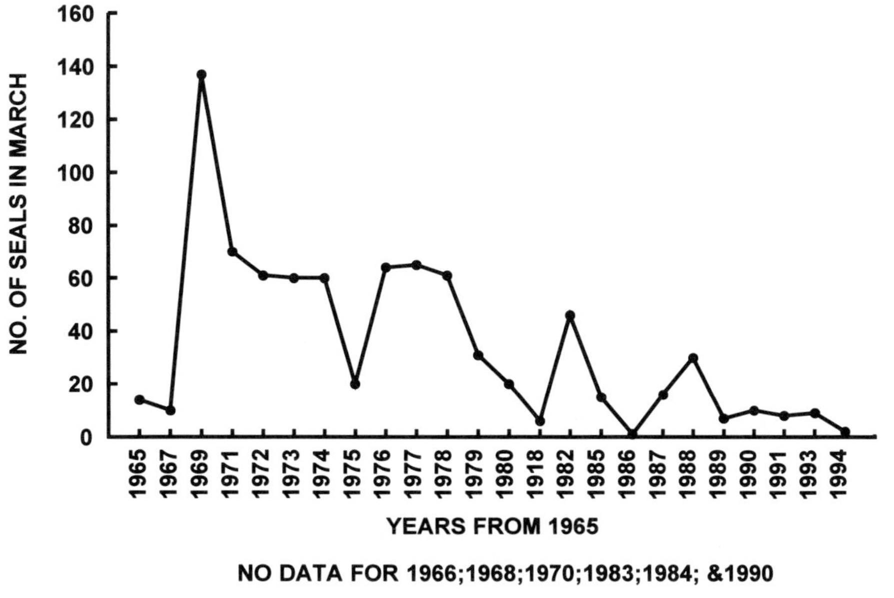

Fig. 3. Numbers of fur seals on Montagu Island in the month of March, 1965 to 1994. (P. Fullagar, pers. comm.).

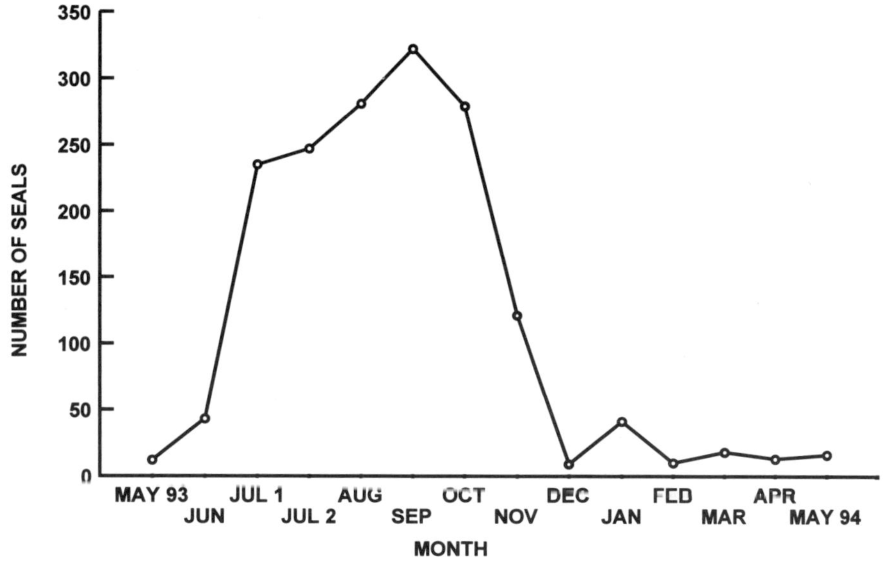

Fig. 4. The maximum number of fur seals (*A. pusillus doriferus* and *A. forsteri*) recorded on Montagu Island from May 1993 to May 1994.

levels in May to reach a maximum in September, then declining to a minimum in December. Sub-adults constituted the most abundant age class, reaching a maximum of 133 animals in July and August. Juveniles like the sub-adults, reached peak numbers in August, decreased slightly in September, then declined rapidly in October to December.

Significant numbers of *A. forsteri* were present only in the months of July to November (Fig. 6).

Two tagged *A. pusillus doriferus* and one *A. forsteri* have been seen at Montagu Island. One of the Australian Fur Seals was tagged in 1989 at Judgement Rock, the other in 1990 at Tenth Island (D. Pemberton, pers. comm.). The New Zealand Fur Seal was tagged at Kangaroo Island in 1992 (Shaughnessy 1993). Tagging sites of all animals are shown in Figure 1.

A male Australian Fur Seal pup was born on 20 October 1993 (Fig. 7). Although the birth was not observed the gathering of seagulls feeding on the placenta drew attention to the pup. It was dead the next day. The earliest previously recorded birth for this species was 26 October at Tenth Island (Fig. 1) (Pemberton and Kirkwood 1994), where median birth date is 26 November with 90% of the pups born between 2 and 20 December (Pemberton and Kirkwood 1994). The median pupping date for the Seal Rocks colony was 1 December (Warneke

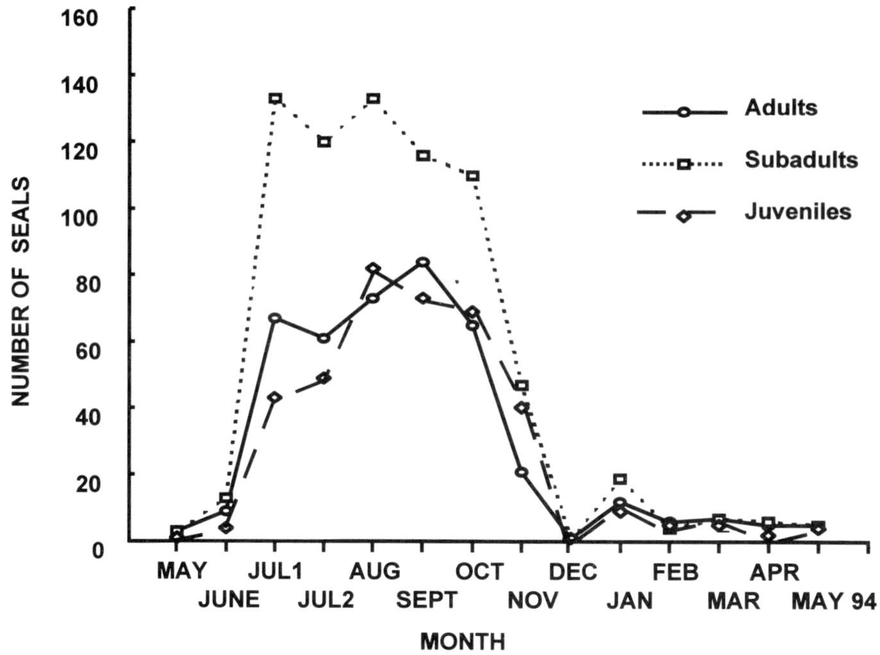

Fig. 5. The age class counts for fur seals on Montagu Island May 1993 to May 1994.

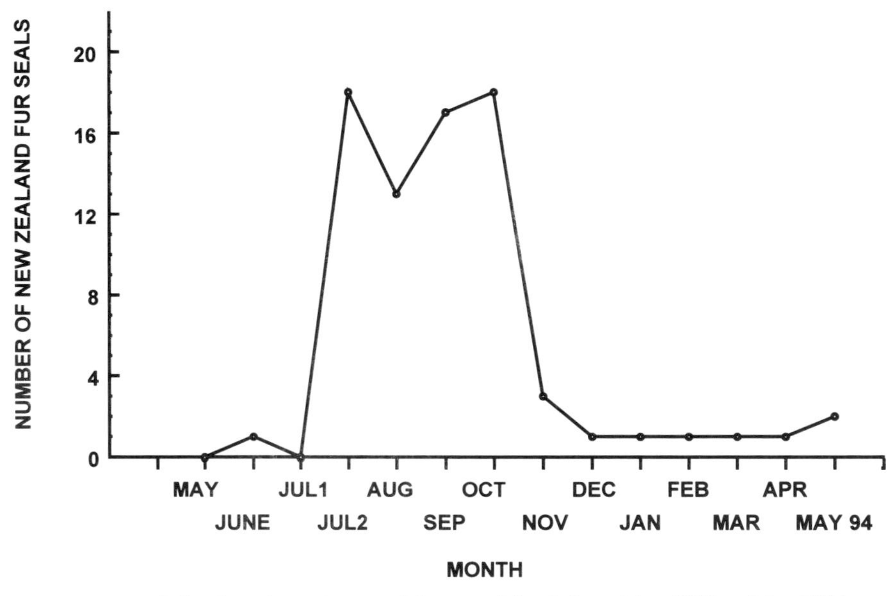

Fig. 6. Arctocephalus forsteri numbers on Montagu Island from May 1993 to May 1994.

and Shaughnessy 1985). These data for normal births at breeding sites suggests that the birth on Montagu Island was premature.

DISCUSSION

Estimates for the Australian Fur Seal population are between 35 000 and 60 000 (Pemberton and Kirkwood 1994), more than two-thirds of which are concentrated at three sites in Bass Strait (Warneke 1982). Breeding only takes place on nine islands within Bass Strait (Warneke and Shaughnessy 1985). The peak of the breeding season is in early December, when the maximum number of seals are found at the breeding sites (Pearse 1978; Pemberton and Kirkwood 1994). Once breeding has concluded the males leave their territories in Bass Strait and do not return until October (Warneke 1979).

Following dispersal from the breeding colonies Australian Fur Seals might seek out specific areas of high food abundance. There are dramatic increases in the number of male *A. pusillus doriferus* in mid-January at Maatsuyker Island (Brothers and Pemberton 1990), their arrival coinciding with large concentrations of Red Bait *Emmelihcthys nitidus* and Jack Mackerel *Trachurus declivis* in the surrounding waters. These fish are

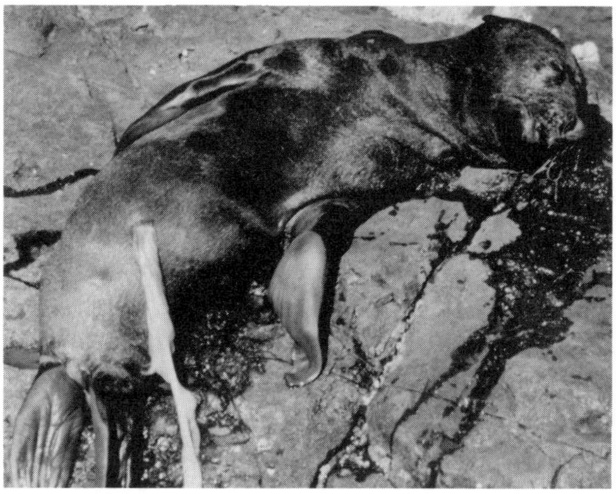

Fig. 7. The male Australian Fur Seal pup born at Montagu Island on 20 October 1993.

known to be the main prey of the seals at this time (Gales and Pemberton 1994). The population then declines steadily and by May and August no Australian Fur Seals are seen (Brothers and Pemberton 1990).

Montagu Island is only 8 km from the continental shelf margin, an area typically high in primary production. Aerial surveys of waters off Tasmania and southeastern Australia in 1936 showed that the greatest concentration of surface swimming fish and birdlife was near Montagu Island (Pacey 1991). Fish was found to be the most important food source (76%) for *A. pusillus doriferus* in Tasmanian waters (Gales and Pemberton 1994). Further, Gales *et al.* (1993), found that a significant component (22%) of the *A. pusillus doriferus* diet in Tasmanian waters was the cephalopod *Sepia apama*. In the last week of July and the last week of August, *Sepia* sp., including *S. apama*, concentrate and spawn within the 36 m depth ranges along the New South Wales south coast (L. Smith, pers. comm.). The peak abundances of Australian Fur Seals at Montagu Island might be related to this concentration of cuttlefish.

The decline in numbers of adult males at Montagu Island during October and November coincides with increasing numbers at the breeding islands, which reach peak numbers in December (Warneke 1975). The adults at Montagu Island are of breeding size and age, and are likely to move to breeding islands during the breeding season. The three re-sightings of tagged animals of known origin at Montagu Island and the recovery of animals tagged at Seal Rocks Victoria, during the period 1966–72 (Warneke 1975), indicates that the range of some animals from their birth sites is considerable.

Although fur seal pups have been reported on Montagu Island prior to this study (Pacey 1991), moulted pups and juveniles 12 to 18 months old are easily confused because of their small body size. The number of females on Montagu Island at any one time is unknown, as it is likely they have been recorded as juveniles. The similarities between juveniles and small adult females make them very difficult to distinguish.

The sub-adult age class is the most abundant on Montagu Island. The sampling technique might have caused some bias, over-estimating the sub-adult age class in the counts. The total was counted first, then the number of adults (as they are the most obvious), then the number of juveniles. The number of sub-adults was obtained by difference, and borderline animals were assigned to this class.

It has long been assumed that all the fur seals at Montagu Island are *A. pusillus doriferus*, however this study has demonstrated the presence of *A. forsteri* (Fig. 8). As the two species can be difficult to distinguish from each other *A. forsteri* may have always been there in the main colony. However Section 5, which appears to be an exclusively *A. forsteri* area, has only been occupied since October 1992 (R. Constable and P. Fullagar, pers. comm.) This might reflect the recent expansion of numbers of *A. forsteri* throughout southern Australian waters documented by Shaughnessy *et al.* (1994).

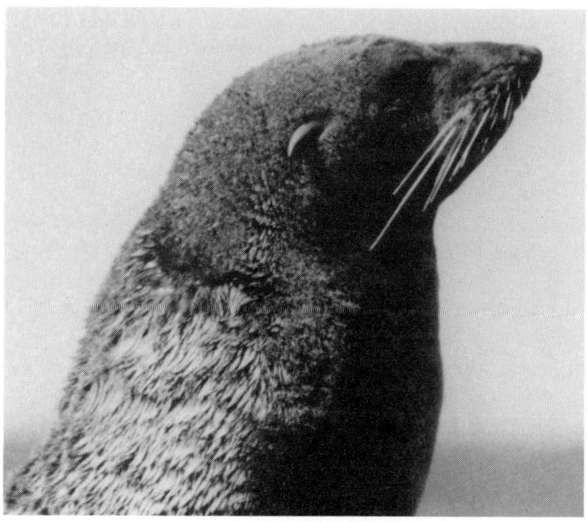

Fig. 8. A New Zealand Fur Seal at Montagu Island.

ACKNOWLEDGEMENTS

Field studies on Montagu Island were made possible by the NSW National Parks and Wildlife Service who provided transport to and from the island and accommodation on the island. We thank the Zoological Parks Board of New South Wales for their support of the project. We are grateful to Ms T. Rogers for reviewing and criticising preliminary drafts of the paper, and two referees for many helpful suggestions. The advice and assistance provided by D. Pemberton and L. Smith was greatly appreciated. Thank you to D. Henderson for assistance with the figures.

REFERENCES

Brothers, N. and Pemberton, D., 1990. Status of Australian and New Zealand fur seals on Maatsuyker Island, Southwestern Tasmania. *Aust. Wildl. Res.* **17:** 563–69.

Fullagar, P. J., 1973. Seabird islands Montagu Island, New South Wales. *Aust. Bird Band.* **2:** 36–39.

Gales, R., Pemberton, D., Lu, C. C. and Clarke, M. R., 1993. Cephalopod diet of the Australian Fur Seal: variation due to location, season and sample type. *Aust. J. Mar. Freshwater Res.* **44:** 657–71.

Gales, R. and Pemberton, D., 1994. Diet of the Australian Fur Seal in Tasmanian. *Aust. J. Mar. Freshwater Res.* **45:** 653–64.

Goldsworthy, S. D., Pemberton, D. and Warneke, R. M., 1997. The field identification of Australian and New Zealand fur seals (*Arctocephalus* spp.). Pp. 63–71 *in* Marine Mammals of the Southern Hemisphere Volume 1: Status, Ecology and Medicine ed by M. Hindell and C. Kemper. Surrey Beatty & Sons: Chipping Norton.

King, J. E., 1969. The identity of the fur seals of Australia. *Aust. J. Zool., Lond.* **17:** 841–53.

King, J. L., 1983. Seals of the World. British Museum (Natural History) Cromwell Road, London SW7 5BD.

Pacey, L., 1991. The Lure of Montague published by Laurella Pacey, P.O. Box 439, Narooma, New South Wales 2546.

Pearse, R. J., 1978. Distribution and conservation of the Australian Fur Seal in Tasmania *Vict. Nat.* **96:** 49–52.

Pemberton, D. and Kirkwood, R. J., 1994. Pup production of Australian Fur Seals In Tasmania. *Aust. Wildl. Res.* **21:** 341–52.

Shaughnessy, P. D., 1993. Abundance of New Zealand fur seals *Arctocephalus forsteri* and Australian sea lion *Neophoca cinerea* on Kangaroo Island and the Neptune Islands South Australia: 1992–93. Report to the South Australian National Parks and Wildlife Service and to Down to Earth Products.

Shaughnessy, P. D., Gales, N. J., Dennis, T. E. and Goldsworthy, S. D., 1994. Distribution and abundance of New Zealand fur seals *Arctocephalus forsteri* in South Australia and Western Australia. *Wildl. Res.* **21:** 667–95.

Stirling, I. and Warneke, R. M., 1971. Implications of comparisons of the airborne vocalizations and some aspects of the behaviour of the two Australian Fur Seals, *Arctocephalus* spp. On the evolution and present taxonomy of the genus. *Aust. J. Zool.* **19:** 227–41.

Warneke, R. M., 1975. Dispersal and Mortality of juvenile fur seals *Arctocephalus pusillus doriferus* in Bass Strait, Southeastern Australia. *Rapp. P.-v Reun. Cons. Int. Explor. Mer.* **169:** 296–302.

Warneke, R. M., 1979. Australian Fur Seal. *FAO Fish. Ser. 5, Vol.* **2:** 41–44.

Warneke, R. M., 1982. The distribution and abundance of seals in the Australasia region, with summaries of biology and current research. *FAO Fish. Ser. 5, Vol.* **4:** 431–75.

Warneke, R. M. and Shaughnessy, P. D., 1985. *Arctocephalus pusillus*, the South African and Australian Fur Seal: Taxonomy, Evolution, Biogeography and life History. Pp. 53–78 in Studies of Sea Mammals in South Latitudes ed by J. K. Ling and M. M. Bryden. South Australian Museum: Adelaide.

Field identification of Australian and New Zealand Fur Seals, *Arctocephalus* spp., based in external characters

S. D. GOLDSWORTHY,[1] D. PEMBERTON[2] and R. M. WARNEKE[3]

Two species of fur seal breed in Australian coastal waters, the Australian *Arctocephalus pusillus doriferus,* and New Zealand Fur Seal *A. forsteri.* The breeding distribution of these species is allopatric, however, recent observations indicate that there is much overlap in their non-breeding distribution. Until recently, identification of species has been largely based on skull morphometrics and an assumed allopatric distribution for the species. Here we provide descriptions of a suite of morphological and behavioural characteristics useful for distinguishing species in the field. Differences in vocalization, locomotion and thigmotactism were found to be the most useful characteristics to separate species. These characteristics also suggest that *A. pusillus* is more closely related to sea lions than other *Arctocephalus* species. Differences in morphology, such as size, head and flipper shape and coloration were less useful in distinguishing species because such features vary markedly with age and sex class, and are only useful in a comparative way at sympatric sites.

INTRODUCTION

Two species of fur seal (*Arctocephalus* spp.) breed around the coast of southern Australia, the Australian Fur Seal *Arctocephalus pusillus doriferus* and the New Zealand Fur Seal *A. forsteri.* Australian Fur Seal breeding colonies are all located in Bass Strait, whilst their non-breeding sites (locations where seals haul-out) extend from Kangaroo Island (South Australia) in the west, south to southern Tasmania, and east to Montagu Island (New South Wales) with occasional records along the the New South Wales coast as far north as the New South Wales/Queensland border (Irvine *et al.* 1997; Goldsworthy, unpubl. data; Llewellyn *et al.* 1994; Shaughnessy and Warneke 1987; Pemberton and Kirkwood 1994) (Fig. 1). The South African Fur Seal *Arctocephalus pusillus pusillus* occurs along the coast of Namibia and west and south coasts of South Africa (Shaughnessy and Warneke 1987). The New Zealand Fur Seal breeds on islands off coastal South Australia, Western Australia and Tasmania (Fig. 1), and on rocky coasts and offshore islands of New Zealand and its subantarctic islands (Shaughnessy *et al.* 1994; Brothers and Pemberton 1990; Mattlin 1987). At present, the breeding distributions of both species in Australia are discrete, however historical information indicates that prior to exploitation by nineteenth century sealers, both species were sympatric in Bass Strait (King 1969; Warneke 1982).

Since the studies of King (1969) and Repenning *et al.* (1971) identification of fur seals in Australia has been based mainly on the geographic location and skull morphometrics of individuals. Problems with identification based on these factors are that (i) dead specimens are required for confirmation of species by skull morphometrics and (ii) although breeding ranges of these species are presently allopatric, recent observations (Brothers and Pemberton 1990; Irvine *et al.* 1997; Goldsworthy, unpubl. data) indicate that these species are regularly seen together at certain sites where their breeding and non-breeding ranges overlap. Further, as populations of each species are increasing, the extent of such overlap is expected to increase also (Shaughnessy *et al.* 1995; Pemberton and Kirkwood 1994).

Because there is limited published material describing the external characteristics of these species, only experienced observers have been able to identify live individuals in the wild. This has led to misidentification of individuals and errors in determining species distribution and breeding status. For example, the small breeding colony of fur seals on Maatsuyker Island was thought to be *A. p. doriferus* until it was recently confirmed that the species was in fact *A. forsteri,* a species not thought to breed in Tasmania (Warneke and Shaughnessy 1985; Brothers and Pemberton 1990). In addition studies examining diet and seal movements are hampered by

[1]Department of Zoology, University of Tasmania, G.P.O. Box 252C, Hobart, Tasmania, Australia 7001.
[2]Department of Parks, Wildlife and Heritage, P.O. Box 44A, Hobart, Tasmania, Australia 7001.
[3]Blackwood Lodge, RSD 273, Mt Hicks Road, Yolla, Tasmania, Australia 7325.

Pages 63–71 *in* MARINE MAMMAL RESEARCH IN THE SOUTHERN HEMISPHERE VOLUME 1: STATUS, ECOLOGY AND MEDICINE ed by M. Hindell and C. Kemper. Surrey Beatty & Sons, Chipping Norton. 1997.

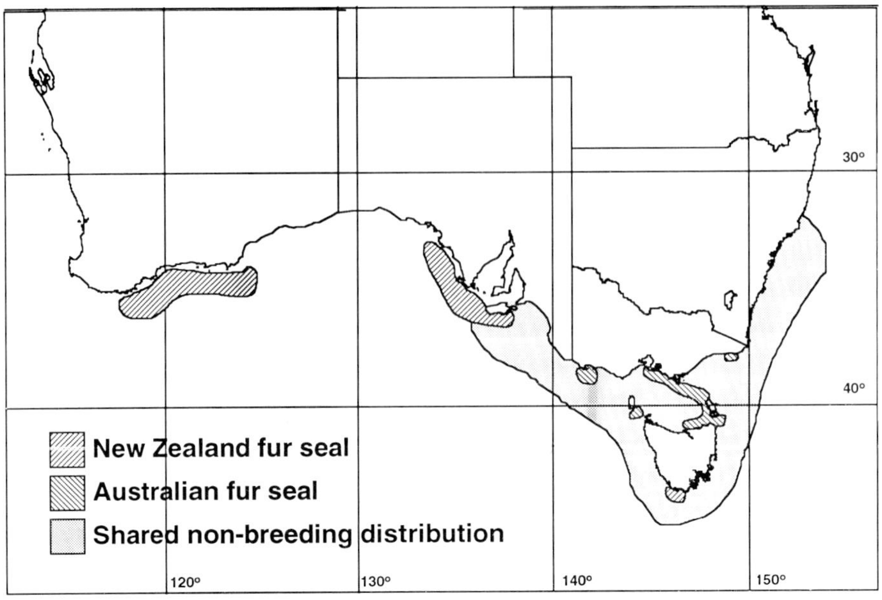

Fig. 1. The breeding distribution of the New Zealand and Australian Fur Seal, within Australia, and the shared non-breeding distribution of both species. Note that distributions encompass islands and other haul-out areas where seals are seen and are not meant to infer at-sea distributions.

identification problems of seals at sympatric sites. Here we provide descriptions of a suite of morphological and behavioural characteristics useful for distinguishing *A. p. doriferus* and *A. forsteri* in the field.

METHODS

Observations on the external morphology and behaviour of *A. p. doriferus* and *A. forsteri* were made at Cape Gantheaume (Kangaroo Island 36°04′S, 137°28′E) (Fig. 1) where groups of mostly male *A. p. doriferus* (juveniles to adults) were observed in the *A. forsteri* breeding colony during frequent visits to Cape Gantheaume between November 1988 and February 1992. At Maatsuyker Island (43°39′S, 146°17′E) (Fig. 1), off the south-west coast of Tasmania, observations on both species were also made over several field trips between 1989 and 1993. Similarly, extensive observations of *A. p. doriferus* were made at breeding colonies in Bass Strait. Seals were observed using binoculars and photographs were taken to help illustrate differences. In addition video recordings were made of each species' terrestrial locomotion in order to analyse interspecific differences.

At Cape Gantheaume and Maatsuyker Island, Australian Fur Seals were distinguished from New Zealand Fur Seals by their highly gregarious and thigmotactic tendencies (Gentry 1975; Warneke and Shaughnessy 1985) and their different vocalizations (Stirling and Warneke 1971). The presence of *A. forsteri* on Maatsuyker Island was verified by the identification of two skulls obtained from illegally shot animals on the island (Brothers and Pemberton 1990).

RESULTS

The external characteristics and behaviours of *A. p. doriferus* and *A. forsteri* are described below, and in the accompanying illustrations:

Coloration of pelage

Coloration can vary markedly between individuals, depending on time of year (in relation to last moult) and age of the animal.

New Zealand Fur Seal

Adult males: Uniform dark grey to brown pelage with pale muzzle fur (Fig. 2).

Adult females: Variable coloured pelage, metallic-grey colour after the moult (February–March) then gradually becoming browner throughout the year until the next moult. Lighter colour ventrally with dark brown abdomen. Lighter ventral pelage also becomes darker post-moult.

Juveniles: Rich dark brown colour (burgundy in yearlings (Goldsworthy and Shaughnessy 1994)), with pale white/cream moustache.

Pups: Dark-brown colouring with grizzled fur (pale tipped guard-hairs) around head and a pale blaze on the top of the head. Muzzle and belly fur is a lighter pale creamy-brown colour. Pups begin to moult into an adult type pelage towards the end of February or early March. Complete moult may take several months; patches of natal pelage may persist until June.

Australian Fur Seal

Adult males: A light sandy brown, slightly paler chest and darker brown belly.

Adult females: Variable coloured, pale fawn (blond) to grey-brown, contrasting pale chest, brown belly.

Juveniles: Coloration similar to adult females, however lighter fur often extends from the jaw below and behind the ears.

Pups: Black at birth with a variable grey ventrum. The natal pelage gradually browns as a result of soiling and fading.

In both species, the colour of the outer pelage changes to dark shades of grey when it is wet. The wet dark pelage of very small juveniles commonly leads to mistaken claims of pups at haul-out sites.

Head shape, rhinarium and vibrissae

New Zealand Fur Seal

Long narrow pointed face, obvious brow and a concave profile (Fig. 2). Adult females have a smaller, narrower skull, compared with males which have a broader and more pronounced bridge. Males also have an enlarged rhinarium which accentuates the pointed snout (Fig. 2a), long pale whiskers which may extend to the shoulders and in some animals there is a small crest of hair on top of the skull between the eyes. The lower jaw and the back of head are curved (Fig. 2e).

Australian Fur Seal

The head is proportionally broader and shorter than in *A. forsteri.* The brow is less obvious, especially in sub-adult animals where the top of the skull slopes evenly down to the tip of the nose without any bridge (Fig. 2). In adult males a small bridge does develop, but it is not as pronounced as in *A. forsteri.* The snout and lower jaw have a more angular profile compared with *A. forsteri,* as does the back of the head. They have fawn coloured whiskers and no crest of fur on the head.

Ears

New Zealand Fur Seal

Ears are dark and lie flat against the side of the head.

Australian Fur Seal

Ears are light in colour and proportionally large and more obvious than in *A. forsteri.* The ears of *A. p. doriferus* also tend to stick out from the side of the head.

Fore-flippers

New Zealand Fur Seal

Very triangular in shape, with little or no thickening of the trailing edge before it joins the body (Fig. 3a).

Australian Fur Seal

Proportionally larger than *A. forsteri* flippers and more paddle-shaped with an obvious thickening of the trailing edge before it joins the body (Fig. 3b). The fore-flipper is broader than in *A. forsteri*, and the distal portion of the flipper has greater flexibility.

In both species, the fore-flippers are proportionally larger and thicker in males compared with females.

Posture/stance

New Zealand Fur Seal

When sitting the neck slopes forward resulting in an angular stance with the head extending at an angle >90 degrees from the base of the neck (Fig. 4). When lying on rocks, fore-flippers are usually held against the body, hind-flippers are usually tucked under the body. *A. forsteri* rarely lie with their fore flippers stretched out either side of their body, a posture common to *A. p. doriferus.*

Australian Fur Seal

Very upright posture when sitting. Neck extends vertically from the back (Fig. 4). The head tends to be held at a right angle from the neck, sometimes forming a double chin. When lying on rocks, fore-flippers and hind-flippers are often stretched out away from the body (Fig. 5). *A. p. doriferus* will also lie with closed flippers (flippers against the body or tucked under) as in *A. forsteri.* These postures are related to thermoregulation.

Terrestrial locomotion

Terrestrial locomotion is best described by illustration, (see Figs 7 and 8). Explanations are given in the figure captions and in the text below.

New Zealand Fur Seal

Hind flippers generally not articulated separately when animals are moving. The head and neck point forward and move up and down (with little lateral movement) because the fore-flippers are moved forward together (Figs 6 and 7). The gait can be described as "bound" or "hop".

Fig. 2 (above and facing page). Comparisons of adult males (a and b), adult females (c and d), Juveniles (e and f) and pups (g and h) of *A. forsteri* and *A. p. doriferus*.

Australian Fur Seal

When moving slowly over rocks, animals sway and rock from side to side. The hind flippers are typically articulated separately (and occasionally dragged as in *Neophoca*). All flippers are moved alternately in sequence like most terrestrial quadrupeds, resulting in a "walking" gait as opposed to the bound of *A. forsteri* (Figs 6 and 7). Their motion was accurately described by the adventurer/sealer John Boultbee, "their motion being so peculiar; they have a sort of a shuffling, uneasy gait, their heads and necks being in continual motion from side to side . . ." (Stark 1986).

Vocalization

A detailed description of the vocalizations of both seal species is given by Stirling and Warneke (1971), who show that *A. forsteri* use two threat calls which are not used by *A. p. doriferus*. It is these vocalizations which are useful for differentiating between the species.

New Zealand Fur Seal

The most distinguishing of the two threat calls is the "broadcast call" produced by adult males. This is a snuffling, nasal whimpering call, which is uttered at regular intervals, especially in the breeding season. The second call, the "full threat vocalization", which is a high-pitched ascending call with a trilling component. This is most commonly used by adult males, but adult females will use the low-intensity threat which also has a high-pitched tone. The calls of *A. forsteri* often have trilling components which are not found in *A. p. doriferus*.

Australian Fur Seal

The most frequent call used by *A. p. doriferus* is the repeated bark (which is equivalent to the broadcast call of *A. forsteri*). This call is typically heard when groups of seals are interacting. Calls tend to be lower pitched than those of New Zealand Fur Seals, with no trilling components.

Thigmotactism

New Zealand Fur Seal

Close physical contact between individuals is avoided in this species, except during mating and between interacting mother/pup pairs. In

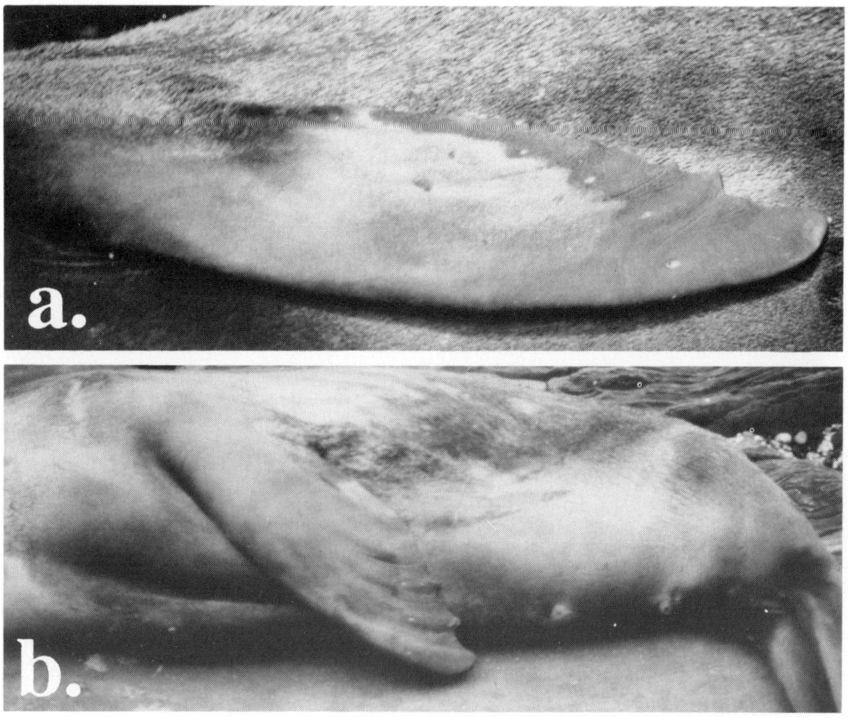

Fig. 3. Comparison of the more triangular shaped fore-flipper of *A. forsteri* (adult female) (a) with the more paddle-shaped and curved fore flipper of *A. p. doriferus* (sub-adult male) (b).

colonies over homogeneous substrate, seals are regularly spaced (Fig. 8a).

Australian Fur Seal

Strongly thigmotactic. *A. p. doriferus* often lie side-by-side or on top of each other (Fig. 8b). Their distribution over homogeneous habitat is often clumped (Fig. 8b).

Habitat

New Zealand Fur Seals

A. forsteri are most often observed on irregular rocky habitat, such as jumbled boulders. They usually avoid open rock platforms and pebble beach habitats.

Australian Fur Seals

A. p. doriferus generally occur on more open flat or sloping rock habitat where many animals can lie together.

Although the terrestrial habitats occupied by these two fur seals are distinctive, we found some intermixing by species. For example at Maatsuyker Island, *A. p. doriferus* do haul-out on to flat boulders located within jumbled boulder habitat where *A. forsteri* breed.

DISCUSSION

The early 18th century sealers distinguished between the "black" and the "brown" fur seal taken from Bass Strait. These probably represent *A. forsteri* and *A. p. doriferus* (Warneke 1975). Whilst the vast experience of the sealers dealing with large numbers of seals at close hand justifies descriptions of two types of seals based on one morphological feature, field identification methods used today should include a range of morphological and behavioural features to increase reliability and confidence of species identification. In this study, differences in

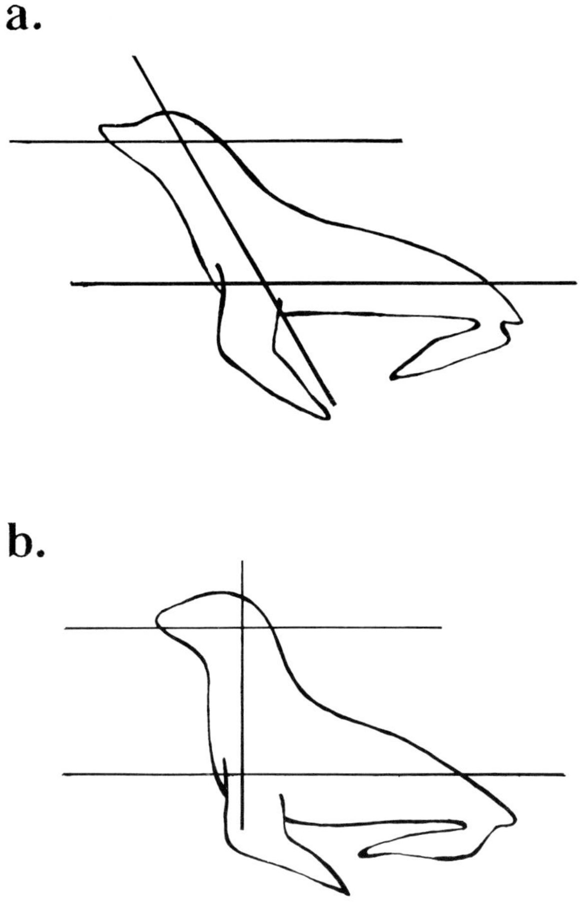

Fig. 4. Typical posture of *A. forsteri* (a) which is more angular than the comparatively erect posture of *A. p. doriferus* (b).

morphology, such as colour and head and flipper shape were less useful in distinguishing species because these features varied markedly with age and sex class, and were found only to be useful for comparative reasons where these species were sympatic. Behavioural differences

Fig. 5. Typical resting posture of *A. p. doriferus* with both fore and hind flippers splayed out on the rocks.

Fig. 6. Comparison of the terrestrial locomotion of *A. forsteri* (a) and *A. p. doriferus* (b) from the side through a complete cycle. Note that in *A. forsteri* hind flippers and fore flippers are moved together, while in *A. p. doriferus* each flipper is moved independently and alternately.

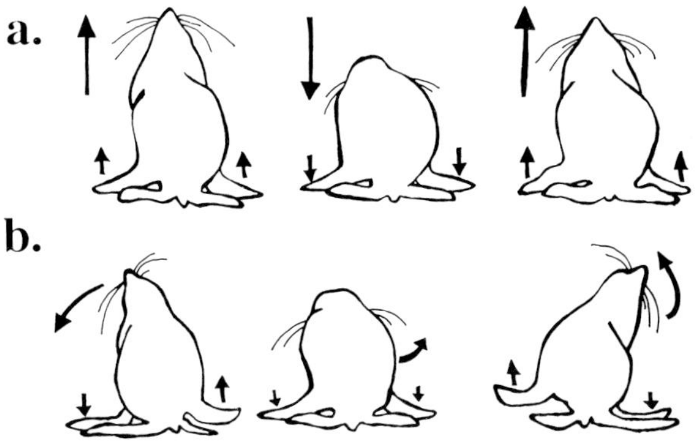

Fig. 7. Comparison of the direction of movement of the head during locomotion in *A. forsteri* and *A. p. doriferus* viewed from the rear. The main difference being that in *A. forsteri* the head and neck move up and down as a result of the fore flippers being moved together, while in *A. p. doriferus* the head and neck move from side to side in a circular motion (both clockwise and counter-clockwise) as a result of the fore flippers being moved alternately.

Fig. 8. Comparison of the degree of spacing between *A. forsteri* (a) (separate) and *A. p. doriferus* (b) (clumped) in typical habitat.

in vocalization, locomotion and thigmotactism were found to be the most useful characteristics to distinguish between species. These features used in conjunction with morphological differences should enable identification of most free ranging individuals.

The three main behavioural differences we observed between *A. forsteri* and *A. pusillus* appear to be general differences between *A. pusillus* (and sea lions) and other *Arctocephalus* species. Whereas male *A. pusillus* produce a stereotyped seal bark, the nearest vocalization we heard in *A. forsteri* was a low frequency, nasal whimpering call, which is also produced in Antarctic, *A. gazella*, and Subantarctic, *A. tropicalis*, Fur Seals (SDG, pers. obs.). These latter calls are analogous vocalizations to the bark used by *A. p. doriferus*, and used in the same behavioural context (Stirling and Warneke 1971). The bark produced by male *A. p. doriferus* appears more similar to the bark produced by the Californian Sea Lion *Zalophus californianus californianus*, than other fur seal species (SDG, pers. obs.). Similarly, the strong thigmotactic tendencies of close body contact and huddling in *A. pusillus* have been noted before as being peculiar, as they are the only fur species which show this type of behaviour (Gentry 1975; Stirling and Warneke 1971). Finally, we also suggest that the terrestrial locomotion of *A. pusillus* to be unique among fur seals. Although Beentjes (1990) described the gait of captive *A. p. doriferus* to be similar to that observed in *A. forsteri*, we suggest otherwise. By comparing the terrestrial locomotion of hundreds of free-ranging *A. forsteri* and *A. p. doriferus* at sympatric locations where both species could be compared, we have been able to show that the gait of *A. p. doriferus* is quite different to that of *A. forsteri*. In fact, the diagonally alternate quadrupedal gait of *A. p. doriferus*, observed in this study appears more similar to that of the Hooker's Sea Lion described by Beentjes (1990), and the Australian Sea Lion *Neophoca cinerea* (SDG, pers. obs.). We have also observed the gaits of *A. gazella* and *A. tropicalis* and they are very similar to that of *A. forsteri*.

The enigmatic taxonomic position of *A. pusillus* between the inferred subfamilies Arctocephalinae (fur seals) and Otariinae (sea lions) has been mentioned previously by Stirling and Warneke (1971) and Kim *et al.* (1975). Repenning *et al.* (1971) also indicated that the large body size relative to head size of *A. pusillus* is strikingly similar to that of sea lions, and that the teeth and skulls of *A. pusillus* are very similar to those of *Neophoca*. Given the morphological and behavioural differences identified between *A. p. doriferus* and other fur seal species in this and other studies, we suggest that the taxonomic status of the *A. pusillus* in the genus *Arctocephalus* may need revision. This suggestion was also implied by Repenning *et al.* (1971), Stirling and Warneke (1971), and Kim *et al.* (1975).

ACKNOWLEDGEMENTS

The occurrence of Australian Fur Seals at Cape Gantheaume was first noticed by Terry Dennis (South Australian National Parks and Wildlife Service). We thank Peter Shaughnessy for informing us of the Boultbee reference, Catherine Bone and Jo Hess for assistance with the figures, and to Peter Shaughnessy for reviewing the manuscript.

REFERENCES

Beentjes, M. P., 1990. Comparative terrestrial locomotion of the Hooker's sea lion *(Phocarctos hookeri)* and the New Zealand Fur Seal *(Arctocephalus forsteri)*: evolutionary and ecological implications. *Zool. J. Linn. Soc.* **98:** 307–25.

Brothers, N. and Pemberton, D., 1990. Status of Australian and New Zealand Fur Seals at Maatsuyker Island, south-west Tasmania. *Aust. Wildl. Res.* **17:** 563–69.

Gentry, R. L., 1975. Comparative social behaviour of eared seals. *Rapp. P.-v. Reun. Cons. int. Explor. Mer.* **169:** 189–94.

Goldsworthy, S. D. and Shaughnessy, P. D., 1994. Breeding biology and haul-out pattern of the New Zealand Fur Seal, *Arctocephalus forsteri*, at Cape Gantheaume, South Australia.*Wildl. Res.* **21:** 365–76.

Irvine, A., Bryden, M. M., Corkeron, P. J. and Warneke, R. M., 1997. A census of fur seals at Montagu Island, New South Wales. Pp. 56–62 *in* Marine Mammal Research in the Southern Hemisphere Volume 1: Status, Ecology and Medicine ed by M. Hindell and C. Kemper. Surrey Beatty & Sons: Chipping Norton.

Kim, K. C., Repenning, C. A. and Morejohn, G. V., 1975. Specific antiquity of the sucking lice and evolution of the Otariid seals. *Rapp. P.-v. Reun. Cons. int. Explor. Mer.* **169:** 544–49.

King, J. E., 1969. The identity of the fur seals of Australia. *Aust. J. Zool.* **17:** 841–53.

Llewellyn, L., Ellis, M., Martin, J. and Ferguson, A., 1994. Atlas of New South Wales wildlife: marine mammals and reptiles. New South Wales NPWS.

Mattlin, R. H., 1987. New Zealand Fur Seal, *Arctocephalus forsteri*, within the New Zealand Region. *In* Status, biology, and ecology of fur seals ed by J. P. Croxall and R. L. Gentry. *NOAA Technical Report. NMFS* **51.**

Pemberton, D. and Kirkwood, R. J., 1994. Pup production and distribution of the Australian Fur Seal *Arctocephalus pusillus doriferus*, in Tasmania. *Wildl. Res.* **21:** 341–52.

Repenning, C. A., Peterson, R. S. and Hubbs, C. L., 1971. Contributions to the systematics of the southern fur seals, with particular reference to the Juan Fernández and Guadalupe species. *In* Antarctic Pinnipedia ed by W. H. Burt. Antarctic Research Series, Vol 18. American Geophysics Union: Washington, D.C.

Shaughnessy, P. D., Gales, N. J., Dennis, T. E. and Goldsworthy, S. D., 1994. Distribution and abundance of New Zealand Fur Seals, *Arctocephalus forsteri*, in South Australia and Western Australia. *Wildl. Res.* **21:** 667–95.

Shaughnessy, P. D., Goldsworthy, S. D. and Libke, J. A., 1995. Changes in the abundance of New Zealand Fur Seals *Arctocephalus forsteri* on Kangaroo Island, South Australia. *Wildl. Res.* **22:** 201–15.

Shaughnessy, P. D. and Warneke, R. M., 1987. Australian Fur Seal, *Arctocephalus pusillus doriferus*. Pp. 73–77 *in* Status, biology, and ecology of fur seals ed by J. P. Croxall and R. L. Gentry. *NOAA Technical Report. NMFS* **51.**

Stark, J., 1986. Journal of a Rambler, The Journal of John Boultbee. Oxford University Press, in association with the Alexander Turnbull Library Endowment Trust, Auckland, New Zealand. Pp. 225.

Stirling, I. and Warneke, R. M., 1971. Implications of a comparison of the airborne vocalizations and some aspects of the behaviour of the two Australian Fur Seals, *Arctocephalus* spp., on the evolution and present taxonomy of the genus. *Aust. J. Zool.* **19:** 277–41.

Warneke, R. M., 1975. Dispersal and mortality of juvenile fur seals *Arctocephalus pusillus doriferus* in Bass Strait, southeastern Australia. *Rapp. P.-v. Reun. Cons. int. Explor. Mer.* **169:** 296–302.

Warneke, R. M., 1982. The distribution and abundance of seals in the Australian region, with summaries of biology and current research. Pp. 431–75 *in* Mammals in the Seas, FAO Fisheries Series No. 5, Volume IV. FOA: Rome.

Warneke, R. M. and Shaughnessy, P. D., 1985. *Arctocephalus pusillus*, the South African and Australian Fur Seal: Taxonomy, Evolution, Biogeography, and Life History. Pp. 53–78 *in* Studies of sea mammals in south latitudes ed by J. K. Ling and M. M. Bryden. South Australian Museum: Adelaide.

Top left: Female Southern Elephant Seal *Mirounga leonina,* with new born pup at Macquarie Island. This female was branded as a pup and the mark was still clear 23 years later. Photo: Mark Hindell.

Top right: Australian Fur Seals *Arctocephalus pusillus doriferus,* at Cape Gantheaume, Kangaroo Island, South Australia. Photo: Simon Goldsworthy.

Bottom left: New Zealand Fur Seal *Artocephalus fosteri,* female and pup, at Cape Gantheaume, Kangaroo Island, South Australia. Photo: Simon Goldsworthy.

Bottom right: Two species of fur seals at Macquarie Island. Left to right: Adult female and adult male Subantarctic Fur Seal *Arctocephalus tropicalis,* and adult female Antarctic Fur Seal *Arctocephalus gazella.* Photo: Simon Goldsworthy.

Right: Adult female Hooker's Sea Lion *Phocarctos hookeri,* Enderby Island. Photo: Mark Hindell.

The importance of being fat: maternal expenditure in the Southern Elephant Seal *Mirounga leonina*

M. A. HINDELL[1] and D. J. SLIP[1]

The energy expenditure of lactating female elephant seals at Macquarie Island was estimated using water turnover and milk composition data. Mothers had a mean daily energy expenditure of 134.5 ± 18.4 MJ/day, and a total expenditure of 3228.1 ± 440.8 MJ over the 24-day duration of the lactation period. Of this 79.6 ± 10.7 MJ/day was lost to pups in the form of milk, with pups ingesting 4.9 ± 0.7 l of milk each day. The remaining 57.6 ± 3.9 MJ/day was energy lost as maternal metabolism. There were no discernible differences between the Macquarie Island population and the South Georgia population with respect to maternal mass at parturition, weaning mass and energy transfer efficiency. This indicates that differences reported between the two populations during the 1960s may no longer exist, or at least are substantially reduced, but greater sample sizes are needed for higher statistical power.

INTRODUCTION

PHOCID seals have a lactation strategy shared by few other groups of mammals. The strategy is characterized by short lactation periods, often constant attendance of a single pup by the mother and very rapid and efficient transfer of energy from mother to pup (Bonner 1984; Oftedal *et al.* 1987). Further, because phocid mothers suckle on land or ice, yet derive their food exclusively from the aquatic environment, the females must fast for much, if not all, of the lactation period (e.g., Costa *et al.* 1986; Kovacs and Lavigne 1992; Bowen *et al.* 1992). Consequently, more than most other mammals, female phocid seals have a marked temporal separation between the acquisition of food and the delivery of energy to their young. This temporal separation between feeding and nursing is most pronounced in the larger species such as the Southern Elephant Seal (Bowen *et al.* 1993).

In the Southern Elephant Seal *Mirounga leonina* there is also considerable spatial separation between where the females feed and where they haul-out to have their pups (Hindell *et al.* 1991; McConnell *et al.* 1992). The temporal and spatial separation of feeding and lactation has several implications for the biology of these animals. In particular, gestating females must acquire sufficient energy to simultaneously supply the growing foetus and her own and the pups energy requirements during the suckling period. How female seals utilize these accumulated adipose reserves to maximize the size of their offspring has been the subject of a number of recent studies (Bowen *et al.* 1993; Lydersen and Hammill 1993; Iverson *et al.* 1993) but has yet to be fully described for Southern Elephant Seals (Fedak *et al.*, in press; Hindell *et al.* 1994a).

A further consequence of the temporal and spatial separation of maternal feeding and suckling is that factors operating thousands of kilometres away from the pupping sites have the potential to influence the outcome of a breeding season. For example, declines in food availability in the feeding areas may influence maternal adipose reserves and ultimately affect the weaning mass of pups at the breeding site. Hindell (1991) proposed that a perturbation in food supply during the 1950s and 1960s was the proximate reason for a dramatic reduction in first year survival and subsequent reduction in the number of Southern Elephant Seals in the Macquarie Island population. The reasons for the reduced food supply are still unclear: one suggestion is that it was a direct effect of the local elephant seal population "overshooting" its pre-exploitation level and reducing the food available to newly weaned pups, or reducing the food available to gestating females who needed to accumulate energy and materials to successfully wean their pups (or a combination of both) (Hindell *et al.* 1994b). An alternative explanation is that variations in oceanographic conditions around important feeding areas was responsible for altering the distribution and abundance of elephant seal prey (Hindell *et al.* 1994b).

The aims of this study were therefore twofold. The first was to quantify the energetic costs of lactation in female Southern Elephant Seals at Macquarie. The second was to use these data to test some of the predictions of hypotheses that have been advanced to explain recent population declines.

METHODS

Field procedures

The study took place on Macquarie Island (54°35′S, 158°55′E) in October 1987 and October 1988. We randomly selected 28 post-partum females and their newborn pups (20 pairs in 1987 and 8 in 1988), anaesthetized each adult using a

[1]Department of Zoology, University of Tasmania, PO Box 252C, Hobart, Tasmania, Australia, 7001.

Pages 72–77 *in* MARINE MAMMAL RESEARCH IN THE SOUTHERN HEMISPHERE VOLUME 1: STATUS, ECOLOGY AND MEDICINE ed by M. Hindell and C. Kemper. Surrey Beatty & Sons, Chipping Norton. 1997.

combination of ketamine HCL and diazepam (Woods *et al.* 1989), then weighed them using a tripod and electronic scales suspended from a one tonne chain block. We injected each female with 370 MBq of tritiated water (HTO) into the lumbar extra-dural intervertebral vein using a 1 ml glass turburculin syringe and an 11 cm spinal needle. Prior to each injection we took a 5 ml blood sample from the same site to evaluate background HTO activity in the animal. The seals were left for six hours after injection to allow full equilibration of the HTO (Costa *et al.* 1986). We then took a second 5 ml blood sample.

We also weighed each pup then injected them with 100 ml of 90 atom per cent deuterated water (HDO) into the lumbar extra-dural intervertebral vein, and them left them for a six-hour equilibration period.

Sixteen of the original 28 mother-pup pairs were still together 20 days later. The other 12 pairs were not relocated due to a variety of causes including post-anaesthetic complications, maternal desertion of pups and early weaning of pups. Towards the end of the lactation period (18–22 days) we anaesthetized and weighed each of the 16 relocated females in the same manner as at the beginning of the study. We took a third 5 ml blood before re-injecting a further 370 MBq of HTO. We reweighed each of the pups, took a third blood sample and re-injected with 100 ml of HDO. A final blood sample was collected after a six-hour equilibration period.

Laboratory procedures

Blood samples were stored in air-tight phials, sealed with parafilm and frozen until analysis in Australia. Water was distilled from the whole blood using rapid vacuum sublimation (Vaughan and Boling 1961). We determined tritium activity from a 200 microlitre aliquot of this extracted water in 2 ml of Emulsifier Safe Scintillation cocktail (Packard) in a Packard Tri-card 4 000 liquid scintillation counter. Each sample was counted three times to less than 0.5% error. Standards were prepared by identical injection of 370 MBq of HTO (using the same syringe and needles as used in the field) into 1 000 ml of distilled water. Dueterium was assayed using mass ratio spectrometry on water distilled from the blood (CSIRO Wildlife Ecology).

Estimation of pup milk consumption and female energy expenditure

The method of estimating pup milk consumption and female energy expenditure followed Costa *et al.* (1986). We derived estimates of milk consumption by pups from water influx during the lactation period, as the only sources of water available to the pups during this time was from their mother's milk and from their own metabolic processes. Metabolic water influx was estimated using figures from Kretzmann *et al.* (1993). The difference of total water influx, derived from the dilution of HDO over time, and metabolic water production was taken to be the rate of milk water ingestion. The total milk input was then estimated from the milk water input and the mean water content of Southern Elephant Seal milk (Hindell *et al.* 1994a).

The total energetic expenditure of the female was derived from the calorific value of the milk consumed by the pup and her own metabolic costs. Calorific value of the milk was estimated from known proportions of fat, protein and water in the milk (Hindell *et al.* 1994a), and published calorific values for each component (Riedman and Ortiz 1979). We estimated the metabolic energy expended by lactating females from water influx derived from dilution of HTO during the experimental period (Nagy and Costa 1980). To estimate the total energy expenditure we assumed that 90% of the total energy was provided by fat catabolism and 10% was derived from protein catabolism (Slip *et al.* 1992). As fat catabolism produces approximately 1.07 g of water and 39.33 kJ of energy and protein produces 0.39 g of water and 17.99 kJ of energy (Slip *et al.* 1992), approximately 2 g of protein must have been used for each 8 g of fat to provide the 10:90 relative energy contributions. The total energy generated for each litre of metabolized water was thus 37.52 MJ.

Results in the text are expressed as mean ± standard deviation.

RESULTS

Consumption of milk and energy by pups

Cross-contamination with tritium of blood samples during storage meant that water influx data were obtained for only five pups and six mothers. During 20 days of sucking these pups increased in mass more than two fold, from a mean of 47.0 ± 5.1 kg to 112.5 ± 16.89 kg. Their average water influx was 2.9 ± 0.3 l/day, of which at least 0.93 ± 0.12 l/day was metabolic water. The remaining 1.99 ± 0.27 l/day of water were therefore acquired from the mother's milk. This translates to an average daily milk intake of 4.86 ± 0.66 l/day for the pups, and a daily energy intake of 79.56 ± 10.72 MJ/day (Table 1).

Changes in body composition and water influx of lactating females

Data on the changes in body composition and water influx during lactation were available for six adult females. These animals lost an average of 161 kg (or 33% of original body mass) during the 20 day experimental period (Table 2). The percentage body fat declined from 36.4 ± 6.1% immediately post-partum to 25.3 ± 3.2% 20 days later. During this time the females had a

Table 1. Estimates of daily milk and energy intake for five Southern Elephant Seal pups. Metabolic water input was estimated from Oritz *et al.* (1978) and the mean water content of Southern Elephant Seal milk was taken to be 41% (derived from Hindell *et al.* (1994a)). Milk fat was assumed to produce 38.1 MJ/kg and milk protein to produce 24.6 MJ/kg.

Pup	Birth Mass (kg)	Final Mass (kg)	Time (days)	Water Influx (l/day)	Metabolic Water (l/day)	Water From Milk (l/day)	Milk Intake (l/day)	Milk Fat (kg/day)	Milk Protein (kg/day)	Energy Intake (MJ/day)
27	38.0	104.5	20	2.87	0.82	2.05	5.00	1.80	0.55	81.80
89	48.5	104.0	20	2.84	0.89	1.95	4.76	1.71	0.52	77.81
90	49.0	101.0	20	2.44	0.88	1.56	3.80	1.37	0.41	62.25
110	49.5	111.0	20	3.20	0.94	2.26	5.51	1.98	0.60	90.18
119	50.0	142.0	20	3.28	1.13	2.15	5.24	1.88	0.57	85.79
Mean	47.0	112.5		2.93	0.93	1.99	4.86	1.75	0.53	79.56
S.D.	5.1	16.9		0.33	0.12	0.27	0.66	0.24	0.07	10.72

Table 2. Estimates of body composition and water influx for six lactating Southern Elephant Seals.

Female	Time (days)	Mass at Birth (kg)	Final Mass (kg)	Fat at Birth (%)	Final Fat (%)	Water Influx (l/day)	Metabolic Rate (MJ/day)
2	20	440	274	47.9	21.9	1.85	69.41
7	20	448	298	30.8	28.8	1.00	37.52
8	20	501	361	36.4	29.5	1.60	60.03
13	20	552	375	36.2	22.8	1.69	63.41
19	20	535	338	35.3	23.5	1.99	74.66
2773	18	489	352	31.8	25.3	1.40	52.53
Mean		494	333	36.4	25.3	1.59	59.59
S.D.		45	39	6.1	3.2	0.35	13.23

Table 3. The average daily energy expenditure (metabolic rate plus milk energy produced), total metabolic rate (MR * average lactation length of 24 days (MacMahon *et al.* 1997)) and total energy expenditure during lactation for four female Southern Elephant Seals.

Female	Metabolic Rate – MR (MJ/day)	Milk Energy (MJ/day)	Average Daily Energy Expenditure (MJ/day)	Total Metabolic Rate (MJ)	Total Energy Expenditure (MJ)
2	69.41	81.80	151.21	1 665.84	3 629.04
7	37.52	77.81	115.33	900.48	2 767.92
8	60.03	62.25	122.28	1 440.72	2 934.72
13	63.41	85.79	149.20	1 521.84	3 580.80
Mean	57.59	76.91	134.50	1 382.22	3 228.12
S.D.	13.93	10.30	18.37	334.38	440.83

Table 4. A comparison of several measures of maternal expenditure at Macquarie Island and South Georgia. The South Georgia data for maternal mass, maternal mass changes and energy transfer efficiency are taken from Arnbom *et al.* (1993) and the pup mass data are taken from Fedak *et al.* (in press). All results are given as means and standard deviations with sample sizes shown in parenthesis.

	Macquarie Island	South Georgia	*t*-test
Post-partum maternal Mass (kg)	507 ± 102 (33)	529 ± 120 (152)	$t_{50.7} = 0.39$, p = 0.69
Maternal Mass at Weaning (kg)	361 ± 82 (29)	357 ± 82 (107)	$t_{42.7} = 1.18$, p = 0.24
Maternal Mass Loss Rate (kg/day)	8.12 ± 1.44 (16)	7.9 ± 1.4 (150)	$t_{165} = 0.09$, $p > 0.05$
Pup Birth Mass (kg)	41.80 ± 6.9 (41)	42.5 ± 1.5 (26)	$t_{49.9} = 0.65$, p = 0.52
Pup Mass at Weaning (kg)	114.0 ± 16.6 (13)	120.8 ± 27.5 (26)	$t_{39.8} = 1.34$, p = 0.19
Energy Transfer Efficiency (%)	42.40 ± 1.0 (16)	45.6 ± 1.5 (130)	$t_{145} = 0.02$, $p > 0.05$

mean water influx of 1.59 l/day. As the animals did not eat or drink, this translates to a metabolic rate of 59.59 ± 13.23 MJ/day.

Energetic cost of lactation

The Average Daily Energy Expenditure (ADEE) of the four females for which there were complete data was 134.5 ± 18.37 MJ/day (Table 3). Consequently, assuming an average lactation length of 24 days (see McMahon *et al.* 1997), these animals would have expended 3 228.12 ± 440.83 MJ to rear their pups, 1 382.22 ± 334.38 MJ (or 42.8%) of which would have been metabolic costs and the remainder lost as the energy in milk.

Comparisons with South Georgia

The energetic costs of lactation for elephant seals at South Georgia were contrasted with similar data from Macquarie Island (Table 4). There were no significant differences in the initial or the final masses of the seals at the two islands. Nor was there any difference, between the populations with respect to the rate of maternal weight loss with mothers losing about 8.0 kg per day. Similarly, there were no differences between the birth mass or the weaning mass of the pups in either study (Table 4), and the energy transfer efficiency was the same for both populations with approximately 45% of the mass lost by mothers being converted to pup tissue.

DISCUSSION

Energetic cost of lactation in phocid seals

Energy expenditure during lactation has so far been estimated for eight species of phocid seals: Northern Elephant Seals (Costa *et al.* 1986), Hooded Seals (Bowen *et al.* 1987), Harp Seals (Stewart 1987). Harbour Seals (Bowen *et al.* 1993), Ringed Seals (Hammill *et al.* 1991), Weddell Seals (Tedman and Green 1987) and Grey Seals (Fedak and Anderson 1982; Iverson *et al.* 1993). Comparisons between Northern and Southern Elephant Seals are particularly interesting as the two species are very closely related taxonomically and morphologically, and yet rear their young in areas of climatic extremes. The initial results of Costa *et al.* (1986) have recently been revised on the basis of improved estimates of metabolic water production in sucking pups (Kretzmann *et al.* 1993). The revised values for Northern Elephant Seals are similar to those derived in the present study for Southern Elephant Seals with respect to average daily energy expenditure (136 MJ/day for *M. angustirostris* and 135 MJ/day for *M. leonina*). However, the longer period of lactation in the northern species results in a greater nett energy cost to the females and larger pups at weaning (Hindell *et al.* 1994a). The small sample sizes in both studies precludes more detailed comparisons, and highlight some apparent anomalies that need to be clarified. For example, the two species appear to be similar in most respects (percentage adipose tissue before and after lactation, daily metabolic rates, rates of weight loss and energy assimilation rates in the pups) and yet *M. angustirostris* females suckle for an average of four days (or 17%) longer *and* have richer milk than *M. leonina* females (i.e., Northern Elephant Seals appear to use similar amounts of energy to produce richer milk for a longer time). Given that the two species also have similar daily energy expenditures, these factors cannot be resolved without more comprehensive studies on both species.

The relative metabolic costs of lactation in harp, hooded, ringed, Weddell and Northern Elephant Seals were compared by Bowen *et al.* (1993); they were found to range between 24.2 and 25.9 $MJ/kg^{0.75}$ for 80% of the lactation period. When expressed in the same fashion, Southern Elephant Seals have a metabolic cost of lactation of 27.2 $MJ/kg^{0.75}$ indicating that the relative costs of lactation are similar for all phocids studied to date.

Bigger seals have more energy reserves relative to their metabolic overheads (Costa 1991), and consequently can transfer more of their nett energy budget to their pups, either as richer milk over a short time or as more, less rich milk over a longer period. Large female elephant seals wean larger pups than do small females, which may be a consequence of differential energy reserves (Arnbom *et al.* 1993). However, larger females also tend to be older, and so it is likely that maternal experience may also play an important role in determining the size of pups at weaning (Iverson *et al.* 1993).

Intra-species comparisons

There were no apparent differences between the Macquarie Island and South Georgia populations of *M. leonina* with respect to (i) initial post-partum mass of mothers, (ii) the mass of mothers towards the end of lactation, (iii) the rate of weight loss and daily energy transfer from mother to pup, and (iv) the duration of lactation (McMahon *et al.* 1997) although this later variable was measured in a different year to present study. The data therefore suggest that females at Macquarie Island and South Georgia are similar with respect to the energy that they have available to deliver to their pups.

However, there is evidence for a significant difference in weaning mass, with the South Georgia animals weaning approximately 6% heavier that those at Macquarie (H. Burton, pers. comm.). This result is based on a substantially bigger sample consisting of several thousand animals, and indicates that large sample sizes are required to resolve the difference between the two populations. It is therefore likely that small differences may exist between the populations for some of the parameters described for mothers in this study, although the ecological significance of these small differences at the population level remains open to conjecture.

Implications for understanding the population decline

A major impetuous for this study was to test the hypothesis that female elephant seals in the declining Macquarie Island population had less energy available, in the form of adipose tissue, to transfer to their pups, which ultimately resulted in lower survivorship of first year seals (Hindell 1991; Hindell *et al.* 1994b) than in the stable South Georgia population (Laws 1994). The results of the present study indicate no apparent differences between the populations, and therefore suggest that differences in maternal energy reserves, and ultimately maternal food resources, were not responsible for the decline in nuumbers at Macquarie Island.

However, these data cannot be regarded as a definitive test of the hypothesis for several reasons:

(a) The major demographic influence for the decline at Macquarie Island seems to have been during the 1960s when juvenile survival was at its lowest (Hindell 1991). Although contemporary survivorship data are not available, there is now evidence to suggest that the Macquarie Island population is presently stable (Antarctic Division, unpubl. data). Ideally therefore, data on maternal expenditure are required from the 1960s. Such data are unavailable, but we do know that females from Macquarie Island were up to 25% smaller (in terms of mass) at puberty than those from South Georgia in the 1950s and 1960s (Carrick *et al.* 1962; Bryden 1968; Laws 1984). This difference declined in older age classes, but Laws (1984) suggested that there would still be a 16% disparity in mass of 20 year old animals. Thus during the 1960s Macquarie Island animals would have had less energy to give to their pups, and this is reflected in the lower birth masses in that population, which were also approximately 16% lighter than from South Georgia (Laws 1984).

(b) There may be differences between the two populations that were not detected in this study, such as the lower weaning masses recorded for Macquarie Island (H. Burton, pers. comm.). However, the difference in weaning mass between contemporary populations (6%) is considerably less than the difference in adult female size (16–25%) in past populations. There is a strong relationship between maternal size and weaning mass (Arnbom *et al.*1993), which suggests that females of the present stable Macquarie Island population are larger than when the population was in decline. This question cannot be completely resolved without comprehensive studies of growth in both populations.

One of the predictions of the "overshoot" hypothesis was that the Macquarie Island females should attain adult body sizes comparable to those at South Georgia, concurrent with stabilization of population size (Hindell *et al.* 1994b). This is based on the premise that the smaller body size during the 1950s was due to reduced food availability to Macquarie Island seals, which was in turn due to the population being at abnormally high levels (Hindell 1991). The results of this study are consistent with this prediction: there are no longer the differences in adult mass of females elephant seals that were described in 1960s (even with larger sample sizes the variation around the population means would decrease, but not the means themselves). A similar result was found when the mass of moulting adult females was compared between Macquarie Island and South Georgia (Hindell *et al.* 1994c).

Deposition of adipose tissue is central to reproductive success in elephant seals, and perhaps even plays an important role in regulation of population size. The amount of fat that a female brings ashore on breeding islands is the result of several factors, including maternal size and foraging success.

ACKNOWLEDGEMENTS

We would like to thank Dr Rupert Woods for his assistance in the field, particularly with the anaesthetics. Also the many other members of the 1987 and 1988 ANARE members at Macquarie Island contributed to the project in many ways. Brian Green and Keith Newgrain from CSIRO Wildlife Ecology analysed the dueterium samples and provided valuable advice throughout. Harry Burton provided unpublished data and much helpful advice.

REFERENCES

Arnbom, T., Fedak, M. A., Boyd, I. L. and McConnell, B. J., 1993. Variation in weaning mass of pups in relation to maternal mass, postweaning fast duration and weaned pup behaviour in southern elephant seals *(Mirounga leonina)* at South Georgia. *Can. J. Zool.* **71:** 1772–781.

Bonner, W. N., 1984. Lactation strategies in pinnipeds: Problems for a marine mammalian group. Pp. 253–72 *in* Physiological Strategis in Lactation ed by M. Peaker, R. G. Vernon and C. H. Knight. Academic Press: London

Bowen, W. D., Boness, D. J. and Oftedal, O. T., 1987. Mass transfer from mother to pup and subsequent mass loss by the weaned pup in the hooded seal, *Cystophora cristata. Can. J. Zool.* **65:** 1–8.

Bowen, W. D., Oftedal, O. T. and Boness, J. B., 1992. Mass and energy transfer during lactation in a small phocid, the harbor seal *(Phoca vitulina). Phys. Zool.* **65:** 844–66.

Bryden, M. M., 1968. Control of growth in two populations of elephant seals. *Nature* **217:** 1106–108.

Carrick, R., Csordas, S. E. and Ingham, S. E., 1962. Studies on the southern elephant seal, *Mirounga leonina* (L.) IV. Breeding and development. *CSIRO Wildl. Res.* **7:** 161–97.

Costa, D. P., 1991. Reproductive and foraging energetics of high latitude penguins, albatrosses and pinnipeds: implications for life history patterns. *Am. Zool.* **31:** 111–30.

Costa, D. P., Le Boeuf, B. J., Huntley, A. C. and Ortiz, C. L., 1986. The energetics of lactation in the northern elephant seal *Mirounga angustirostris*. *J. Zool., Lond.* **209:** 21–33.

Fedak, M. A. and Anderson, S. S., 1982. The energetics of lactation: accurate measurements from a large wild mammal, the grey seal *(Halichoerus grypus)*. *J. Zool., Lond.* **198:** 473–79.

Fedak, M. A., Arnbom, T. and Boyd, I. L., in press. The relation between size of southern elephant seal mothers, the growth of their pups and the use of maternal energy, fat and protein during lactation. *Physiol. Zool.*

Hammill, M. O., Lydersen, C., Ryg, M. and Smith, T. G., 1991. Lactation in the ringed seal *(Phoca hispida)* based on cross sectional sampling. *Can. J. Fish. Aq. Sci* **48:** 2471–476.

Hindell, M. A., 1991. Some life history parameters of a declining population of southern elephant seals. *J. Anim. Ecol.* **60:** 119–34.

Hindell, M. A., Bryden, M. M. and Burton, H. R., 1994a. Early growth and milk composition in southern elephant seals *(Mirounga leonina)*. *Aust. J. Zool.* **42:** 723–32.

Hindell, M. A., Burton, H. R. and Slip, D. J., 1991. Foraging grounds of southern elephant seals, *Mirounga leonina*, as inferred from water temperature data. *Aust. J. Mar. Freshwater Res.* **42:** 115–28.

Hindell, M. A., Slip, D. J. and Burton, H. R., 1994b. Possible causes of the decline of southern elephant seal populations in the Southern Pacific and Southern Indian Oceans. Pp. 66–84 *in* Elephant Seals: Population Ecology, Behaviour and Physiology ed by B. J. LeBoeuf and R. M. Laws. University of California Press: Berkley.

Hindell, M. A., Slip, D. J. and Burton, H. R., 1994c. Body mass loss in moulting female southern elephant seals at Macquarie Island. *Polar Biology* **14:** 275–78.

Iverson, S. J., Bowen, W. D., Boness, D. J. and Oftedal, O. T., 1993. The effect of maternal size and milk enegy output on pup growth in grey seals *(Halichoerus grypus)*. *Physiol. Zool.* **66:** 61–88.

Kovacs, K. M. and Lavigne, D. M., 1992. Maternal investment in otariid seals and walruses. *Can. J. Zool.* **70:** 1953–964.

Kretzmann, M. B., Costa, D. P. and Le Boeuf, B. J., 1993. Maternal energy investment in elephant seal pups: evidence for sexual equality? *Am. Nat.* **141:** 466–80.

Laws, R. M., 1984. Seals. Pp. 621–713 *in* Antarctic Ecology ed by R. M. Laws. Academic Press: London.

Laws, R. M., 1994. History and present status of southern elephant seal populations. Pp. 49–65 *in* Elephant Seals: Population Ecology, Behaviour and Physiology ed by B. J. LeBoeuf and R. M. Laws. University of California Press: Berkley.

Lydersen, C. and Hammill, M. O., 1993. Activity, milk intake and energy consumption in free-living ringed seal *(Phoca hispida)* pups. *J. Comp. Physiol. B* **163:** 433–38.

McConnell, B. J., Chambers, C. and Fedak, M. A., 1992. Foraging ecology of southern elephant seals in relation to the bathymetry and productivity of the Southern Ocean. *Antarct. Sci.* **4:** 393–98.

McMahon, C. R, van den Hoff, J., Burton, H. R. and Davis, P. D., 1997. Evidence for precocious development in female pups of the Southern Elephant Seal *Mirounga leonina* at Macquarie Island. Pp. 92–96 *in* Marine Mammal Research in the Southern Hemisphere Vol. 1: Status, Ecology and Medicine ed by M. Hindell and C. Kemper. Surrey Beatty & Sons: Chipping Norton.

Nagy, K. A. and Costa, D. P., 1980. Water flux in animals: analysis of potential errors in the tritiated water method. *Am. J. Physiol.* **238:** R454–R465.

Oftedal, O. T., Boness, D. J. and Tedman, R. A., 1987. The behaviour, physiology, and anatomy of lactation in the pinnipedia. *Curr. Mamm.* **1:** 175–245.

Riedman, M. and Ortiz, C. L., 1979. Changes in milk composition during lactation in the northern elephant seal. *Physiol. Zool.* **52:** 240–49.

Slip, D. J., Gales, N. J. and Burton, H. R., 1992. Body mass loss, utilization of blubber and fat, and energetic requirments of male southern elephant seals, *Mirounga leonina*, during the moulting fast. *Aust. J. Zool.* **40:** 235–43.

Stewart, R. E. A., 1987. Behavioural reproductive effort of nursing harp seals, *Phoca groenlandica*. *J. Mamm.* **68:** 348–58.

Tedman, R. and Green, B., 1987. Water and sodium fluxes and lactational energetics in suckling pups of Weddell seals *(Leptonychotes weddelli)*. *J. Zool., Lond.* **212:** 29–42.

Vaughan, B. E. and Boling, E. A., 1961. Rapid assay procedures for tritium-labeled water in body fluids. *J. Lab. Clinical Med.* **57:** 159–64.

Woods, R., Hindell, M. A. and Slip, D. J., 1989. The effects of physiological state on sensitivity to ketamine anaesthesia in the southern elephant seal. *J. Wildl. Dis.* **25:** 586.

The Australian Sea Lion: a review of an unusual life history

N. J. GALES[1] and D. P. COSTA[2]

The supra-annual (17.6 month), non-seasonal breeding cycle of the Australian Sea Lion is uncharacteristic of pinnipeds. This endemic Australian species numbers only 9 300–11 700 animals and has been shown to (i) have a varied diet, (ii) work hard during foraging, (iii) produce relatively low energy milk, and (iv) control its unique reproductive timetable endogenously. It is sparsely distributed, with less than 50 pups born each season at most breeding sites. The reproductive behaviour of *N. cinerea* is unusual among otariids, as males engage in sequential polygyny (sequential defence of single mates) and do not form harems. We review the current knowledge of this species' life history and propose that a prime determinant for its reproductive cycle and behaviour is a nutrient poor, stable marine environment in which seasonality offers no advantage to the temporal targeting of energetically expensive lactation. Rather, *N. cinerea* has opted to spread its anabolic effort over a greater period, during which the dependent young supplement their nutritional needs and learn a wide variety of foraging skills necessary for an animal with such opportunistic feeding habits. Concomitant with the extended lactation is an extended gestation, facilitated by an increase in the placental phase of pregnancy.

INTRODUCTION

MAMMALIAN reproductive parameters generally follow an annual timetable in response to predictable fluctuations in resources. The family Pinnipedia has evolved particular life-history patterns that enable it to exploit aquatic food resources whilst still using terrestrial habitats for whelping and nurturing its young. Within the Pinnipedia, two fundamentally different reproductive systems have evolved. The Phocidae (true seals) give birth on ice or land, have a brief lactation during which the females usually fast, produce an extremely lipid-rich, energy dense milk, fuelled primarily from stored blubber, the pups gain weight rapidly and weaning is abrupt. The Otariidae (sea lions and fur seals) also give birth on land, but differ from phocids by producing milk of lower lipid content, having a more protracted lactation, during which the females alternate feeding the pup with replenishing their food reserves at sea, having slower pup growth rates and weaning less abruptly. The extent of habitats utilized by the otariids (from equatorial with little seasonal variation to sub-polar with high seasonal variation) has led to the development of a range of reproductive and foraging strategies that compensate for the particular proximate environmental influences. Despite these environmentally driven behavioural modifications, most otariids still have in common a life history characterized by synchronous, annual reproduction and the behavioural and physiological mechanisms that drive this. The exception is the Australian Sea Lion *Neophoca cinerea*.

N. cinerea is an Australian endemic which has been shown to have a non-annual, non-seasonal breeding cycle of 17–18 months (Ling and Walker 1978; Higgins 1993; Gales *et al.* 1994), to be temporally asynchronous between breeding sites (Gales *et al.* 1994), and to have a non-typical otariid mating strategy (Higgins 1990). Other studies have elucidated this species' maternal behaviour (Higgins 1990), its at sea behaviour (Costa *et al.* 1989; Costa *et al.* 1991) and diet (Richardson and Gales 1987; Gales and Cheal 1992), as well as described aspects of proximate milk composition (Kretzmann *et al.* 1991; Gales *et al.*, in press) and reproductive physiology (Gales *et al.*, in prep). In this paper we summarize these studies and speculate on the reasons why this sea lion has evolved this unique life history.

ABUNDANCE, DISTRIBUTION AND POPULATION TRENDS

The Australian Sea Lion is distributed from Houtman Abrolhos, Western Australia to The Pages in South Australia (Fig. 1) (Gales *et al.* 1994). Within this range it utilizes about two thirds of the available islands as breeding or resting sites. Mainland habitat is far less extensively used with only a few, isolated locations on the Australian mainland hosting year-round aggregations of sea lions (Gales *et al.* 1994). Breeding has been recorded on 51 islands, 28 of which are in Western Australia and 23 in South Australia (Fig. 2) (Gales *et al.* 1994). A further six islands in Western Australia and 13 islands in South Australia where sea lions were seen are possible breeding locations (Gales *et al.* 1994).

The population of the Australian Sea Lion is estimated at 9 300–11 700 (Gales *et al.* 1994). Previous estimates were 2 300–5 000 (Walker and Ling 1981; Inns *et al.* 1979). This increase in estimated population size does not imply

[1]School of Veterinary Studies, Murdoch University, Murdoch, Western Australia, Australia, 6150.
Present Address: Department of Conservation, PO Box 10–420, Wellington, New Zealand.
[2]Institute of Marine Sciences and Department of Biology, University of California, Santa Cruz, CA 95064, USA
Pages 78–87 *in* MARINE MAMMAL RESEARCH IN THE SOUTHERN HEMISPHERE VOLUME 1: STATUS, ECOLOGY AND MEDICINE ed by M. Hindell and C. Kemper. Surrey Beatty & Sons, Chipping Norton. 1997.

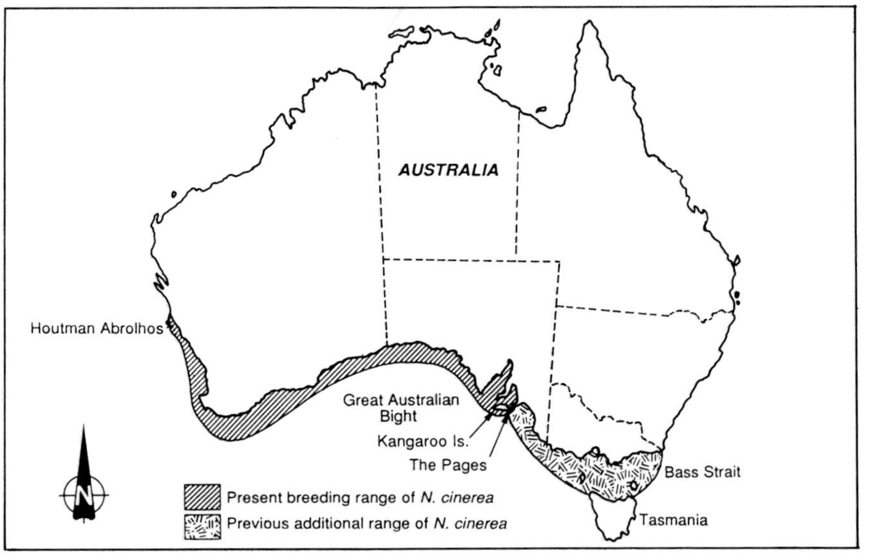

Fig. 1. Current range of *Neophoca cinera* and range prior to commercial sealing.

population growth; rather, the study by Gales *et al.* (1994) is the first systematic and comprehensive survey of the species to date.

There are few data from which to assess the status of *N. cinerea* populations. Based on counts of pups and the total number of animals ashore, the colony at Seal Bay, Kangaroo Island is reported to have been stable over the past decade (Robinson and Dennis 1988). No equivalent data exist for other colonies. Pup production on three small colonies off the west coast of Western Australia varied markedly in pup production and pup mortality over three consecutive breeding seasons (Gales *et al.* 1992). This variability highlights the need for long-term monitoring programmes at pre-determined colonies if population status is to be determined accurately.

The average size of *N. cinerea* breeding populations is among the smallest of any otariid. Only five of the known 51 breeding populations produce more than 100 pups per season. The two islands of The Pages are the sites of the largest colonies, producing about 300 pups each per season, which accounts for almost a quarter of the species' total pup production. The other three large colonies are all in South Australia: Dangerous Reef (275 pups/season), Kangaroo Island (180 pups/season) and Purdie Island (120 pups/season). The remaining 46 colonies produce approximately 30 pups each per season, and account for a little more than half of the species' pup production (Gales *et al.* 1994).

There is an eastern bias in the population distribution of *N. cinerea*. Despite approximately 55% of the breeding locations being reported in Western Australia, the sea lion colonies in South Australia account for 71% of the estimated population. This difference is primarily because the five largest colonies are in South Australia. The remaining small colonies in South Australia have an estimated mean seasonal pup production of 32, compared with a pup production of 28 at all Western Australia (Gales *et al.* 1994).

There are no records of the pristine population size of *N. cinerea*. Although this species was less commercially valuable than the sympatric New Zealand Fur Seal *Arctocephalus forsteri* it was taken in large numbers throughout its range for food and skins. Sealing in Australia was largely unregulated and it is not possible to estimate the extent of the take of sea lions from sources such as cargo manifests. Descriptions from early explorers provide strong evidence that *N. cinerea* was considerably more abundant in some parts of its range than today (e.g., Vancouver 1801; Freycinet 1807; Flinders 1814), as well as being more widespread with populations previously existing in Bass Strait (Flinders 1814) and Tasmania (Wood Jones 1925).

REPRODUCTIVE BEHAVIOUR

Seasonality and duration of pupping cycle

The most atypical aspect of the reproductive behaviour of *N. cinerea* is its supra-annual breeding cycle. The most detailed elucidation of this cycle is provided by Higgins (1990) and Higgins (1993) who studied individual sea lions at Seal Bay, Kangaroo Island. She described 33 interbirth intervals of 22 individual females and provided further data from median dates of birth of unmarked females to conclude that the mean interbirth interval was 17.6 ± 0.3 months. Although no such data from marked females have been collected elsewhere for this species, a 17–18 month interval between pupping seasons has been demonstrated at 14 other sites, throughout its range (Gales *et al.* 1992; Gales *et al.* 1994). Despite extensive survey effort, there are no data to indicate that intervals between breeding seasons are other than 17–18 months at any Australian Sea Lion colony.

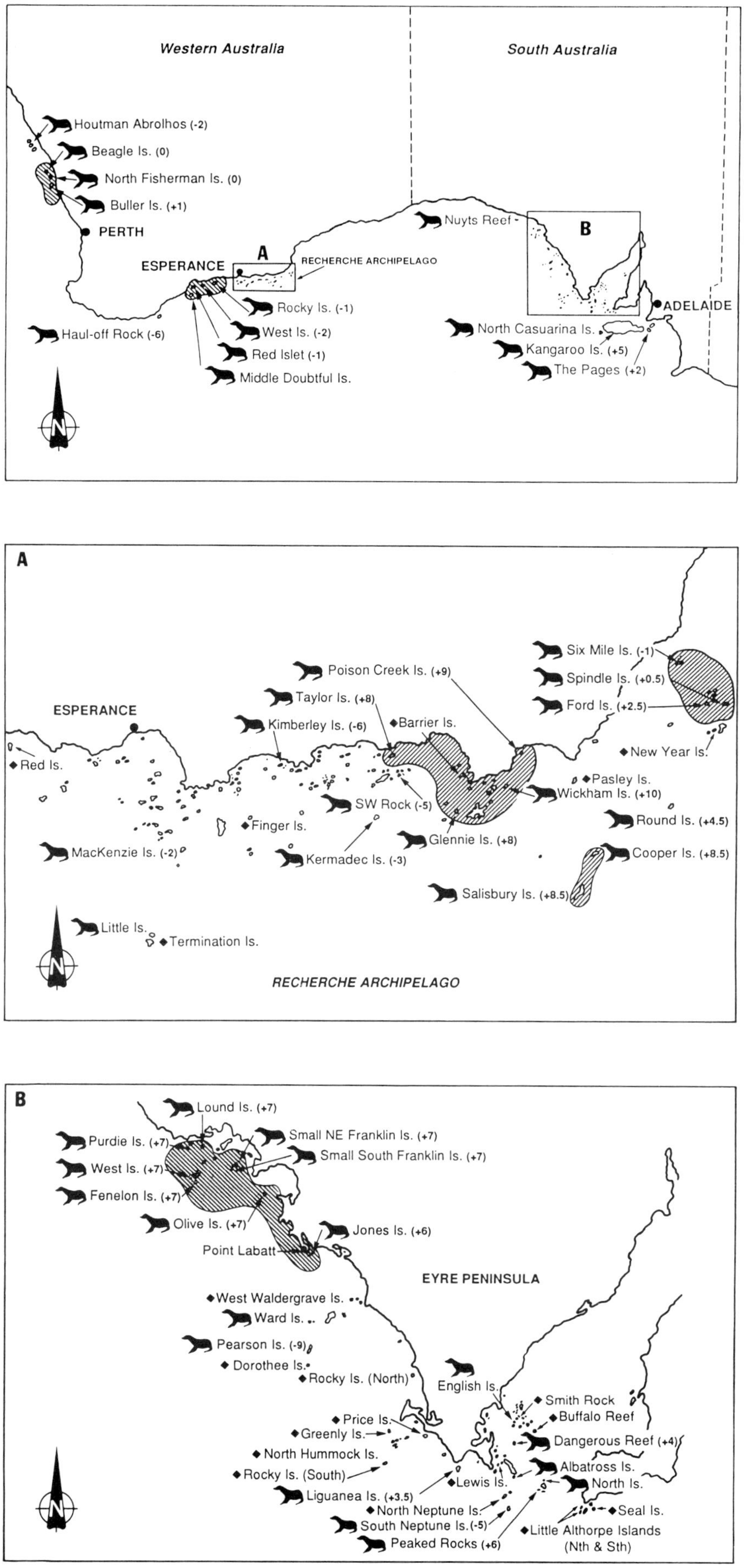

Fig. 2. Breeding colonies and haul-out sites of *Neophoca cinera*. Known breeding islands are marked with a positive or negative number which relates to the timing (months) of their breeding season relative to that of the reference colony, North Fisherman Island. Hatching demarcates groups of proximate islands with breeding seasons in relatively close synchrony and which may represent sub-populations. Unconfirmed, or possible breeding locations are marked with diamond icons.

A clear consequence of this supra-annual breeding cycle is a lack of seasonality. If the period of the breeding cycle was fixed at 18 months, a system of alternating seasons could be expected with, perhaps, the development of alternating strategies appropriate for each season. Assuming the period of slightly less than 18 months is common to all sites, then there will be a seasonal drift over time, with pupping seasons eventually occuring at all times of year at any one colony. Consequently, no seasonally based reproductive behaviour could operate. Such seasonal drifts in pupping seasons have been described for Kangaroo Island (Higgins 1990) and colonies on the west coast of Western Australia (Gales *et al.* 1992).

A further intriguing aspect of the breeding cycle of *N. cinerea* is the lack of synchrony of pupping seasons between sites. Gales *et al.*(1994) documented the pupping seasons of 37 *N. cinerea* breeding sites and found no pattern that could be used to predict the timing of one colony in relation to others. Some colonies in close proximity to one another shared, or were within a month of, the timing of pupping seasons, but generally the pattern was random (Fig. 2). Such asynchrony between sites differs from the types of clinal variation in pupping dates reported for Weddell Seals *Leptonychotes weddellii* (Stirling 1969) and Harbour Seals *Phoca vitulina* (Temte *et al.* 1991) which follow a predictable pattern and can be explained by variations in photoperiod.

A pupping season of five months has been described for *N. cinerea* at several locations (Gales *et al.* 1992; Higgins 1993; Gales *et al.* 1994) and is likely to occur throughout its range. This is the most protracted pupping season reported for any otariid.

Mating strategy

Neophoca cinerea demonstrated a typical otariid pattern of polygyny with the formation of small harems of females attended by a single male (Marlow 1975). It was however, unclear if the polygyny was based on female defence, resource (or habitat) defence, male dominance, or scramble competition (Boness 1991). A more detailed investigation of reproductive behaviour of *N. cinerea* at Seal Bay, Kangaroo Island described male breeding behaviour as a combination of territoriality (resource defence) and female defence (Higgins 1990). Contrary to the findings of Marlow (1975), who described males attending up to four females at one time, Higgins (1990) described a system of sequential polygyny where males typically attended only one female at a time. Mating followed the post-partum oestrus, after which the female made her first foraging trip and the male would seek another pre-oestrus female. As the pupping season is protracted (up to five months), this system allows males to achieve more than one mating during a single season. Furthermore, as the energy reserves of males are insufficient to allow them to fast throughout the pupping season, periods of time ashore can be interspersed with foraging trips. Such a mating system leads to a high rate of turnover of territories, with as many as five males occupying a single territory during a breeding season at Seal Bay (Higgins 1990). The territories themselves were plastic in size and duration of male tenure, with flexible boundaries and no ritualized boundary displays by males. Higgins (1990) believed that the existence of "pup holes" (an area in which pups could be secreted and protected from aggression from bulls) may be a determinant of territoriality; such environments being attractive to the females. Gales *et al.* (1994) also noted the utilization of "pup holes" at most breeding sites, but commented upon the paucity of such habitat at the islands with the highest pup production and most dense colonies (viz. The Pages and Dangerous Reef). The habitat at these colonies is characterized by rocky substratum with little vegetation cover and few ledges or holes. Consequently, pups are mostly born and nurtured on open ground. These types of density dependent factors are clearly important when investigating reproductive behaviour. The site of Higgins' (1990) study is an area of low female density, and indeed low pup production. Most pup production at Seal Bay occurs in Pup Cove, a cliff-bound bay to the west of Higgins' study site. There, female density is higher and slightly different mating strategies may be utilized. Similarly, Marlow's (1975) study on Dangerous reef was of a colony where density is high and resources (in terms of area, and possibly "pup holes") are more limited. Sea lions at such colonies may well utilize strategies at variance with those described by Higgins (1990) for colonies of low female density. However, based on observations by one of us (NG) from survey work during the breeding seasons at the majority of the small colonies where *N. cinerea* breed (and indeed all those in Western Australia), a system of sequential, single mate defence operates.

Neophoca cinerea is a highly sexually dimorphic species with males weighing up to 300 kg and females up to 120 kg. Consequently, we might expect a more polygynous breeding system than has been reported. However, Boness (1991) notes that the use of sexual dimorphism as an indicator of degree of polygyny should be done with caution and, further, that phylogenetic and ecological factors can constrain the ability to capitalize on the polygamy potential. Indeed, Higgins (1990) noted that the differential reproductive success of males at Seal Bay was low, and that a high proportion of them had access to at least one female during each breeding season.

Maternal investment: attendance cycles and duration of lactation

The only intensive investigation of maternal investment in *N. cinerea* was conducted on

Kangaroo Island (Higgins 1990; Higgins and Gass 1993). Many aspects of female behaviour were typical of other otariids: females usually exhibit site fidelity in their choice of birth sites, they arrive a mean of 1.8 days prior to birth and the first foraging trip begins a mean of 9.8 days after birth (oestrus occurs at about 7 days post-partum) (Higgins and Gass 1993). During the first five months of lactation female attendance cycles averaged 33.0 ± 7.3 hrs ashore and 48.5 ± 12.6 hrs at sea; foraging trips during the first month tended to be the shortest, levelling off thereafter (Higgins and Gass 1993). Duration of time at sea and ashore were independent of the sex of the pup.

In contrast to the annual breeding otariids, *N. cinerea* females continue to suckle their pups for 15–18 months. Pups are typically weaned one month prior to the cow giving birth (Higgins and Gass 1993). Clearly, this extended lactation is related to the protracted breeding cycle. The level of nutritional and developmental dependence of the pup throughout lactation has not been quantified, and warrants special attention.

At Seal Bay, an estimated 71% of females gave birth to pups in consecutive seasons (Higgins and Gass 1993). Cows that do not give birth in consecutive seasons and whose pups survived the lactation period, continued to suckle their yearlings for a further 15–18 months. Occasionally, females suckle their offspring for three complete lactational cycles. Similarly, failure to wean a previous offspring at the time of parturition will often lead to the death of the new pup by starvation.

REPRODUCTIVE PHYSIOLOGY

Duration of embryonic diapause and placental gestation

The mechanism by which *N. cinerea* achieves its extended gestation is intriguing. The confirmation of a short post-partum oestrus and the documentation of inter-birth intervals in individual females demonstrated that their extended reproductive cycle was not achieved by extending the temporal separation of birth and mating (such as in the walrus), but through an extended gestation (Higgins 1993). Gestation in pinnipeds is characterized by a period of embryonic diapause, followed by post-implantation gestation. Studies in many species have shown some variation in the duration of diapause, but a consistency in duration of placental gestation (Boyd 1991). It may therefore have been expected that *N. cinerea* had achieved a gestation of between 17 to 18 months by extending embryonic diapause to about nine months, thus maintaining a placental gestation of some nine months.

Embryonic diapause was first reported in *N. cinerea* by Tedman (1991) who recovered a blastocyst from a female know to have given birth about eight weeks earlier. In a more recent study, blood samples were collected from longitudinally sampled *N. cinerea*. One hundred and fifty-four blood samples were collected from cows at various stages of gestation and were analysed for oestradiol and progesterone (Gales *et al.*, unpubl. data). There was a significant increase in the concentrations of these hormones during the fourth month of pregnancy, peaking in the sixth month of gestation. Oestradiol concentrations then decreased over the next few months before plateauing at nine months after parturition of the previous pup. This plateau level was higher than for the first three months of gestation. Circulating progesterone concentrations remained high during the sixth and seventh months after parturition, decreasing during the eighth month to levels approximating those of the first three months of pregnancy. Concentrations of both progesterone and oestradiol showed a discernible rise during the thirteenth month of pregnancy, then returned to plateau levels (Gales *et al.*, unpubl. data) (Fig. 3). The changes in these two circulating hormones at around 4–5 mnths provided strong evidence that reactivation and implantation of the blastocyst in *N. cinerea* occurs at this time, indicating that embryonic diapause is of similar duration to that in other pinnipeds. This would lead to a prolonged active (placental) gestation of up to 14 months, the longest recorded for any pinniped.

The extended placental gestation suggested for *N. cinerea* should perhaps result in the production of a more developmentally advanced pup. Assessing this in terms of pup birth mass as a percentage of maternal mass suggests this is not the case. Indeed, *N. cinerea* produce a pup of only 9% maternal body mass compared to 10.7 ± 2.58% (mean ± sd) for ten other otariids (see summary in Costa 1991). Accordingly, it seems likely that foetal development in *N. cinerea* is slower than that of other pinnipeds, except perhaps the walrus which, in common with the Australian Sea Lion, has an extended placental gestation. Such a reduced growth rate may represent an energetic advantage to female *N. cinerea* which can spread the energetic cost of gestation over a longer period.

Proximate composition of milk

The Australian Sea Lion produces milk that contains less energy (fat) than other temperate or high latitude otariids. The mean proximate composition of *N. cinerea* milk during lactation is 31.1% lipid, 56.8% water, 9.8% protein and 0.9% ash; fat content increases during lactation and is inversely proportional to water content (Gales *et al.*, in press). Of the otariids for which proximate milk composition has been studied throughout lactation (n = 6 species), only the Galapagos fur seal has a lower lipid content than that of *N. cinerea* (Trillmich and Lechner 1986).

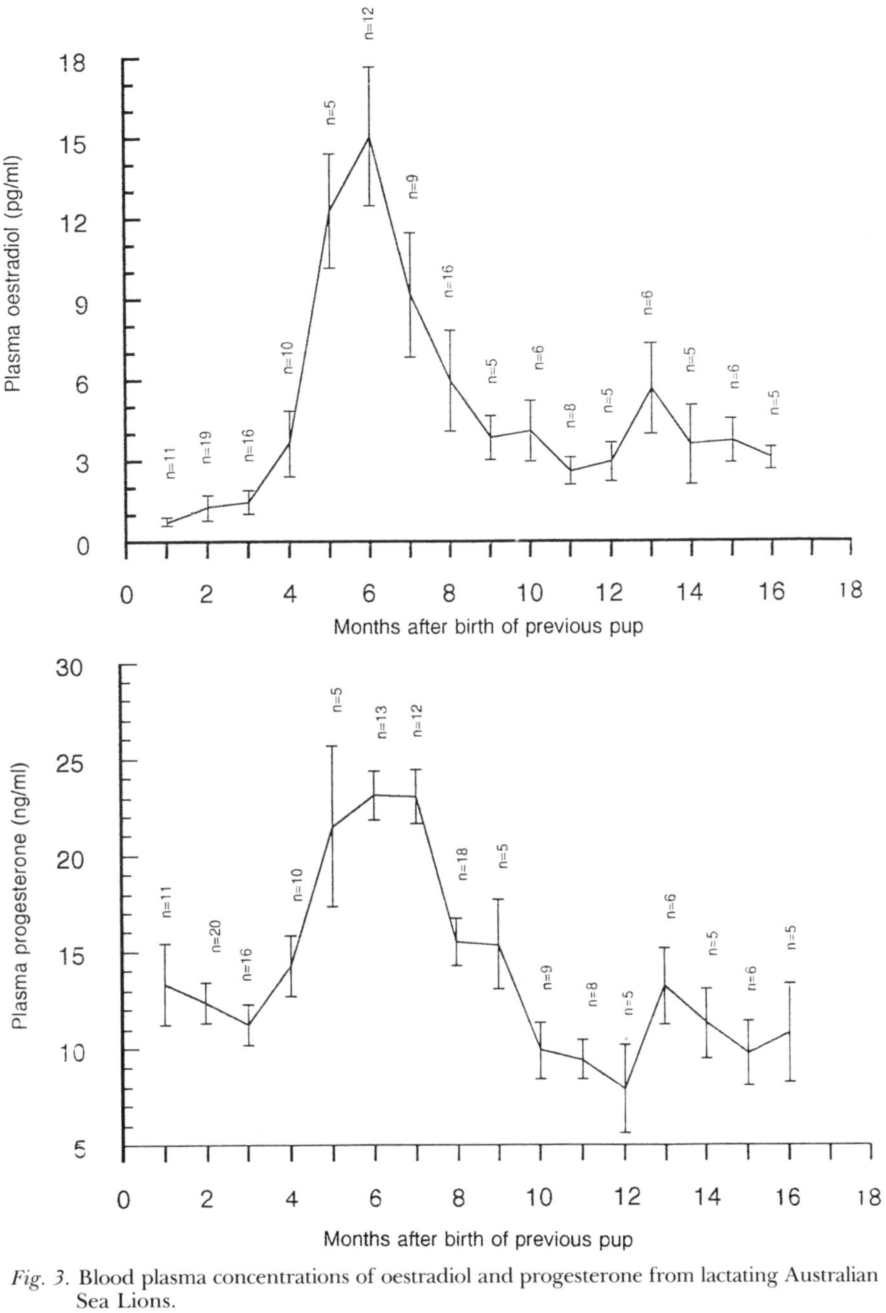

Fig. 3. Blood plasma concentrations of oestradiol and progesterone from lactating Australian Sea Lions.

A further feature of milk from *N. cinerea* is its high degree of variability between samples. This variability is in part associated with pup age, but did not correlate with cow size, foraging trip duration, or time spent ashore prior to milk sample collection (Kretzmann *et al.* 1991; Gales *et al.*, in press). Possible explanations for the variability in lipid content between samples are that it reflects variable foraging success (patchy prey distribution) or variable foraging strategies (diet).

The correlation between pup age and lipid content of the milk (lipid content increased during lactation) suggests that some variation in resource availability may occur over the duration of lactation. However, this explanation seems unlikely as breeding in this species is non-annual, non-seasonal and asynchronous across its range. As such, the different stages of lactation will occur during different seasons, and these stages will shift during subsequent lactations. The reasons for the temporal changes in lipid content are unknown and may be controlled by a wide variety of systems such as a feedback mechanism based on pup demand (as the pup begins to forage for itself, and attendance patterns change), a genetic coding, or changes in maternal foraging strategy. Further research on maternal foraging skills and attendance patterns throughout lactation, as well as ontogeny of foraging skills of juveniles is required to elucidate the temporal changes in lipid content of the milk.

FORAGING ECOLOGY

Diet

The diet of free ranging *N. cinerea* has not been well studied. Anecdotal accounts have indicated a broad diet with prey species including teleost fish, elasmobranchs, cephalopods, crustaceans, penguins and other sea birds, and even turtles (Marlow 1975; Walker and Ling 1981; King 1983; Richardson and Gales 1987; Ling 1992). In a study of the efficacy of using scats as an indicator of prey consumption in *N. cinerea* Gales and Cheal (1992) determined that fewer than 2% of ingested otoliths were excreted, and excretion was biased to larger otoliths. Similarly, with cephalopod mouth parts there was differential excretion of beaks of different sizes in vomits and scats. Analysis of prey remains in scats of *N. cinerea* was thus described as an unreliable indicator of diet and other techniques such as gastric lavage, emesis, or perhaps immunology need to be utilized to quantify diet more adequately. Nevertheless, it is clear that the Australian Sea Lion has a broad, perhaps opportunistic diet, with benthic organisms being well represented.

Diving behaviour

The diving pattern of Australian Sea Lions is distinct from that reported for most other otariids. Other otariids typically feed on midwater prey and dive deeper during the day and shallower at night as they follow vertically migrating prey (Gentry and Kooyman 1986). In contrast, the diving behaviour of 17 female lactating Australian Sea Lions measured over three field seasons at Seal Bay, Kangaroo Island (Costa *et al.* 1991) indicates that they feed predominantly on the bottom with no diurnal pattern. While at sea 50–60% of the time was spent underwater, with only short periods at the surface between dives. Animals leave the beach and begin diving immediately, apparently following the bottom contour until they reach the feeding grounds. While feeding on the bottom, their dives are shallow, reaching an average maximum depth of 92 m with a mean depth of 67 m. Dives last an average of 3.3 min, with the mean maximum dive duration of 5.74 min, the longest dive recorded being 8.3 min. Their dives are on average longer than other otariids, especially the similar sized California Sea Lion, which has a mean dive duration of about 2 min (Feldkamp *et al.* 1989)(Fig. 4). Diving behaviour appears to change with season; in winter dives were exclusively flat bottomed, whereas in summer they consisted of some flat bottomed dives interspersed with bouts of shallower V and U shaped dives. The flat bottomed dives were interpreted as being benthic dives, as the depth range was that of the shelf waters adjacent to the sea lion colony. The shallower V and U shaped dives were interpreted to be mid-water

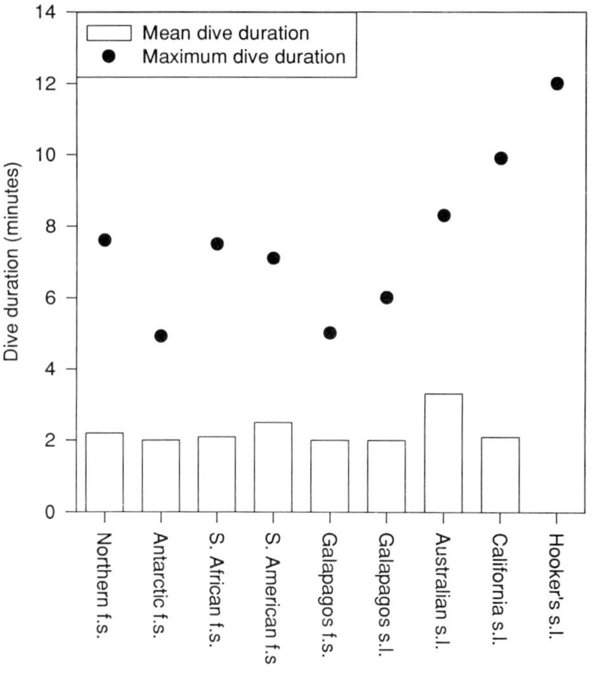

Fig. 4. Maximum (dots) and mean (histogram) dive duration for 11 pinnipeds. Body mass of pinnipeds increases from left to right. Data are from Galapagos Fur Seal (Trillmich *et al.* 1986); Antarctic Fur Seal (Kooyman *et al.* 1986); Cape Fur Seal (Kooyman and Gentry 1986); Northern Fur Seal (Gentry *et al.* 1986); California Sea Lion (Feldkamp *et al.* 1989); Australian Sea Lion (Costa *et al.* 1989); Hooker's Sea Lion (Gentry *et al.* 1987).

dives. During both seasons the sea lion's diving effort (ratio of time spent diving to time spent on surface) was high, inferring a need for similar effort in acquiring resources year round (Costa, unpubl. data).

Foraging energetics

Australian Sea Lions have at-sea metabolic rates that are comparable with those measured on other otariids (Costa 1991). Metabolic rates have been measured using the O_{18} doubly labelled water method (Costa *et al.* 1989) on 16 lactating female *N. cinerea*. The overall metabolic rate of females was 435 W with a mean mass of 79 kg, or 4.8 times the predicted basal metabolic rate (BMR) for a terrestrial animal of equal size. However, when these rates are corrected for the various periods of time the animals spent onshore and at sea, the at-sea metabolic rate is 616 W, or 6.8 times the predicted BMR. These results indicate that Australian Sea Lions expend energy faster than equivalently sized, deeper mid-water feeding California Sea Lions, which expend energy at 4.8 times the predicted BMR (Feldkamp *et al.* 1989; Costa *et al.* 1991). The higher at-sea metabolic rate of Australian Sea Lions is surprising given that they exhibit dives of longer duration than most other otariids (Fig. 4). If *N. cinerea* have a similar oxygen storage capacity to California Sea Lions (Feldkamp *et al.* 1989), they should have utilized their oxygen reserves during the first two minutes of their 3.3 min dives. If the animals utilize aerobic metabolism we might expect prolonged surface intervals following their dives. However, the surface interval following a dive of an Australian Sea Lion is typically less than the preceding dive (Costa *et al.* 1991). The implication of these comparisons is that adult female *N. cinerea* are working hard during foraging, at least compared to similar sized California Sea Lions. As larger animals have disproportionately greater diving capability than smaller animals (Costa 1993) it is interesting to consider what foraging behaviour might be adopted by immature (smaller) *N. cinerea*. Studies of diving behaviour of juvenile *N. cinerea* could be most instructive.

DISCUSSION

Previous studies of *N. cinerea* create a picture of a most unusual pinniped. The supra-annual life cycle, lack of temporal synchrony between breeding populations, widespread and dispersed distribution and its opportunistic diet all suggest an animal adapted to a low energy, stable, environment. Indeed, the marine environment inhabited by *N. cinerea* is characterized by waters of low productivity (Pearce 1991) and has been described as one of the most nutrient poor marine environments in the world (Rochford 1980). This is in contrast to most continental west and (in the southern hemisphere) south coasts that are influenced by nutrient-rich, cold currents flowing towards the equator which are part of the general ocean circulation. These run clockwise in the northern hemisphere and anticlockwise in the southern hemisphere. These waters run westward along the equator, until they collide with the east coasts of the various continents where they are diverted back towards the poles. In Australia some of these nutrient depleted, warm waters pass through the Indonesian archipelago and flow south along the west coast of Australia as the Leeuwin Current. This flow down the west coast has the effect of preventing the normal, northward flow of cold, high energy water and extends a tropical marine influence well into temperate latitudes. This influence gradually dilutes as the Leeuwin current flows down the west coast of Australia and then eastward along the south coast. The diminution of its influence is likewise reflected in the eastward distribution of *N. cinerea,* and also in the eastward distribution of another south Australian pinniped, the New Zealand Fur Seal *Arctocephalus forsteri* (Shaughnessy *et al.* 1994).

This nutrient poor, relatively stable marine environment may explain some of the adaptive strategies utilized by *N. cinerea*. A reproductive rate that is set endogenously may be assumed to be most adaptive to non-variable trophic and climatic influences (Murton and Westwood 1977). In Western Australia there are examples of other marine predators (birds) that have evolved non-annual reproductive patterns (Wooller *et al.* 1991). The establishment of the supra-annual cycle of *N. cinerea* offers the advantage of spreading the energetic demands of lactation over a longer time, as well as enabling dependent young more time to acquire foraging skills (Gales *et al.* 1994). Similarly, the protracted placental gestation spreads the energetic cost of reproduction over a longer period. The widespread, small, temporally asynchronous breeding colonies dilute any concentration of resource requirements over their range.

The mating strategy of *N. cinerea* is similarly well adapted for its environment. The protracted pupping season precludes the option of males maintaining tenure of territories for the entire season, indeed the changeover of territory tenure in *N. cinerea* is known to be high (Higgins 1990). As such, a system of sequential polygyny, operating perhaps as a combination of resource and female defence, allows the males access to more than one female during the season, as well as enabling them to feed periodically. The lower level of polygyny reported for *N. cinerea* compared to other otariids (Higgins 1990) has the likely consequence of decreasing the intensity of sexual selection. Indeed the ratio of adult female mass to adult male mass is less for *N. cinerea* than for most other otariids. A further consequence of the low variance in male reproductive success is that any absolute decline in population size must be greater before a genetic bottleneck is reached than would be the case if the variance in reproductive success was high.

The flow chart (Fig. 5) summarizes our suggested determinants of the life history traits of *N. cinerea.*

If the reproductive cycle evolved by *N. cinerea* is optimal for its low energy environment, why is it that similar life history patterns have not been adopted by other otariids living in stable, low energy habitats such as the tropics (e.g., the Galapagos Fur Seal and Galapagos Sea Lion). The most likely explanation for this is that the Galapagos Islands are non-typical for tropical environments as they experience annual upwellings of nutrient rich waters, and thus offer a clear advantage to focus reproduction to coincide with this trophic window. This discrete period of food availability is reflected in the variations in milk fat (energy) content of milk in the Galapagos Fur Seal *Arctocephalus galapagoensis* (Trillmich and Lechner 1986). Contrary to the trend in other pinnipeds to have an increase in milk fat over the duration of lactation, this species has its highest fat content during the period of maximum food availability. Later in lactation the content is the lowest recorded for any pinniped. Indeed, in the absence of the annual upwellings (as occurs during El Niño events), *A. galapagoensis* behaves much like *N. cinerea* in extending its lactation (albeit via the neglect and loss of the ensuing offspring).

The Australian Sea Lion shares a significant portion of its range with the New Zealand Fur Seal which breeds annually and is undergoing a

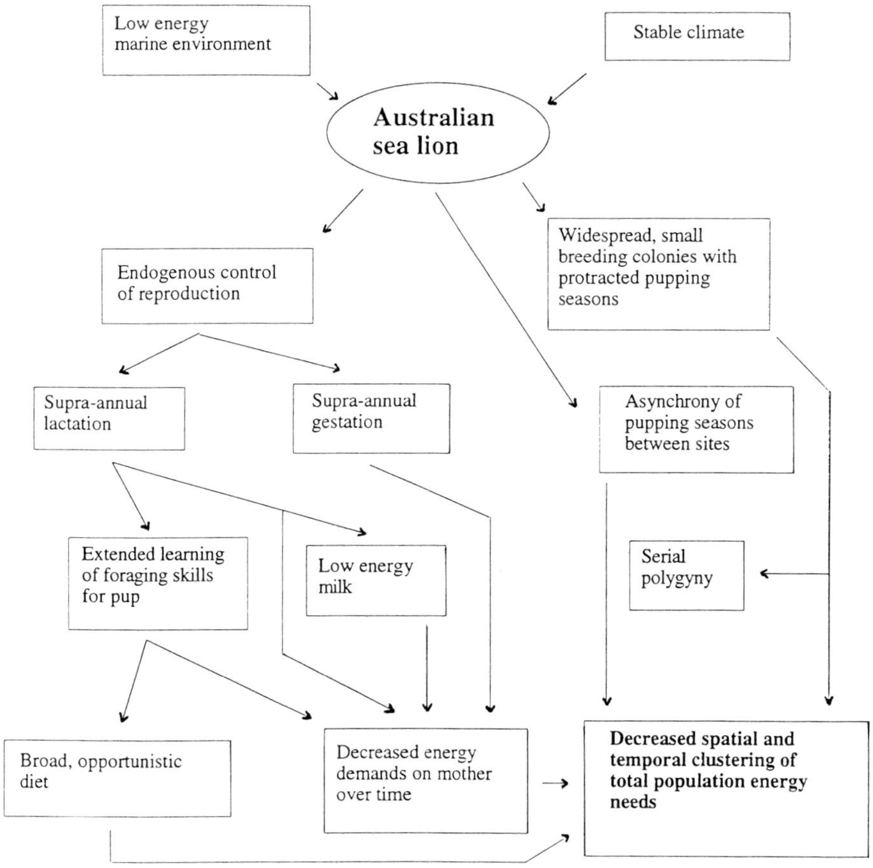

Fig. 5. Flow chart summary of determinants of life history traits of the Australian Sea Lion.

population increase (Shaughnessy *et al.* 1994). The apparent success of the fur seals in this region suggests that they are better adapted to the marine environment of southern Australia than *N. cinerea*, and that consequently non-annual seasons could be regarded as mal-adaptive. This apparent contradiction of our hypothesis for the evolution of *N. cinerea's* reproductive life cycle may be explained when the entire range of these species is taken into account. The New Zealand Fur Seals inhabiting the south coast of Australia are at the western limit of their distribution; their eastern limit being the islands of New Zealand where they are far more abundant. Indeed, within the range of their distribution they share with *N. cinerea* there is a marked eastern bias. *N. cinerea*, on the other hand, are distributed eastwards only as far as the islands off South Australia, the area about which the effects of the Leeuwin current are at their weakest. The western and northern distribution of *N. cinerea* continues into less productive waters. It is possible that *N. cinerea* have evolved in these nutrient limited waters of Western Australia and moved eastwards to the limit of their competitive capacity. *A. forsteri* on the other hand have evolved the more typically otariid strategies that suit them well over their range, the west coast of Western Australia representing the western limits of their competitive (or adaptive) capacity.

Future research of *N. cinerea* should perhaps occur in the western half of this species' range, where it is probable that the primary determinants of its life history are most potent.

ACKNOWLEDGEMENTS

This review was considerably improved by the constructive criticisms of Drs D. Boness, F. Trillmich and P. Shaughnessy.

REFERENCES

Boness, D. J., 1991. Determinants of mating systems in the Otariidae (Pinnipedia). Pp. 1–44 *in* Pinniped Behaviour ed by D. Renouf. Chapman and Hall: London.

Boyd, I. L., 1991. Environmental and physiological factors controlling the reproductive cycles of pinnipeds. *Can. J. Zool.* **69:** 1135–148.

Costa, D. P., 1991. Reproductive and foraging energetics of pinnipeds: Implications for life history patterns. Pp. 300–44 *in* Pinniped Behaviour ed by D. Renouf. Chapman and Hall: London.

Costa, D. P., 1993. The relationship between reproductive and foraging energetics and the evolution of the Pinnipedia. *Symp. zool. Soc. Lond.* **66:** 293–314.

Costa, D. P., Kretzmann, M. and Thorson, P. H., 1989. Diving pattern and energetics of the Australian sea lion, *Neophoca cinerea*. *Am. Zool.* **29:** 71A (Abstr.).

Costa, D. P., Swain, U. and Gales, N. J., 1991. Seasonal changes in the diving pattern and energetics of the Australian sea lion, *Neophoca cinerea* P. 14 *in* Proceedings of 9th Biennial Conference on the Biology of Marine Mammals, Chicago, Illinois, December 5–9, 1991. (Abstr.)

Feldkamp, S. D., DeLong, R. L. and Antonelis, G. A., 1989. Diving patterns of California sea lions, *Zalophus californianus. Can. J. Zool.* **67:** 872–83.

Flinders, M., 1814. A voyage to Terra Australis. Vol. 1. G. and W. Nicol: London.

Freycinet, L. C. D., 1807. Voyage de découvertes aux Terres Australes. Vol 1. Péron, F. Paris: Imprimerie Impériale.

Gales, N. J. and Cheal, A. J., 1992. Estimating diet composition of the Australian sea lion *(Neophoca cinerea)* from scat analysis: an unreliable technique. *Wildl. Res.* **19:** 447–56.

Gales, N. J., Cheal, A. J., Pobar, G. J. and Williamson, P., 1992. Breeding biology and movements of Australian sea-lions, *Neophoca cinerea,* off the West coast of Western Australia. *Wildl. Res.* **19:** 405–16.

Gales, N. J., Shaughnessy, P. D. and Dennis, T. E., 1994. Distribution, abundance and breeding cycle of the Australian sea lion *Neophoca cinerea* (Mammalia: Pinnipedia). *J. Zool., Lond.* **234:** 353–70.

Gales, N. J., Costa, D. P. and Kretzmann, M. B. Proximate composition of Australian sea lion milk throughout the entire supra-annual lactation period. *Wildl. Res.* In press.

Gentry, R. L. and Kooyman, G. L., 1986. Fur seals: maternal strategies on land and at sea. Pinceton University Press: Princeton.

Higgins, L. V., 1990. Reproductive behavior and maternal investment of Australian sea lions. Ph.D. Thesis, University of California: Santa Cruz.

Higgins, L. V., 1993. The nonannual, nonseasonal breeding cycle of the Australian sea lion, *Neophoca cinerea. J. Mamm.* **74:** 270–74.

Higgins, L. V. and Gass, L., 1993. Birth to weaning parturition, duration of lactation, and attendance cycles of Australian sea lions *(Neophoca cinera) Can. J. Zool.* **71:** 2047–055

Inns, R. W., Aitken, P. F. and Ling, J. K., 1979. Mammals. Pp. 91–102 *in* Natural history of Kangaroo Island ed by M. J. Tyler, C. R. Twidale, and J. R. Ling. Royal Society of South Australia: Adelaide.

King, J. E., 1983. Seals of the World. British Museum, Natural History: London.

Kooyman, G. L. and Gentry, R. L., 1986. Diving behaviour o south African fur seals. Pp. 142–52 *in* Fur Seals: Maternal Strategies on Land and at sea ed by R. L. Gentry and G. L. Kooyman. Princeton University Press: Princeton.

Kooyman, G. L., Davis, R. W. and Croxall, J. P., 1986. Diving behaviour of Antarctic fur seals. Pp. 115–25 *in* Fur Seals: Maternal Strategies on Land and at sea ed by R. L. Gentry and G. L. Kooyman. Princeton University Press: Princeton.

Kretzmann, M. B., Costa, D. P., Higgins, L. V. and Needham, D. J., 1991. Milk composition of Australian sea lions, *Neophoca cinerea:* variability in lipid content. *Can. J. Zool.* **69:** 2556–561.

Ling, J. K., 1992. *Neophoca cinerea.* Mammalian species No. 392. *Am. Soc. Mamm.* **392:** 1–7.

Ling, J. K. and Walker, G. E., 1978. An 18-month breeding cycle in the Australian sea lion? *Search* **9:** 464–65.

Marlow, B. J., 1975. The comparative behaviour of the Australasian sea lions *Neophoca cinera* and *Phocarctos hookeri* (Pinnipedia: Otariidae). *Mammalia* **39:** 159–230.

Murton, R. K. and Westwood, N. J., 1977. Avian breeding cycles. Clarendon Press: Oxford.

Pearce, A. F., 1991. Eastern Boundary currents of the southern hemisphere. *J. Roy. Soc. West. Aust.* **74:** 35–45.

Richardson, K. C. and Gales, N. J., 1987. Functional morphology of the alimentary tract of the Australian sea lion, *Neophoca cinera. Aust. J. Zool.* **35:** 219–26.

Robinson, A. C. and Dennis, T. E., 1988. The status and management of seal populations in South Australia. Pp. 87–110 *in* Marine Mammals of Australasia; field biology and captive management ed by M. L. Augee. The Royal Zoological Society of New South Wales: Sydney.

Rochford, D. J., 1980. Nutrient status of the oceans around Australia. *CSIRO Division of Fisheries and Oceanography Report* **1977–79:** 9–20.

Shaughnessy, P. D., Gales, N. J. and Dennis, T. E., 1994. Distribution and abundance of New Zealand fur seals *Arctocephalus forsteri* in South Australia and Western Australia. *Wildl. Res.* **21:** 667–95.

Stirling, I., 1969. Ecology of the Weddell seal in McMurdo Sound, Antarctica. *Ecology.* **50:** 573–86.

Tedman, R. A., 1991. The female reproductive tract of the Australian sea lion, *Neophoca cinerea* (Peron, 1816) (Carnivora: Otariidae). *Aust. J. Zool.* **39:** 351–72.

Temte, J. L., Bigg, M. A. and Wiig, Ø., 1991. Clines revisited: the timing of pupping in the harbour seal *(Phoca vitulina). J. Zool., Lond.* **224:** 617–32.

Trillmich, F. and Lechner, E., 1986. Milk of the Galapagos fur seal and sea lion, with a comparison of the milk of eared seals (Otariidae). *J. Zool., Lond.* **206:** 271–77.

Trillmich, F., Kooyman, G. L., Majiluf, P. and Sanchez-Grinan, M., 1986. Attendance and diving behaviour of south American fur seals during El Niño in 1983. Pp. 153–67 *in* Fur seals: maternal strategies on land and at sea ed by R. L. Gentry and G. L. Kooyman. Princeton University Press: Princeton.

Vancouver, G., 1801. A voyage of discovery to the North Pacific Ocean and round the World. Vol. 1. J. Stockdale: London.

Walker, G. E. and Ling, J. K., 1981. Australian Sea Lion *Neophoca cinerea* (Péron, 1816). Pp. 99–118 *in* Handbook of Marine Mammals, Volume 1. The Walrus, Sea Lions, Fur Seals and Sea Otter ed by S. H. Ridgway and R. J. Harrison. Academic Press: London.

Wood Jones, F., 1925. The eared seals of South Australia. *Rec. South Aust. Mus.* **3:** 9–16.

Wooller, R. D., Dunlop, J. N., Klomp, N. I., Meathrel, C. E. and Wienecke, B. C., 1991. Seabird abundance, distribution and breeding patterns in relation to the Leeuwin Current. *J. Roy. Soc. West. Aust.* **74:** 129–32.

Growth in Subantarctic Fur Seal *Arctocephalus tropicalis* pups as a possible indicator of offshore food availability

M. N. BESTER[1] and A. S. VAN JAARSVELD[1]

Both male and female pups at Marion Island and Amsterdam Island were significantly heavier in 1981/82 than their counterparts at Gough Island at 30–36 days (in 1984/85) and 287–320 days (1979/80) of age. There were negligible mass differences between Marion and Amsterdam Island populations. This may suggest consistent conditions between populations, although differences in pup growth rates linked to differences in food availability to their mothers may only be detected in unusual seasons. Comparison with the much larger Gough Island Fur Seal population, where pups grow poorly, suggests that there are differences in the *per capita* prey abundance/availability to lactating females. In addition, disparate growth rates were observed in male and female pups at Gough Island which may reflect the inability of mothers to meet the milk demands of male pups.

INTRODUCTION

ENVIRONMENT and genotype interact to produce phenotypic parameters such as growth rate and final body size of individuals. Body size can be considered to be a measure of nutrient intake (Fitzhugh 1975) which, in turn, varies with changes in population size (Innes *et al.* 1981). Adult size in pinnipeds is largely influenced by early growth (Bryden 1968), and therefore investigations of growth to weaning may give insight into large scale environmental influences on seal populations through the energetic bond between mothers and their offspring (McLaren and Smith 1985).

In otariids, maternal care typically consists of a relatively short peri-natal period ashore with the single pup, followed by a long period of alternating onshore activity (predominantly suckling) and foraging at sea. Attendance behaviour (i.e., the pattern in which fur seal females deliver nourishment to their growing young) has been examined as a possible indicator of food availability and environmental change, with longer trips indicating scarcer resources (McCann 1987). One of the end-products of attendance behaviour, pup growth, has been used in the Antarctic Fur Seal *Arctocephalus gazella*, the New Zealand Fur Seal *A. forsteri*, and the Subantarctic Fur Seal *A. tropicalis* to investigate possible seasonal variations in local food availability within a population (Doidge *et al.* 1984; Mattlin 1981) and for comparison between populations (Kerley 1985; Bester and Van Jaarsveld 1994).

This paper compares data on pup mass of Subantarctic Fur Seal *Arctocephalus tropicalis* populations from Amsterdam Island (37°50′S, 77°35′E), Gough Island (40°20′S, 09°54′W) and Marion Island (46°54′S, 37°45′E) as a measure of early growth to weaning in different environments.

METHODS

Live pups were caught, restrained by hand, and weighed to the nearest 100 g three times; (i) within a week of birth, (ii) at 30–36 d after the median birthdate, and (iii) around weaning at 287–320 days in different years (Table 1). Inter-island comparisons are made amongst the *A. tropicalis* populations from Amsterdam Island (median birthdate = 11 December — Roux (1987)), Gough Island (median birthdate = 11 December — Bester and Van Jaarsveld (1994)), and Marion Island (median birthdate = 17 December — Kerley (1987)) using the original data collected between 1979/80 and 1984/85. All pup ages were calculated from the appropriate median date of birth, and it was assumed that the mean birthdate did not vary in a significant way between years (*vide* Bester 1987). Student's *t*-tests were used to test for differences in measurements between the samples with significance set at the 1% level ($P < 0.01$). All means are given together with one standard deviation.

The power coefficient c which scales post-natal growth between birth and weaning (Charnov 1991), such that growth is proportional to W^c where W is body mass (Berrigan *et al.* 1993), was estimated for each population and separately for the sexes. These relationships are expressed as: $Y = aW^c$. Estimates of c were carried out using regression analysis following log-transformation of data. Log-transformation allows the estimation of c without reference to whether fur seal growth

[1]Department of Zoology and Entomology, University of Pretoria, Pretoria 0002, South Africa.
Pages 88–91 *in* MARINE MAMMAL RESEARCH IN THE SOUTHERN HEMISPHERE VOLUME 1: STATUS, ECOLOGY AND MEDICINE ed by M. Hindell and C. Kemper. Surrey Beatty & Sons, Chipping Norton. 1997.

Table 1. Body mass of *A. tropicalis* pups at comparative ages (days) from birth to weaning in different island populations (taken from Bester and Van Jaarsveld 1994).

Island	Sex	(Days)	Age (kg)	Mass sd	Range	n	Year	Source
1. Amsterdam	M	birth	4.9		3.6–6.1	25	55/56	1
	M	31–36	8.3	1.4	6.0–11.9	100	81/82	3
	M	≥298	18.2	2.3	13.5–23.0	55	81/82	2
2. Marion	M	<1	4.5	0.0		2	81/82	5,6
	M	30	7.9	1.4	4.8–9.8	15	81/82	5,6
	M	287	16.0	3.6	10.5–21.5	18	81/82	5,6
3. Gough	M	<7	4.4	0.9	3.5–6.3	7	75/76	4
	M	35	6.8	1.0	4.3–8.5	25	84/85	6
	M	320	12.9	2.2	9.4–15.8	12	79/80	4
1. Amsterdam	F	birth	4.0		3.3–4.5	29	55/56	1
	F	31–36	7.5	1.2	4.4–11.7	88	81/82	3
	F	≥298	15.5	2.5	11.0–22.5	46	81/82	2
2. Marion	F	<1	3.9	0.5	3.6–4.3	2	81/82	5,6
	F	30	7.0	0.9	4.8–8.4	20	81/82	5,6
	F	287	13.7	2.1	10.6–16.4	14	81/82	5,6
3. Gough	F	<7	4.0	0.8	2.6–4.8	6	75/76	4
	F	35	6.9	1.4	3.6–10.5	25	84/85	6
	F	320	9.5	1.8	6.2–11.7	8	79/80	4

Sources: 1 = Paulian (1964), 2 = Roux (1987), 3 = Roux (pers. comm.), 4 = Bester (1987), 5 = Kerley (1985), 6 = Bester and Van Jaarsveld (1994).

from birth to weaning is linear (Kerley 1985; Doidge *et al.* 1984; Doidge and Croxall 1989) or sigmoid (Trites 1993).

RESULTS

On Amsterdam Island, male pups were significantly heavier than female pups at approximately 34 days (t = 4.32, d.f. = 186, $P < 0.001$) and at weaning (approximately 300 days) (t = 5.17, d.f. = 99, $P < 0.001$). The mean mass difference at birth was 0.9 kg, but could not be tested for significance because the original data are not available. At Gough Island significant sexual dimorphism in pup mass was only manifested at weaning (t = 3.68, d.f. = 18, $P < 0.01$). Sexual dimorphism for all recorded ages of Marion Island pups (Table 1) failed to reach significance.

Inter-island comparisons showed little difference in mass amongst 30–36 days old females from Amsterdam, Gough and Marion Islands ($P > 0.02$) and none between Amsterdam and Marion males (t = 0.32, d.f. = 113, $P > 0.05$). Gough Island male pups weighed significantly less than male pups from Amsterdam Island (t = 5.29, d.f. = 123, $P < 0.001$) and Marion Island (t = 2.99, d.f. = 38, $P < 0.01$). Towards the end of lactation there was little difference in mass between Amsterdam Island and Marion Island pups of both sexes (t = 2.56, d.f. = 71, $P > 0.01$ and t = 2.45, d.f. = 58, $P > 0.01$ for males and females respectively), but a significant difference was observed between Amsterdam and Gough Islands (t = 6.41, d.f. = 65, $P < 0.001$ and t = 6.78, d.f. = 52, $P < 0.001$ for males and females respectively), as well as for females between Marion and Gough Islands (t = 4.89, d.f. = 20, $P < 0.001$).

The estimated power coefficients c for the three populations are shown in Table 2. The power coefficients represent the slope of a regression line fitted to the log-transformed data (age in years *v* mass in kg) depicting growth between birth and weaning. Male and female pups from Gough Island displayed disparate growth rates, when compared with the other populations (Table 2). However, the growth rate of female pups from Gough Island lies between that of pups of both sexes at Amsterdam Island (lowest growth rates) and Marion Island (highest growth rates).

Table 2. Estimated power coefficients c for different island populations with sexes analysed separately.

Island	Sex	c
1. Amsterdam	M	4.02
	F	4.08
2. Gough	M	3.54
	F	4.23
3. Marion	M	4.34
	F	4.49

DISCUSSION

Meaningful age-specific comparison of pup masses between island populations of *A. tropicalis* (Table 1) is confounded by small sample sizes for the Gough Island and Marion Island populations. Small sample sizes probably also obscured likely significant mass differences between sexes which are manifested in the Amsterdam Island population (Roux 1987; Bester and Van Jaarsveld 1994) and in other fur seal species (Mattlin 1981; Doidge *et al.* 1984; Kerley 1985; Trillmich 1986; Trites 1991) where male pups are larger and grow faster than female pups. Since both male and female pups at Amsterdam Island and at Marion Island are significantly heavier than their counterparts at Gough Island, the recorded difference in pup masses may suggest intra-specific variations in growth to weaning between populations as a result of differences in prey abundance/availability to their mothers. All three populations are increasing rapidly, but the Amsterdam Island and Marion Island populations were relatively small at 35 000 and 19 857 respectively in 1981/82 (Kerley 1987; Roux 1987) when most of the pup measurements were taken, while the large Gough Island population (estimated at over 200 000 in 1978 — Bester (1980)) presumably removed more from its environment in the period 1979/80–1984/85 when the population continued to increase (Bester 1990).

The low mean mass of the few Gough Island pups weighed at 320 days (Table 1) may have resulted from the inclusion of individuals during their early post-weaning dispersion phase (Bester 1981), as none of those weighed had their mothers in attendance and they were found on a part of the coastline where few births occurred during the austral summer. However, the low mean pup mass of 9.6 ± 3.0 kg (n = 50) and 8.9 ± 2.1 kg (n = 50) for males and females respectively at 296 days of age in 1993/94 at Gough Island (unpubl. data) would indicate that the values obtained in 1979/80 are likely to be a true reflection of the weaning mass of the pups (this study).

The similarity in age specific mass of different cohorts of pups both within and between Amsterdam Island and Marion Island (Table 1) suggests consistent environmental conditions. However, differences in pup growth rates linked to changes/differences in food to their mothers, may only be detected in unusual seasons (Doidge *et al.* 1984). Furthermore, variability in pup mass increases as fur seal pups grow older and increased sample sizes are needed (for Marion Island in particular) to detect statistically significant differences in mean body mass of older pups (Trites 1993).

A latitudinal graded difference in adult size seems to exist in *A. tropicalis* with the largest animals occurring at low latitude Amsterdam Island, followed by Gough Island, and finally Marion Island, at the southern most extension of the breeding range of the species (Bester and Van Jaarsveld 1994). Older, larger Northern Fur Seal females *Callorhinus ursinus*, produce larger foetuses than younger, smaller ones (Trites 1991). In *A. gazella*, maternal size and birth mass of male pups were highly correlated (Boyd and McCann 1989), as were birth mass and weaning mass of female pups (Doidge and Croxall 1989). The graded difference in adult size according to latitude did not, however, extend to age specific pup mass for the different *A. tropicalis* populations in the present study. Maternal and offspring body mass and condition might therefore be unrelated as in *A. gazella* where high weaning mass was associated with those pups (females) whose mothers spent more time ashore attending their offspring (Doidge and Croxall 1989).

We therefore hypothesize that despite the biases that may have been introduced (*vide* Trites 1993; this study) pups from Gough Island showed the poorest growth which is linked to a lower *per capita* prey abundance within the foraging range of the large population of lactating females at this locality. By contrast, within the small population of *A. tropicalis* at Macquarie Island, with only 68 pups of three species born in 1990/91 (Shaughnessy and Goldsworthy 1993), male pups were weaned at a mean mass of 23.3 kg at 281 days of age (Goldsworthy 1992), an approximately twofold increase over that recorded for Gough Island (this study). This hypothesis is further supported by the large difference in growth rates between male and female pups at Gough Island (Table 2), suggesting that although mass at weaning (or early post-weaning) was probably compromised in both sexes, female pups managed to maintain their growth rate whereas male pups could not do so. This appears to support findings of Boyd and McCann (1989) which indicated that maternal size and the birth mass of male pups (but not female pups) is correlated, which suggests that maternal size may be limiting male, but not female pup growth. The inability of lactating females at Gough Island to produce heavier offspring of both sexes therefore raises questions about female condition and the provisioning of their pups, and consequently *per capita* prey abundance/availability.

The results obtained in the present study are of direct relevance to the CCAMLR Ecosystem Monitoring Program (Anonymous 1988), and has a bearing on studies on Australian pinnipeds (Shaughnessy 1992) at Heard and Macquarie Islands (Shaughnessy and Goldsworthy 1993).

ACKNOWLEDGEMENTS

The South African Department of Environmental Affairs and Tourism (DEAT) provided logistical and financial support on the advice of the South African Committee for Antarctic Research (SACAR). Financial support from the Foundation for Research Development to ASvJ is gratefully acknowledged. The DEAT provided additional funding which made the attendance of MNB at this symposium possible. We thank Allan Seabrook, Steve Atkinson, James Gleeson, Lex Hes, Jean-Paul Roux, Pat Condy and Graham Kerley for their inputs during previous studies. Pieter Greyling and members of the Gough 39 team recorded pup masses in 1993/94, while the work at Amsterdam Island in 1981/82 formed part of a co-operative research programme under the auspices of Terres Australes et Antarctiques Francaises (TAAF) and the then South African Scientific Committee for Antarctic Research.

REFERENCES

Anonymous, 1988. Standard methods for monitoring parameters of predatory species. Scientific Committee for the Conservation of Antarctic Marine Living Resources: Hobart, Australia.

Berrigan, D., Charnov, E. L., Purvis, A. and Harvey, P. H., 1993. Phylogenetic contrasts and the evolution of mammalian life histories. *Evol. Ecol.* **7:** 270–78.

Bester, M. N., 1980. Population increase in the Amsterdam Island fur seal *Arctocephalus tropicalis* at Gough Island. *S. Afr. J. Zool.* **15:** 229–34.

Bester, M. N., 1981. Seasonal changes in the population composition of the fur seal *Arctocephalus tropicalis* at Gough Island. *S. Afr. J. Wildl. Res.* **11:** 49–55.

Bester, M. N., 1987. Subantarctic fur seal, *Arctocephalus tropicalis*, at Gough Island (Tristan da Cunha Group). *NOAA Techn. Rep. NMFS No.* **51:** 57–60.

Bester, M. N., 1990. Population trends of Subantarctic fur seals and southern elephant seals at Gough Island. *S. Afr. J. Antarct. Res.* **20:** 9–12.

Bester, M. N. and Van Jaarsveld, A. S., 1994. Sex-specific and latitudinal variance in postnatal growth of the Subantarctic fur seal *(Arctocephalus tropicalis). Can. J. Zool.* **72:** 1126–133.

Boyd, I. L. and McCann, T. S., 1989. Pre-natal investment in reproduction by female Antarctic fur seals. *Behav. Ecol. Sociobiol.* **24:** 377–85.

Bryden, M. M., 1968. Control of growth in two populations of elephant seals. *Nature, Lond.* **217:** 1106–108.

Charnov, E. L., 1991. Evolution of life history variation among female mammals. *Proc. Natl Acad. Sci., USA.* **88:** 1134–137.

Doidge, D. W. and Croxall, J. P., 1989. Factors affecting weaning weight in Antarctic fur seals *Arctocephalus gazella* at South Georgia. *Polar Biol.* **9:** 155–60.

Doidge, D. W., Croxall, J. P. and Ricketts, C., 1984. Growth rates of Antarctic fur seal *Arctocephalus gazella* pups at South Georgia. *J. Zool., Lond.* **203:** 87–93

Fitzhugh, H. A., 1975. Analysis of growth curves and strategies for altering their shape. *J. Anim. Sci.* **42:** 1036–051.

Goldsworthy, S. D., 1992. Maternal care in three species of southern fur seal (*Arctocephalus* spp.). PhD thesis, Monash University: Clayton.

Innes, R. E., Stewart, R. E. A. and Lavigne, D. M., 1981. Growth in Northwest Atlantic harp seals *Phoca groenlandica*. *J. Zool., Lond.* **194:** 11–24.

Kerley, G. I. H., 1985. Pup growth in the fur seals *Arctocephalus tropicalis* and *A. gazella* on Marion Island. *J. Zool., Lond.* **205:** 315–24.

Kerley, G. I. H., 1987. *Arctocephalus tropicalis* on the Prince Edward Islands. *NOAA Techn. Rep. NMFS No.* **51:** 61–64.

Mattlin, R. H., 1981. Pup growth of the New Zealand fur seal *Arctocephalus forsteri* on the Open Bay Islands, New Zealand. *J. Zool., Lond.* **193:** 305–14.

McCann, T. S., 1987. Female fur seal attendance behaviour. *NOAA Techn. Rep. NMFS.* **51:** 199–200.

McLaren, I. A. and Smith, T. G., 1985. Population ecology of seals: retrospective and prospective views. *Mar. Mamm. Sci.* **1:** 54–83.

Paulian, P., 1964. Contribution a l'etude de l'otarie de l'Ile Amsterdam. *Mammalia* **28:** 3–146.

Roux, J.-P., 1987. Subantarctic fur seal, *Arctocephalus tropicalis*, in French Subantarctic territories. *NOAA Techn. Rep. NMFS No.* **51:** 79–81.

Shaughnessy, P. D., 1992. New mammals recognized for Australia — Antarctic and Subantarctic fur seals *Arctocephalus* species. *Aust. Mammal.* **15:** 77–80.

Shaughnessy, P. D. and Goldsworthy, S., 1993. Feeding ecology of southern fur seals (*Arctocephalus* spp.) and their management at Heard and Macquarie Islands (extended abstract). *Proc. NIPR Symp. Polar Biol.* **6:** 173–75.

Trillmich, F., 1986. Maternal investment and sex-allocation in the Galapagos fur seal, *Arctocephalus galapagoensis*. *Behav. Ecol. Sociobiol.* **19:** 157–64.

Trites, A. W., 1991. Foetal growth of northern fur seals: life-history strategy and sources of variation. *Can. J. Zool.* **69:** 2608–617.

Trites, A. W., 1993. Biased estimates of fur seal pup mass: origins and implications. *J. Zool., Lond.* **229:** 515–25.

Evidence for precocious development in female pups of the Southern Elephant Seal *Mirounga leonina* at Macquarie Island

C. R. McMAHON[1], J. van den HOFF[1], H. R. BURTON[1] and P. D. DAVIS[1]

Data on birth date, birth mass, weaning mass, growth rate, nursing period, mass ratio and tooth development were collected from 974 Southern Elephant Seal *Mirounga leonina* pups on Macquarie Island from September to November 1993. On average, male pups outweighed females by 4 kg (10%) at birth and 4.5 kg (4%) at weaning. Female pups were nursed for significantly longer (24.75 ± 2.8 days) than males (24.30 ± 2.63 days). At weaning, 80.5% of the male pups lacked teeth altogether, while 88.1% of females had one or more teeth. Male and female weaners with teeth weighed more and were nursed for significantly longer than those without teeth. The lack of teeth appears to hold no advantage for pups of either sex in terms of a prolonged nursing period. Rather the presence or absence of teeth at weaning appears to be related to pre and post-partum development and to the pup's sex.

INTRODUCTION

THE Southern Elephant Seal *Mirounga leonina* is among the most sexually dimorphic of all mammals, with mature males weighing up to ten times more than adult females (McCann *et al.* 1989). However, this dimorphism is almost indistinguishable at birth and weaning with males weighing only 7% and 4% more, respectively, than females (Carrick *et al.* 1962a; Little *et al.* 1987).

Size is one of many variables that determines the social and subsequent reproductive success in male elephant seals (Bartholomew 1970; Le Boeuf and Reiter 1988). Reiter *et al.* (1978) considered that pups of both sexes embark on their respective reproductive "strategies" early in life, and that adult males and females maximize their reproductive success differently; dominant males strive to inseminate as many cows as possible over a short period of time, while females distribute their efforts throughout their lives. The nursing period is an ideal time for males and females to gain a competitive edge over their cohort members, as this is the time of most rapid growth (Reiter *et al.* 1978; Clutton-Brock *et al.* 1981).

Reiter *et al.* (1978), in a study of the Northern Elephant Seal *M. angustirostris*, showed that male pups not only weighed more than females at birth and at weaning, but that males also weaned a day later. However, Kretzmann *et al.* (1993) did not observe any differences in the nursing duration between the sexes. Reiter *et al.* (1978) also found that male pups moult and develop canines later than the females, and concluded that these characteristics may be a "strategy" employed by males to facilitate milk stealing, thus prolonging the active growth period. Stealing milk has been suggested as a possible "strategy" by which a male pup may gain a weight advantage over his cohort members and indeed Reiter *et al.* (1978) found all exceptionally large weaners were male and that males were the most successful at milk stealing. Although risky and often dangerous, the advantages of such a strategy may be of great value in this polygynous mammal (Bartholomew 1970; Reiter *et al.* 1978). Other authors (Clinton 1990; Haley 1990), however, dispute the importance of size, and the advantage of milk stealing.

Dental development in *M. leonina* post-partum is known to be extremely variable (Laws 1953). The maxillary canines are the first to appear, followed by the mandibulary canines and later by the maxillary and mandibulary incisors (Laws 1953). The average number of days from birth to eruption of the canines is 8–10 days (Briggs 1974). The typical weaning dentition, which did not differentiate between male and female weaners, was then described by the formula: $(I\,\frac{0}{1}, C\,\frac{1}{1}$ (Laws 1953). Laws (1953) also suggested that pups born later in the season develop teeth earlier.

This study describes the dentition of male and female Southern Elephant Seal pups on Macquarie Island and relates dental development with various birth and weaning parameters.

METHODS

This study was conducted on the isthmus of Macquarie Island (54°30′S, 158°57′E), from

[1]Australian Antarctic Division, Channel Highway, Kingston, Tasmania, Australia 7050.

Pages 92–96 *in* MARINE MAMMAL RESEARCH IN THE SOUTHERN HEMISPHERE VOLUME 1: STATUS, ECOLOGY AND MEDICINE ed by M. Hindell and C. Kemper. Surrey Beatty & Sons, Chipping Norton. 1997.

September through to the end of November 1993. The isthmus study area is described in detail by Carrick *et al.* (1962a).

Thirteen harems on the isthmus beaches were searched each day in the early morning for new born pups. New born pups (black pups) were weighed to the nearest kilogram within 24 hours of birth, using a 200 kg dial face, spring balance (Salters). The pups were tagged individually in the inter digital webbing of the hind flippers with a single blue "Jumbo Roto" tag (Dalton Supplies, Ltd). Pups that were not handled were marked with white oil based paint applied by means of a brush attached to a pole. This prevented pups from being classified as new born on future occasions. Thus, all new born pups were marked daily, either by tag or paint.

Three weeks after the first pup was born, daily searches were made of both beach and tussock areas for weaned pups. Pups were considered weaned once they were observed outside the breeding harems, even though their mothers may still have been present in the harems. The previously tagged pups that had left the harems were re-captured, re-weighed and permanently marked by hot iron brands on each flank. On the day of weaning, all the teeth that had erupted through the gums during the nursing period were recorded using the classification of Laws (1953).

Relationships between tooth development and the various parameters were examined using the General Linear Models in the SAS statistical package. The mean values were compared using Tukey's studentized range test. A value with $P \leq 0.05$ denotes statistically significant differences throughout. Mean values are given ± SD. Growth rates were calculated as the average mass gain during the nursing period. The nursing period for each pup was taken as the period between the date of first weighing (birth) and date of re-capture (weaning).

RESULTS

A total of 974 Southern Elephant Seal pups were weighed twice during the 1993 pupping season. Of these, 960 pups were checked at weaning for any dentition, and 49.4% (474) were found to be toothless. The number of individuals weighed at birth (n = 1 172) was divided evenly between males (605) and females (567) (Chi-square = 1.22, df = 1, $0.50 > P > 0.25$). At weaning the sex ratio remained at unity, males (536) and females (500) (Chi-squared = 1.25, df = 1, $0.50 > P > 0.25$).

At birth, males weighed significantly more than females ($P < 0.05$, *t*-test). The mean mass for male pups at birth was 42.0 ± 6.12 kg (n = 523), while the mean female mass was 38.0 ± 5.11 kg (n = 449, Table 1a), a difference of 4 kg or 10% of body weight. This significant difference was maintained until weaning (Table 1e) when males weighed 120.2 ± 26.87 kg (n = 523), and females 115.7 ± 24.48 kg (n = 450, $P < 0.05$, *t*-test). The mass of female pups at weaning was only 4% less than that of male weaners. Female pups therefore gained proportionately more weight during the nursing period than did males. This proportional weight gain is exemplified by the significant difference in mass ratio where female pups gained on average 3.05 ± 0.56 (n = 450) times their birth weight at weaning and males gained 2.87 ± 0.53 (n = 524, Table 1d).

There was no significant difference in the growth rates of male and female pups nor did the presence or absence of teeth have any statistically significant effect on growth rates over the nursing period. Male pups gained 3.26 ± 1.04 kg. day^{-1} (n = 522), females gained 3.18 ± 0.98 kg. day^{-1} (n = 449) over the nursing period (Table 1c).

The nursing period for female pups (24.75 ± 2.8 days, n = 450) was significantly ($P < 0.05$, *t*-test) longer than that of males (24.30 ± 2.63 days, n = 524) (Table 1b). At the above growth rates, this translates to approximately 1.4 kg of weight for the extra 0.45 days of nursing.

Tooth development at weaning

The development of the dentition was highly variable, with 39 unique eruption patterns (Table 2). Males and females differed in the degree of eruption, with 80.5% of males lacking teeth compared to 11.9% of females. The typical dental formula for male and female weaners respectively was: $I\,\frac{0}{0},\ C\,\frac{0}{0}$ and $I\,\frac{0}{1},\ C\,\frac{1}{1}$.

Weaning dentition and birth mass

Weaned female pups with teeth weighed (38.2 kg ± 4.86, n = 396) significantly ($P < 0.05$, *t*-test) more at birth than female pups without teeth (36.3 kg ± 6.51 kg, n = 53, Table 1a). A similar difference was noted for males, where weaned pups with teeth weighed more at birth (43.5 kg ± 5.79, n = 102) than their toothless counterparts (41.6 kg ± 6.14, n = 420). Weaned pups with teeth consistently weighed approximately 5% more at birth than those without teeth.

Weaning dentition and nursing period

There were significant differences between the nursing periods of female pups with and without teeth ($P < 0.05$, *t*-test) and between male pups with and without teeth, (Table 1b). Weaned female pups with teeth were nursed for

Table 1. Means, standard deviations and *t*-test results for factors associated with dental development in weaned Elephant Seal pups at Macquarie Island. Comparisons are presented for animals with and without teeth for each sex (denoted by [1]), and for overall comparisons between sexes (denoted by [2]) for each of five parameters. The five parameters are (a) birth mass, (b) nursing period, (c) growth rate, (d) mass ratio and (e) weaning mass.

		Mean	St. Dev	n	*t*-test[1]
(a) BIRTH MASS (kg)	Teeth	38.18	4.86	396	
Female (37.96 ± 5.11, n = 449)					$P < 0.05$
	No teeth	36.28	6.51	53	
t-test[2] $P < 0.05$					
	Teeth	43.53	5.79	102	
Male (41.99 ± 6.12, n = 523)					$P < 0.05$
	No teeth	41.62	6.14	420	
(b) NURSING PERIOD (days)					
	Teeth	24.85	2.73	396	
Female (24.75 ± 2.80, n = 450)					$P < 0.05$
	No teeth	23.96	3.26	53	
t-test[2] $P < 0.05$					
	Teeth	24.80	2.61	102	
Male (24.30 ± 2.63, n = 524)					$P < 0.05$
	No teeth	24.18	2.62	421	
(c) GROWTH RATE (kg day^{-1})					
	Teeth	3.17	0.98	396	
Female (3.18 ± 0.98, n = 449)					$P > 0.05$
	No teeth	3.26	0.91	53	
t-test[2] $P > 0.05$					
	Teeth	3.21	1.05	101	
Male (3.26 ± 1.04, n = 522)					$P >> 0.05$
	No teeth	3.27	1.04	421	
(d) MASS RATIO					
	Teeth	3.04	0.58	394	
Female (3.05 ± 0.56, n = 450)					$P > 0.05$
	No teeth	3.12	0.46	53	
t-test[2] $P < 0.05$					
	Teeth	2.82	0.54	101	
Male (2.87 ± 0.53, n = 524)					$P > 0.05$
	No teeth	2.88	0.52	420	
(e) WEANING MASS (kg)					
	Teeth	116.10	24.71	396	
Female (115.70 ± 24.48, n = 450)					$P > 0.05$
	No teeth	112.58	22.71	53	
t-test[2] $P < 0.05$					
	Teeth	122.32	27.06	101	
Male (120.17 ± 26.87, n = 523)					$P > 0.05$
	No teeth	119.71	26.83	421	

24.9 ± 2.73 days (n = 396), while toothless weaned female pups were nursed for only 24.0 ± 3.26 days (n = 53). Males with teeth were nursed for 24.8 ± 2.61 days (n = 102), and toothless males for 24.2 ± 2.62 days (n = 421).

Weaning dentition and mass ratio

The differences in mass ratio of males and females, with or without teeth was not statistically different. Male and female pups without teeth did, however, gain proportionally more weight than those with teeth (Table 1d).

Weaning dentition and weaning mass

Within male and female pups, the presence or absence of teeth at weaning had no statistically significant effect on weaning mass. Weaners with teeth, whether male or female, out-weighed toothless pups (Table 1e).

Weaning dentition and birth date

There were no statistical differences between the dentition of animals born early in the season and those born later.

DISCUSSION

It appears from these data that the presence and number of teeth provides a measure of the degree of development of weaned Southern Elephant Seal pups from Macquarie Island. Whilst male pups weighed significantly more at birth and weaning than females their tooth development was far behind, with 80.5% of males lacking teeth at weaning. There appears to be precocious development with respect to weaning dentition in female pups during the nursing period, and perhaps even during gestation.

Table 2. Variation in tooth eruption patterns of Southern Elephant Seal *Mirounga leonina* pups. Decimals refer to that fraction of the normal number of teeth, e.g., I $\frac{1.5}{0.5,}$ equates to three (3) maxillary incisors and one (1) mandibulary incisor. Teeth were not classified according to the side of the mouth in which they were present.

Category	Tooth combination			Frequency
One	I $\frac{0}{0,}$	C $\frac{0}{0}$		474
Two	I $\frac{0}{0.5,}$	C $\frac{0}{0}$		1
Three	I $\frac{0.5}{0,}$	C $\frac{0}{0}$		17
Four	I $\frac{0}{0,}$	C $\frac{0}{0.5}$		2
Five	I $\frac{0}{0,}$	C $\frac{0.5}{0}$		2
Six	I $\frac{0.5}{0,}$	C $\frac{0.5}{0}$		1
Seven	I $\frac{0}{0,}$	C $\frac{1}{0}$		13
Eight	I $\frac{1}{0,}$	C $\frac{0}{0}$		22
Nine	I $\frac{0.5}{0.5,}$	C $\frac{0}{0}$		1
Ten	I $\frac{0}{0,}$	C $\frac{0}{1}$		15
Eleven	I $\frac{0}{1,}$	C $\frac{0}{0}$		17
Twelve	I $\frac{1.5}{0,}$	C $\frac{0}{0}$		3
Thirteen	I $\frac{0}{0.5,}$	C $\frac{0}{1}$		1
Fourteen	I $\frac{0}{1,}$	C $\frac{0}{0}$	PM $\frac{0}{0.5}$	1
Fifteen	I $\frac{0}{0,}$	C $\frac{0.5}{1}$		1
Sixteen	I $\frac{0}{1,}$	C $\frac{0}{0.5}$		1
Seventeen	I $\frac{0}{1,}$	C $\frac{0.5}{0}$		2
Eighteen	I $\frac{1}{1}$	C $\frac{0}{0}$		5
Nineteen	I $\frac{2}{0,}$	C $\frac{0}{0}$		4
Twenty	I $\frac{0}{0,}$	C $\frac{1}{1}$		194
Twenty-one	I $\frac{1}{0,}$	C $\frac{1}{0}$		1
Twenty-two	I $\frac{0}{0.5}$	C $\frac{0.5}{1}$		1
Twenty-three	I $\frac{0}{0.5}$	C $\frac{1}{1}$		4
Twenty-four	I $\frac{0.5}{0}$	C $\frac{1}{1}$		16
Twenty-five	I $\frac{0.5}{1}$	C $\frac{1}{0}$		1
Twenty-six	I $\frac{0}{0}$	C $\frac{1}{1}$	PM $\frac{0.5}{0}$	1
Twenty-seven	I $\frac{1.5}{0}$	C $\frac{0.5}{1}$		1
Twenty-eight	I $\frac{2}{1}$	C $\frac{0}{0}$		9
Twenty-nine	I $\frac{0}{1}$	C $\frac{1}{1}$		13
Thirty	I $\frac{1}{0}$	C $\frac{1}{1}$		51
Thirty-one	I $\frac{1}{1}$	C $\frac{1}{0}$		1
Thirty-two	I $\frac{1}{1}$	C $\frac{0}{0}$	PM $\frac{0}{1}$	1
Thirty-three	I $\frac{0.5}{0.5}$	C $\frac{1}{1}$		1
Thirty-four	I $\frac{1.5}{0}$	C $\frac{1}{1}$		4
Thirty-five	I $\frac{1}{0.5}$	C $\frac{1}{1}$		4
Thirty-six	I $\frac{1}{1}$	C $\frac{1}{1}$		48
Thirty-seven	I $\frac{1.5}{1}$	C $\frac{1}{1}$		6
Thirty-eight	I $\frac{2}{1}$	C $\frac{1}{1}$		19
Thirty-nine	I $\frac{2}{1}$	C $\frac{1}{1}$	PM $\frac{1}{0}$	1
Total				960

Our data show there was no statistically significant sex difference in the growth rates of pups during the nursing period, whether they lacked teeth or not (Table 1c). Pre-partum maternal investment, expressed as birth mass (males 4 kg heavier), was carried through to weaning when males were still 4 kg heavier. Therefore, post-partum maternal investment was equal for male and female offspring. Growth rates (post partum maternal investment) did not differ between sexes for Northern Elephant Seal *M. angustirostris* pups (Kretzmann *et al.* 1993) and Southern Elephant Seal pups at South Georgia (McCann *et al.* 1989).

The nursing period was significantly longer (0.45 days or 1.4 kg mass gain) for female pups than for males, at Macquarie Island. This difference was also apparent within the sexes where male and female weaners with teeth were nursed for significantly longer than those without teeth. The reverse has been reported for Northern Elephant Seal pups, where males suckled one full day longer than females (Reiter *et al.* 1978), however, this difference was not found by Kretzmann *et al.* (1993).

Reiter *et al.* (1978) speculated that the absence of teeth in weaned male Northern Elephant Seal pups may allow them to "steal milk" more successfully than females and therefore extend their nursing period beyond their filial mother for a gain in size and condition. Milk stealing is dangerous to the weaned pup and Reiter *et al.* (1978) suggest that males are selected to pursue this growth strategy through delays in molting and canine teeth eruption. Whilst our data does not allow comment on this point, it does show that the presence of teeth in a pup does not reduce the nursing period with its mother (Table 1b). From this, there appears to be no adverse stimulus to the cow from her toothsome pup, at least during the nursing period.

Bonner (1984) and Fogden (1971), believe milk stealing to be of no value as the thieves gain very little from their efforts. Fogden (1971) showed that Grey Seal *Halichoerus grypus* pups who steal milk do not thrive as well as pups who suckled from one cow only. There are however possible maternal advantages, cows without pups benefited from adopting pups as it has been shown that Northern Elephant Seal cows who failed to suckle also failed to copulate (Le Boeuf *et al.* 1972).

Mass differences of pups at birth and at weaning have been related to the relative condition of the cow (Reiter *et al.* 1981; Le Boeuf *et al.* 1989; Campagna *et al.* 1992; Wilkinson 1992; Arnbom *et al.* 1993). Wilkinson (1992) and Arnbom *et al.* (1993) have shown that post partum mass of mothers is linearly related to the mass of their weaned pup, the heaviest mothers producing the heaviest pups. Here we show the heaviest pups at birth and weaning are those with teeth, regardless of sex. Weight and tooth development therefore appear to be related.

Heavier pups remain on the beaches for longer and may therefore have more time to develop social and diving skills (Arnbom *et al.* 1993). Larger pups are also more successful at surviving the first year of life (Campagna *et al.* 1992; Arnbom *et al.* 1993). Pups with well developed teeth may gain an advantage over their cohorts who have less well developed or no teeth in their feeding at sea. However, the presence or absence of teeth can be of no benefit to the weaned elephant seal pup during the post-weaning fast as the weaners are still lying on a beach, without feeding, while tissue differentiation occurrs. Of absolute importance is that the pups have developed teeth by the time they leave their natal beaches.

There was a significant sex difference in mass ratio as female pups gained proportionally more weight than male pups over the nursing period. This decrease in relative mass difference between male and female pups over the nursing period (10% to 4%) indicates that female pups may be more successful at extracting and/or assimilating resources from their mothers than are male pups. If this so, female pups may have the capacity to develop post weaning skeletal characteristics faster than male pups. However, Kretzmann *et al.* (1993) showed female Northern Elephant Seal pups were slightly, though not significantly, less metabolically efficient than males.

Female Southern Elephant Seals at Macquarie Island have, depending on the cohort, the ability to breed at age 5.2 ± 1.8 years (Hindell 1991), males may breed as late as 14 years of age (Carrick *et al.* 1962b). Female pups could therefore be expected to develop earlier and more rapidly than males. This study shows that sex differences in development, as indicated by the presence of teeth, may occur during gestation and are carried through the nursing period to weaning and, perhaps, furthur still to age at first production. The degree to which parental investment, in male and female Southern Elephant Seal pups, is related to each sexes respective long term sexual strategies needs further study.

ACKNOWLEDGEMENTS

We are indebted to Dr M. N. Bester and Dr M. Hindell who commented and improved earlier drafts of the paper, and to J. Easson, D. Glaken, P. Croswaite and D. Hudspeth for valuable assistance in collecting field data. Doctors Tom Arnbom and R. L. De Long provided helpful comments on the manuscript.

REFERENCES

Arnbom, T., Fedak, M. A., Boyd, I. L. and McConnell, B. J., 1993. Variation in weaning mass of pups in relation to maternal mass, post weaning fast duration, and weaned pup behaviour in southern elephant seals *(Mirounga leonina)* at South Georgia. *Can. J. Zool.* **71:** 1772–781.

Bartholomew, G. A., 1970. A model for the evolution of pinniped polygyny. *Evol.* **24:** 546–59.

Bonner, W. N., 1984. Lactation strategies in pinnipeds: problems for a marine mammalian group. *Symp. zool. Soc. Lond.* **51:** 253–72.

Briggs, K.T., 1974. Dentition of the northern elephant seal. *J. Mamm.* **55:** 158–71.

Campagna, C., Le Boeuf, B. J., Lewis, M. and Bisioli, C., 1992. Equal investment in male and female offspring in the southern elephant seals. *J. Zool., Lond.* **226:** 551–61.

Carrick, R., Csordas, S. E., Ingham, S. E. and Keith, K., 1962a. Studies on the southern elephant seal, *Mirounga leonina* (L) III. The annual cycle in relation to age and sex. *CSIRO Wildl. Res.* **7:** 119–60.

Carrick, R., Csordas, S. E. and Ingham, S. E., 1962b. Studies on the southern elephant seal, *Mirounga leonina* (L) IV. Breeding and Development. *CSIRO Wildl. Res.* **7:** 119–60.

Clinton, W. L., 1990. Sexual selection and the life-history of male northern elephant seals. PhD thesis, University of California, Santa Cruz.

Clutton-Brock, T. H., Albon, S. D. and Guinnes, F. E., 1981. Parental investment in male and female offspring in polygynous mammals. *Nature* **289:** 487–89.

Fogden, S. C. L., 1971. Mother-young behaviour at grey seal breeding beaches. *J. Zool., Lond.* **164:** 69–92.

Haley, M., 1990. Fighting and dominance in male northern elephant seals. PhD dissertation, University of California, Santa Cruz.

Hindell, M., 1991. Some life-history parameters of a declining population of southern elephant seals, *Mirounga leonina. J. Anim. Ecol.* **60:** 119–34.

Kretzmann, M. B., Costa, D. P. and Le Boeuf, B. J., 1993. Maternal energy investment in elephant seal pups: evidence for sexual equality? *Am. Nat.* **141:** 466–80.

Laws, R. M., 1953. The elephant seal *Mirounga leonina.* 1. Growth and age. *Scient. Rep. Falkld. Isl. Depend. Surv.* **8:** 1–61.

Le Boeuf, B. J. and Reiter, J., 1988. Lifetime reproductive success in northern elephant seals. Pp. 344–62 *in* Reproductive success, studies of individual variation in contrasting breeding systems ed by T. H Clutton-Brock. University of Chicago Press: Chicago and London.

Le Boeuf, B. J., Whiting, R. J. and Gannt, R. F., 1972. Perinatal behaviour of northern elephant seal females and their young. *Behaviour* **43:** 121–56.

Le Boeuf, B. J., Condit, R. and Reiter, J., 1989. Parental investment and secondary sex ratio in northern elephant seals. *Behav. Ecol. Sociobiol.* **25:** 109–17.

Little, G. J., Bryden, M. M. and Barns, A., 1987. Growth from birth to 20 days in the elephant seal, *Mirounga leonina*, at Macquarie Island. *Aust. J. Zool.* **35:** 307–12.

McCann, T. S., Fedak, M. A. and Harwood, J., 1989. Parental investment in southern elephant seals, *Mirounga leonina. Behav. Ecol. Sociobiol.* **25:** 513–14.

Reiter, J., Stinson, N. L. and Le Boeuf, B. J., 1978. Northern elephant seal development: the transition from weaning to nutritional independence. *Behav. Ecol. Sociobiol.* **3:** 337–67.

Reiter, J., Panken, K. J. and Le Boeuf, B. J., 1981. Female competition and reproductive success in northern elephant seals. *Anim. Behav.* **29:** 670–87.

Wilkinson, I. S., 1992. Factors affecting reproductive success of southern elephant seals, *Mirounga leonina,* at Marion Island. Ph.D dissertation, University of Pretoria, 185 pp.

Diving behaviour of Antarctic Fur Seals *Arctocephalus gazella* Peters around Heard Island

K. GREEN[1,2]

The diving behaviour of Antarctic Fur Seals around Heard Island is characterized by different modes of foraging in different seasons and by different sexes. Differences in foraging ecology between the Heard Island region and other areas where the species has been studied are not reflected in intra-sex differences in diving behaviour. Females foraged to the north-east of the island at night and at shallow depths on schooling fish, resulting in a similar diving pattern to krill-feeding seals elsewhere. However, males generally fed on benthic and bentho-pelagic fish on the shelf around Heard Island or foraged to the south of Heard Island in deep water on the deep scattering layer at depths greater than 200 m with some travelling to Antarctic waters in winter to feed at shallow depths, presumably on krill.

INTRODUCTION

THIS study of diving behaviour of Antarctic Fur Seals *Arctocephalus gazella* around Heard Island was aimed at assessing their diving depth and locality as a part of a larger study of foraging ecology. Studies of the diet at Heard Island indicated that fur seals fed mainly on fish in spring and summer (Green *et al.* 1989) and on fish and squid in the autumn/winter period (Green *et al.* 1991). During the late spring and early summer of 1987, fish from the Heard Island continental shelf area comprised the bulk of the diet. However, earlier in the spring and later in the summer pelagic fish (mainly myctophids) constituted more than 50% of the prey taken by fur seals. In autumn/winter 1990 this figure rose to 96%. A greater level of erosion of otoliths from a key species *(Gymnoscopelus nicholsi)* suggested that the seals were foraging further from Heard Island in winter (Green *et al.* 1991). These data were used to model prey consumption in the waters surrounding Heard Island (Woehler and Green 1992). However inferences from diet and sightings from ships were not sufficient to determine with any certainty the foraging areas used by fur seals.

Most of the diving data on Antarctic Fur Seals come from the major population on South Georgia. However, the diet of Antarctic Fur Seals from South Georgia differs markedly from that at Heard Island, consisting mainly of Krill *Euphausia superba* (Bonner 1968; North *et al.* 1983). The diving behaviour at South Georgia can therefore be related to the concentrations and movements of krill (Croxall *et al.* 1985). In addition, studies around South Georgia have been confined to female seals and then only during lactation (Boyd and Croxall 1992).

Previous studies of pinnipeds (Hindell *et al.* 1991; Green and Burton 1993) have commented on the difficulty in investigating their foraging ecology from dietary samples taken whilst the animal is ashore, maybe thousands of kilometres from the main feeding areas. The present study therefore attempted to extend our knowledge of the types of prey fur seals might be taking by inferences drawn from diving behaviour, times and depths. It also arose out of the need to understand more fully the diving behaviour of Antarctic Fur Seals in the complexity of the waters of the Heard Island shelf, with a diversity of prey types in different seasons and with differing requirements of the sexes.

METHODS

Antarctic Fur Seals were captured with a long-handled net, the net being detachable from the support arms for ease of handling the seal (Furman Diversified Inc. USA). After capture, males were immobilized either physically or chemically. In the former instances, the seals were positioned in sternal recumbency on a restraint board (112 cm × 70 cm) modified from Gentry and Kooyman (1986). Webbing belts 4 cm in width were tightened over the neck and back with ratchet buckles. Once the seal was secured to the restraint board, the netting was drawn forward to expose the dorsal midline and front flippers. The head and shoulders remained within the net. For females, physical restraint only was used and the board was a smaller version of that used for males. Chemical restraint was used only on males. Ketamine hydrochloride and diazepam were injected at a ratio of 1000:1 at a dose rate of approximately 6 mg/kg into the lumbar muscles using a hand-held syringe.

[1]Antarctic Division, Channel Highway, Kingston, Tasmania, Australia 7050.
[2]Present address: NPWS, Snowy Mountains Region, PMB Cooma, New South Wales, Australia 2630.
Pages 97–104 *in* MARINE MAMMAL RESEARCH IN THE SOUTHERN HEMISPHERE VOLUME 1: STATUS, ECOLOGY AND MEDICINE ed by M. Hindell and C. Kemper. Surrey Beatty & Sons, Chipping Norton. 1997.

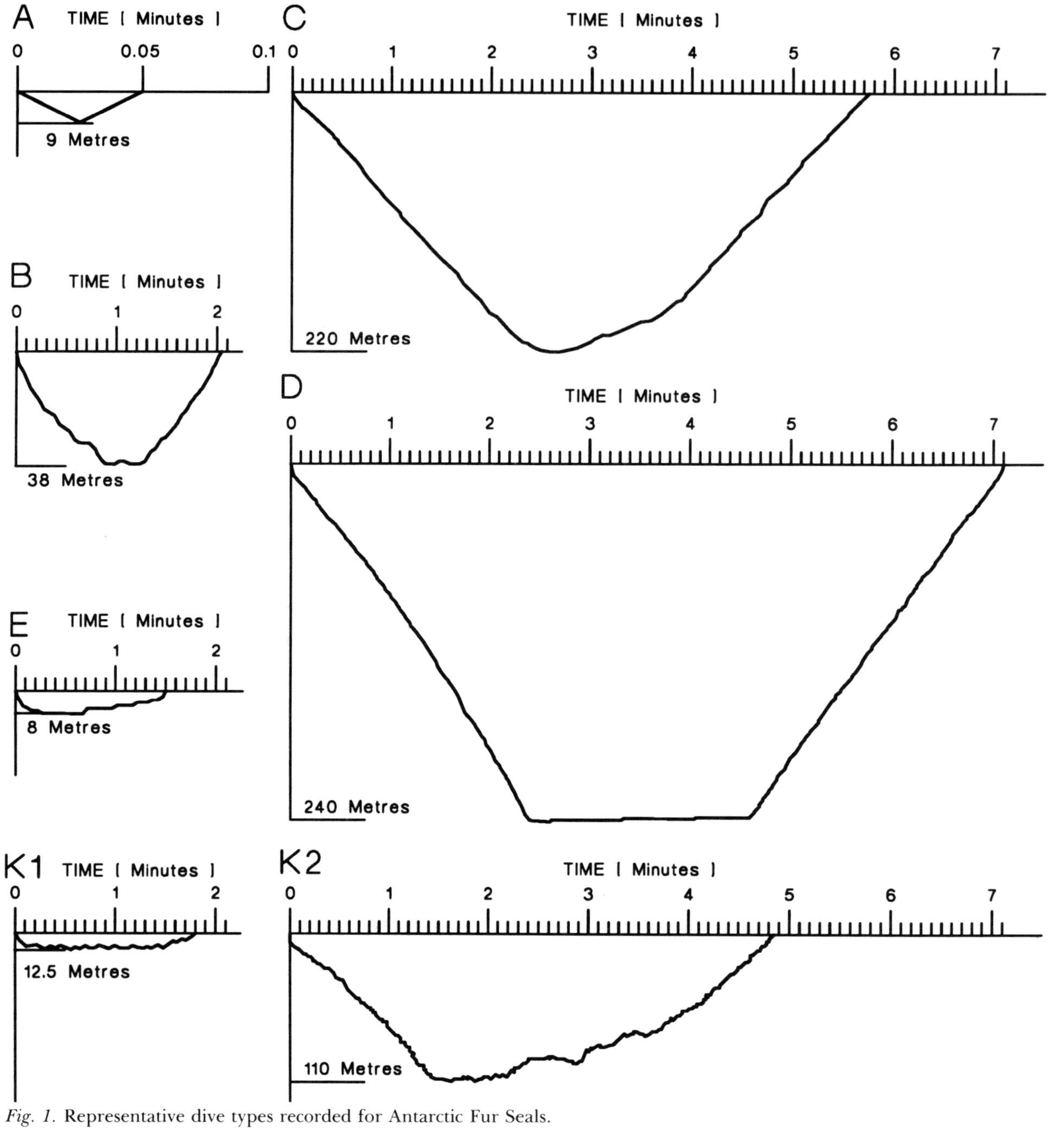

Fig. 1. Representative dive types recorded for Antarctic Fur Seals.

A Time Depth Recorder (Mk 3, Wildlife Computers) was glued to the dorsal midline of each seal with Araldite K268 (Ciba Geigy). The recorders were programmed to record depth every three seconds, and temperature every 240 seconds. A delay until recording began was programmed for units carried by males consisting of 20, nine, five and six days for the four deployments. No delays were programmed for females. Numbered plastic tags were placed in the skin fold caudal to the fifth phalanges of each front flipper.

For adult males there were three winter deployments: (i) Nine units on 23 and 24 April, (ii) five units on 6 July, and (iii) five units on 4 and 5 August. There was one spring deployment of five units on 9 and 10 December. Three females were fitted with Mark 3e units on 9 March 1992. Nine Mark 3e units were deployed on females between parturition and the first post-parturition foraging trip; these were deployed on 30/11/92 (three units) and 8/12/1992 (six units). Two further Mark 3e units were deployed on 17 January 1993. Units were retrieved and redeployed opportunistically on females in the summer of 1992/93.

On recapture the units were retrieved by cutting the fur beneath the glue and returning the units to the laboratory for down-loading. Software supplied by Wildlife Computers was used to analyse the dive data.

The data produced by the program "Dive Analysis" (Wildlife Computers) were sorted by depth and then duration. Dives shallower than four metres were generally simple in profile and were therefore considered to be travelling dives (Type A), and were discarded from further analysis. The remaining dives were then sorted in their depth categories by duration. After examination of some hundreds of short duration, shallow dives (<10 m) all dives of less than 30 seconds duration and shallower than 5 m were allocated to dive type A. The traces of all other dives were then examined. Dives (Fig. 1) were classified as:

- A Shallow, short duration dives possibly only used for travel
- E Shallow, <20 m, dives of varying duration (*cf* K)
- B Mid-level dives (20 metres to 175 metres)
- C Deep dives without flat bottom
- D Deep dives with flat bottom
- K Shallow, short duration dives with characteristic "saw tooth" profile
- K2 Deeper, longer duration dives with characteristic "saw tooth" profile

Latitude and longitude were calculated using the "Geolocation" program (from Wildlife Computers) which determines the time of dawn and dusk then calculates the location using standard navigational formulae. The latitude and longitude which the program produces is a convenient centre of the likely location along a line between an estimated maximum and minimum. The location was checked for correctness using water temperature data collected by the TDR at the surface and at 100 m depth compared with sea temperature charts from Gordon and Molinelli (1982).

RESULTS

Of the 24 TDRs deployed on male fur seals, eight yielded useable data. TDRs were retrieved from all females with only one recorder from a female not yielding any data.

The majority of dives recorded were type A dives, but the proportion of these was affected by the time at which the recorder turned on. If the recorder turned on immediately prior to the seal departing for sea, then travel dives assumed a greater overall importance. To restrict this analysis to presumed foraging dives, type A dives were removed from the analysis and the distribution of presumed foraging dives was recalculated (Table 1).

Seal 572 foraged in waters around Heard Island in May, and Antarctic waters in June. These two foraging periods for this animal are henceforth referred to as 572H and 572A respectively. The least common dive type for six non-breeding males around Heard Island was the shallow type E, averaging 15.0% of dives;

Table 1. Basic dive statistics for adult male and female Antarctic Fur Seals from Heard Island. Also listed are the percentage number of dives for each of six dive types. Type A dives are excluded from these analyses.

Seal Number	No. of dives	Foraging dives/day	Max depth (metres)	Mean Max depth (metres)	Mean duration (min.)	Dive Type B	C	D	E	K	K2
Males											
565	467	46	232	139.6	4.41	8.5	38.0	26.1	27.4	0.0	0.0
587	347	37	232	187.3	5.60	8.4	7.8	74.3	9.5	0.0	0.0
572H	1 232	41	234	130.7	4.38	26.7	14.3	33.4	25.8	0.0	0.0
572A	3 008	207	62	17.1	1.65	2.5	0.0	0.0	30.8	66.7	0.0
601	66	55	264	126.4	4.22	47.3	18.2	29.1	5.5	0.0	0.0
602	266	36	240	153.6	3.85	28.5	11.4	48.4	11.4	0.0	0.0
608	291	37	130	82.8	4.50	0.0	0.0	0.0	7.7	13.5	78.7
004	678	121	214	73.5	2.92	70.0	26.4	0.1	3.4	0.0	0.0
597	2 758	36	444	216.3	6.54	4.8	18.1	66.4	10.6	0.0	0.0
Females											
1991/92											
564	1 729	110	50	10.3	1.03	2.2	0.0	0.0	97.9	0.0	0.0
566	1 745	123	49	10.0	0.95	1.6	0.0	0.0	98.4	0.0	0.0
568	2 983	112	88	19.3	1.75	31.7	0.0	0.0	68.3	0.0	0.0
1992/93											
626	1 612	185	109	36.2	1.39	56.4	0.6	0.0	43.0	0.0	0.0
628.1	3 865	186	120	31.9	1.38	60.2	0.6	0.0	39.2	0.0	0.0
628.4	4 459	–	–	–	–	61.5	0.0	0.0	38.5	0.0	0.0
623.1	3 249	179	91	20.0	1.09	30.2	0.0	0.0	69.8	0.0	0.0
623.2	2 253	233	138	36.5	1.30	65.7	2.6	0.0	31.7	0.0	0.0
630	3 871	129	192	46.5	1.83	91.0	5.8	0.0	3.3	0.0	0.0
614	4 632	279	125	29.9	1.34	57.3	1.5	0.0	41.2	0.0	0.0
Hole 2	3 504	130	99	37.6	1.37	65.6	0.0	0.0	34.4	0.0	0.0
619	329	–	85	38.8	1.44	66.4	0.0	0.0	33.5	0.0	0.0
621.1	3 842	–	–	–	–	95.1	0.6	0.0	4.3	0.0	0.0
613	7 474	–	–	–	–	47.4	0.9	0.0	51.6	0.0	0.0
618	10 881	–	–	–	–	54.1	1.2	0.0	33.3	0.0	0.0
627.2	4 254	231	136	47.4	1.23	85.5	1.0	0.0	13.5	0.0	0.0

mid level dives (type B) averaged 20.7% and deep dives (type C and type D) averaged 61.3% of dives (Table 1). Three individuals (572A, 608 and 004) could not be fitted into any generalization. Except for these three individuals, the average maximum foraging dive depth for males was between 139.6 m and 216.3 m (Table 1) with only seal 572A having a shallower average than for most female seals. For 16 females during the breeding season, 43.9% of dives were shallow E type; mid level dives (type B) averaged 54.5% and deep dives (type C generally with only one type D) averaged 1.5% of dives (Table 1). The average maximum foraging depth for females lay between 10.0 m and 47.4 m (Table 1)

Diving by males tended to be during the day around Heard Island and at night in Antarctic waters. Diving of animals between May and August started in the early morning (the latest being 0826 h) and generally finished by 1700 h (Fig. 2). Seal 572H was an exception to this, starting diving earlier in the day and finishing later. Seal 597 (October) started earlier but, except on one day finished at about the same time. Seal 004 was a male that was peripheral to a breeding harem and whose diving patterns (Table 1) were similar to those for breeding females. Seals in Antarctic waters (572A and 608) dived at all hours but with rest periods concentrated during the day rather than the night (Fig. 2). Seal 572A (in June) concentrated its foraging dives (type K) overnight whereas seal 608, whose dives in October/November were generally deeper, foraged at all times with dive type K also being conducted at night.

Diving sessions by lactating females tended to occur at night with little diving by day (Fig. 2). There was no significant difference between the number of dives per day by seals in December/January 1992/93 and March 1992, however, there was a significant difference ($t = 3.369$, $p < .01$) between the time submerged, with animals in March spending less time underwater and at generally shallower depths.

All female seals travelled to the north-east of Heard Island where most diving was over an area of raised sea floor extending out to the north-east of the Kerguelen/Heard plateau (Fig. 3). Males generally foraged in waters to the south of Heard Island around a submarine hole in the shelf, or further to the south in Antarctic waters (Fig. 4). Seal 004, an animal that was peripheral to a breeding harem, travelled to the north to forage in areas similar to those used by females.

Sea surface temperature data obtained from TDRs were compared with those from sea surface temperature charts contained in the atlas by Gordon and Molinelli (1982). These charts provided accurate predictions of seasonal sea temperature at Heard Island in 1992/93 (Green 1993). For all females for which data were available, there was good accord between

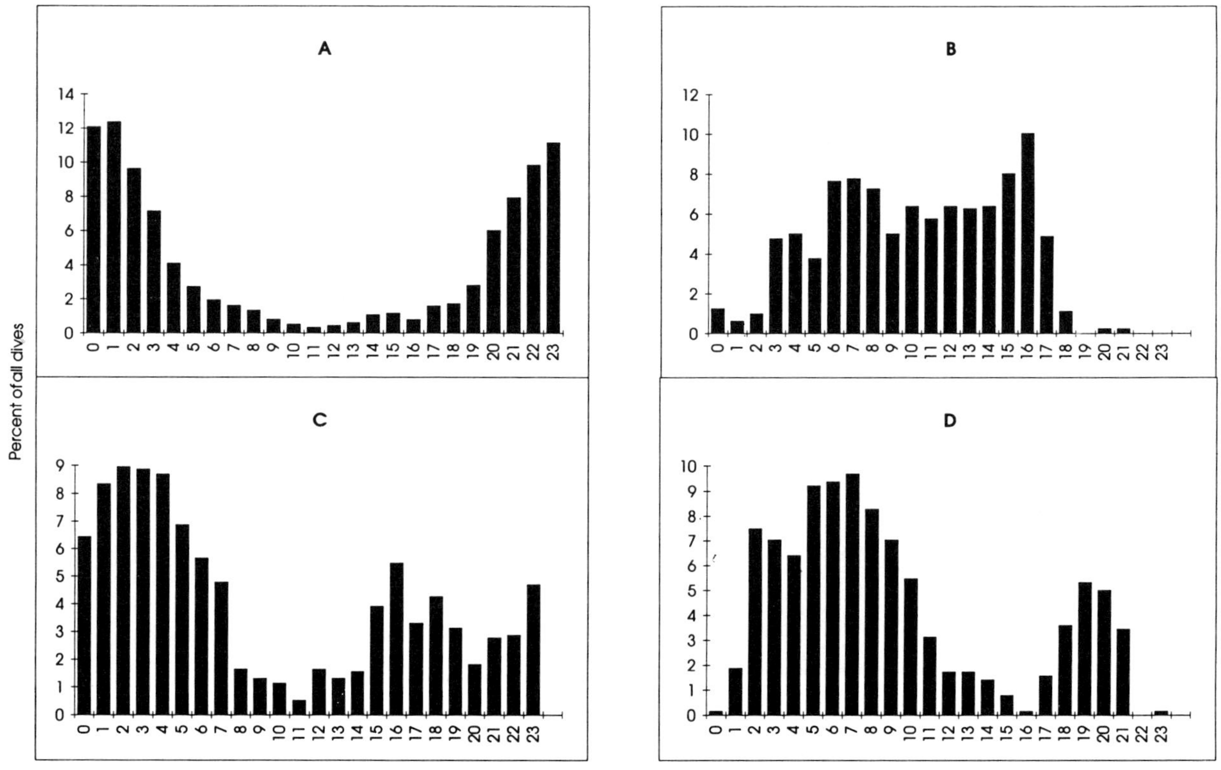

Fig. 2. Diurnal distribution of dives for (a) females, (b) males around Heard Island in winter, (c) males in Antarctic waters, and (d) one male in summer.

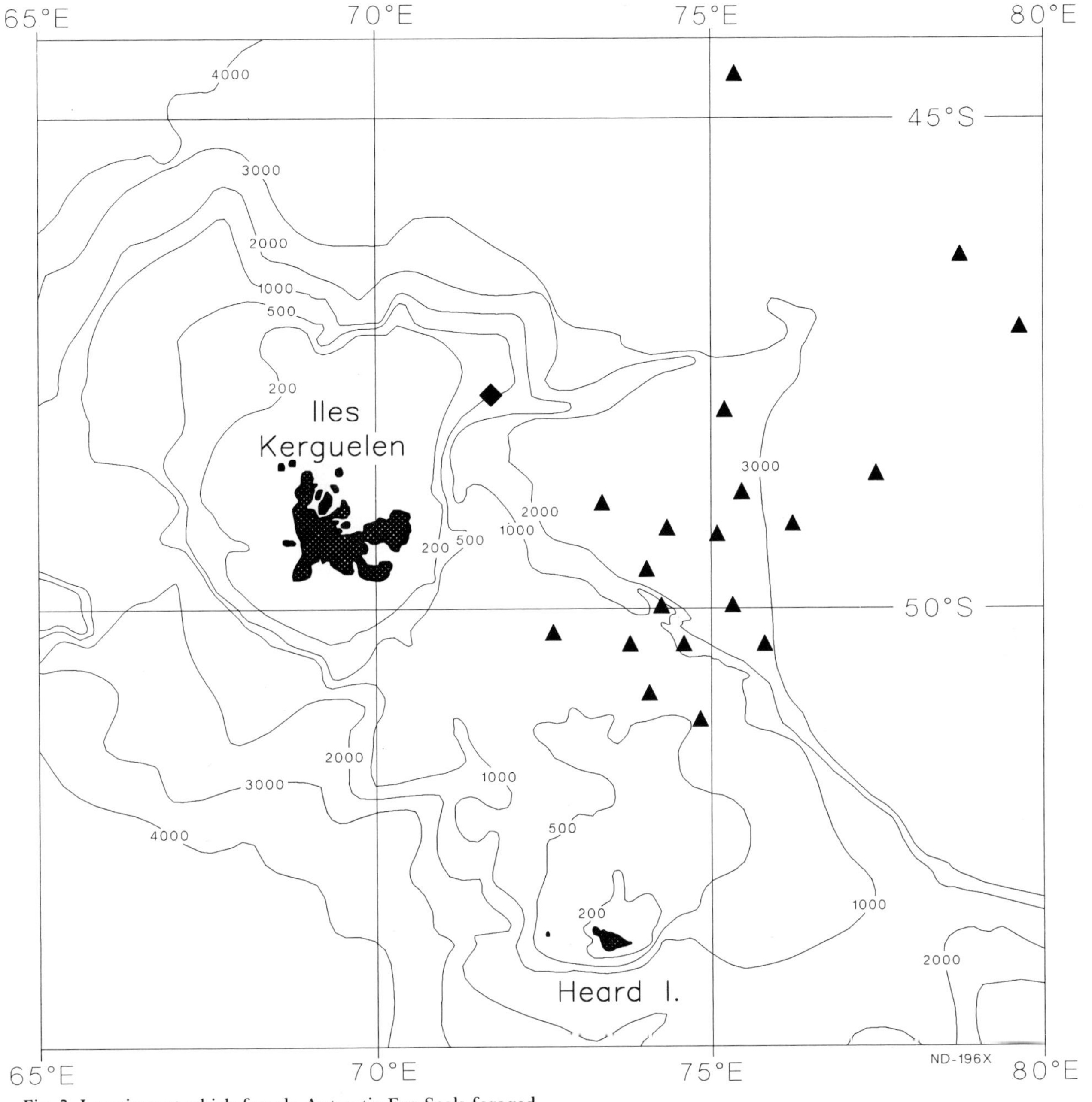

Fig. 3. Locations at which female Antarctic Fur Seals foraged.

geolocation and temperature data. Recorded temperatures of −1.8° for male 572 confirmed that it had been in Antarctic waters. The temperature data for most other males did not provide good supportive evidence of the geolocation result and for most males in subantarctic waters the recorded temperatures were slightly too high for their southerly position. The exception was for data from seal 004 which indicated a position further south than that obtained from geolocation data. However, evidence for its geolocation position (similar to that for the females) was reinforced by diving behaviour and dive times which were similar to those for females.

DISCUSSION

Although there have been a number of studies of diving behaviour of fur seals in general, and of Antarctic Fur Seals in particular (Boyd and Croxall 1992), the present study of diving behaviour around Heard Island differs from all others in that males as well as females were instrumented. The concentration on females is a result of both the types of questions being posed, and the practical necessity of retrieving recorders, an event more likely with females returning to pups. However, site fidelity of males in the present study, although not quantified, was high, with many males returning after an absence of from days to months to the same or a nearby tussock. The few failures to recover TDRs from males in this study may have been a result of the deployment of recorders late in the moult and the early departure of the expedition from Heard Island in relation to the peak in fur seal numbers when they would have

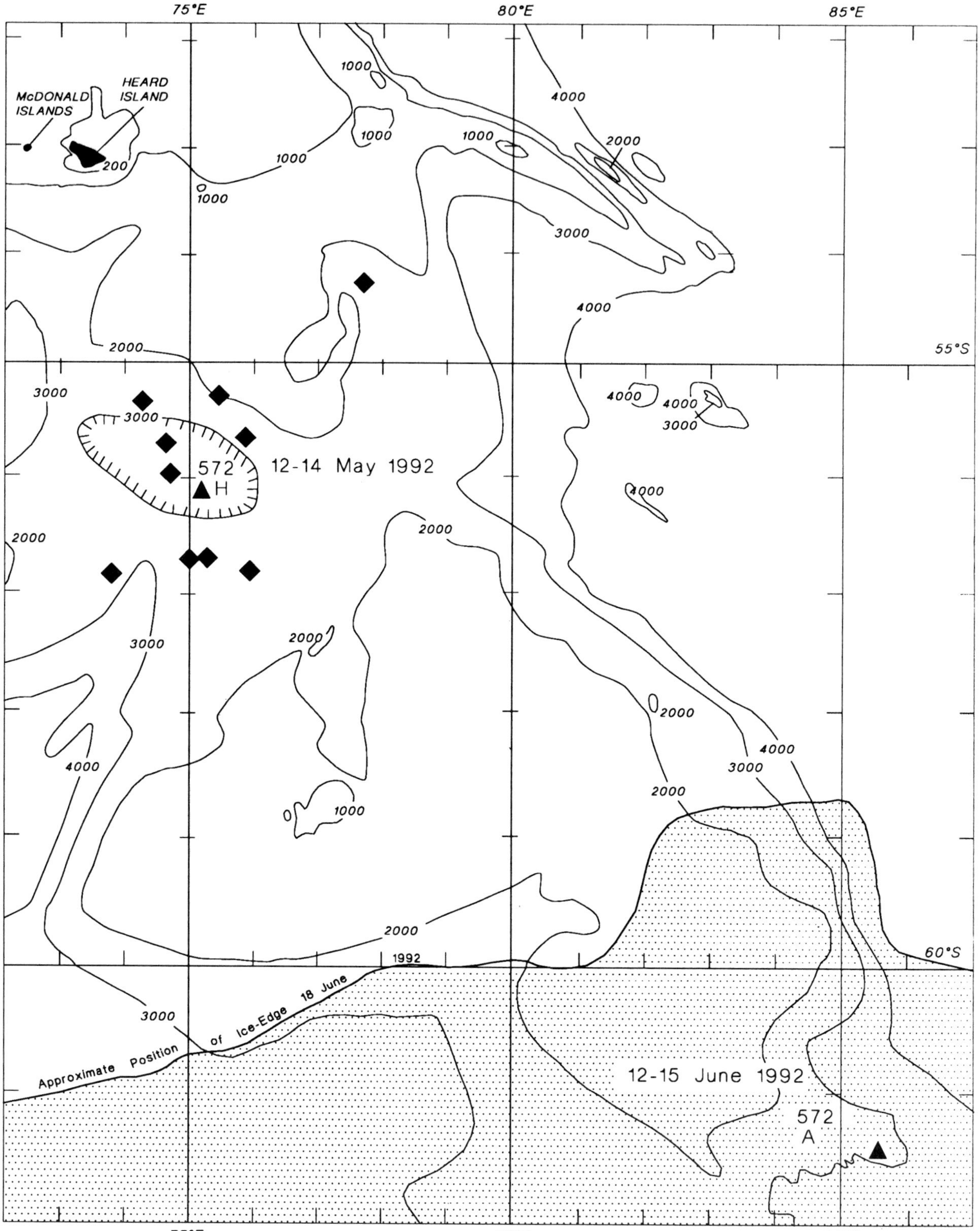

Fig. 4. Locations at which male Antarctic Fur Seals foraged.

been expected to return. The decision to examine males was necessary because at Heard Island they constituted the majority of the 21 500 seals ashore in the summer (Green *et al.*, unpubl. data) compared with about 250 females (Shaughnessy and Goldsworthy 1990) and therefore their impact on fish stocks would be greater than for females.

Male diving behaviour consisted of making a few relatively deep dives per day. Males generally made less than 50 foraging dives per day during daylight hours compared with the 100 or more dives for most females. Two males were exceptions to this, one of which followed an essentially "female" diving pattern (seal 004) and one which foraged in

Antarctic waters. This male (seal 572 A — Table 1) swam into Antarctic waters and exhibited diving behaviour that very much fits the pattern of krill feeding (Croxall *et al.* 1985) with shallow diving occurring mainly at night. This was in contrast to the foraging of the same animal in Heard Island waters where nearly 50% of its dives were the deep C and D type dives and most were during daylight hours. Female seals, though feeding on fish rather than krill, had a similar diving pattern to females at South Georgia in that the majority of dives occurred at night (Fig. 2) and many were type E dives within the top 20 m of the water column (Table 1 *cf* Boyd and Croxall 1992). The differences in foraging between males and females probably indicates both the constraints on females which have to obtain high energy food to produce milk for suckling their pups and their shorter aerobic dive limit that would limit them to dives of about 160 m (Boyd and Croxall 1992). This is too shallow for them to reach the scattering layers and the bentho-pelagic fish apparently targeted by males (Green *et al.* 1997). The limit of 160 m was exceeded by only one female, which did so on ten occasions, eight within the same day, with a maximum depth of 192 m. The maximum depth recorded previously for an Antarctic Fur Seal (female) was 181 m from around South Georgia (Boyd and Croxall 1992). In the present study that figure was exceeded by all males except one. The maximum recorded depth from one male was 444 m. The maximum time submerged also exceeded the ten minutes recorded by Boyd and Croxall (1992), but only by one male that dived for 11.65 minutes.

The dive types documented here probably reflect different diving strategies depending upon the targeted prey. The type A dives were of short duration and occurred mainly at the beginning and at the end of diving sessions and were possibly only used for travel. Observations from shore and ships indicate that such depths may regularly be achieved by Antarctic Fur Seals diving through breaking swells. The type E dives which were shallow, and of variable duration, occurred particularly at night. These dives possibly occurred when seals were feeding on myctophids. Around Marion Island, the only potential prey species found at these shallow depths was *Krefftichthys anderssoni*, which was recorded moving up to within 10 m of the surface at night (Perissinotto and McQuaid 1992). The purpose of the mid-level type B dives may also have been to feed on these myctophids as densities of *K. anderssoni* were found to increase tenfold to depths of around 150 m (Perissinotto and McQuaid 1992). The type C and D dives were deep enough to reach the bentho-pelagic *Champsocephalus gunnari* on the shelf, or pelagic myctophids off it. These fish, essentially *Electrona, Gymnoscopelus, Protomyctophum* and *Krefftichthys,* are characteristic of deep, off-shelf water and usually constitute the deep scattering layers at about 200–500 m (Green *et al.* 1997). During the late winter/spring/summer *C. gunnari* dominated the diet (Green *et al.* 1997) but no record was found of type C or D dives on the shelf at this time. Sightings of Antarctic Fur Seals around Heard Island have been mainly on banks where the predominant fish species has been *C. gunnari.* To prey on these would have required dives by the males to about 200 m on the shelf. The fact that these were not documented relates mainly to the difficulty in instrumenting animals at about the time of the moult (their major reason for being ashore). The shallow, short duration K dives were possibly a result of foraging for krill but the associated K2 dives also with the characteristic "saw tooth" profile but deeper and of longer duration could not be ascribed to any foraging type. In these dive types the dive record both expanded and confirmed information previously gathered on the foraging ecology from simple dietary studies.

The closest record for foraging by females around Heard Island was at about 200 km on the eastern side of the shelf, extending out to the north-east in deep water (Fig. 2). This is in contrast to around South Georgia where krill remain readily available within 50–150 km (Kooyman *et al.* 1986). At South Georgia the mean duration of foraging trips was recorded as 4.3 ± 2.1 days (McCann and Croxall 1986) which gives an attendance time of about 2.3 days per trip. The time mothers are available ashore for nursing is the major criterion in weaning mass at least for female pups (Doidge and Croxall 1989). In this regard the approximately 1.8 days ashore for a mean trip duration of 5.2 ± 2.4 days recorded by Bester and Bartlett (1990) for Marion Island is very similar. Maternal attendance at Heard Island is approximately 1.5 days ashore for five to seven days at sea (Shaughnessy and Goldsworthy 1993). The difference, if it is significant, may be explained by the greater distance that seals have to travel to forage but the effects might be mitigated by the rich food source in term of the high oil content of *K. anderssoni,* the major pelagic prey of the summer (Green *et al.* 1997).

The present study has shown that the diving behaviour of fur seals may be more complex than first thought, and that there are a variety of diving patterns used by fur seals around Heard Island. In Antarctic waters male seals dived to shallow depths at night presumably to feed on krill, or deeper to feed on an unknown food source. Around Heard Island, they foraged by night at shallow depths, presumably on pelagic myctophids, or deeper by day to feed on deep shoaling fish. These four types of behaviour are documented, in the present study. One type of diving behaviour not identified from dive profiles was that necessary to prey on benthic and bentho-pelagic fishes, as there were no deep diving profiles from over shelf waters. There were fewer behavioural types for females, possibly because they were only studied at one time of the year when they were constrained by their need to suckle pups. The paucity of sightings of females on land during the winter indicates that their foraging areas may not necessarily be the same as for the males. The situation around Heard Island is therefore quite complex and cannot be fully understood in a single year's study. The attachment of further recorders, possibly linked to satellite transmitters, will be necessary to obtain better detail on diving behaviour, especially through the winter.

The diving behaviour of males at Heard Island differed markedly from the summary of previous studies (Boyd and Croxall 1992) in that in addition to pelagic foraging the animals fed in shallow coastal waters (from dietary studies), foraging dives normally exceeded 100 m and were generally longer than three minutes. For females around Heard Island, however, the general summary by Boyd and Croxall (1992) held true and the only time 181 m was exceeded was by one female which dived to 192 m. Therefore, although diving by female fur seals may be remarkably similar inter-specifically and intra-specifically, regardless of food type (krill at South Georgia, fish at Heard Island), the difference between the sexes may be profound. This has important ramifications for modelling of resource use as males can obviously target a greater variety of prey types and in locations inaccessible to females.

ACKNOWLEDGEMENTS

Thanks to Dave Slip who gave much useful advice concerning Time Depth Recorders and, together with Mark Hindell, commented on the manuscript. Dave Slip, Geoff Moore and Erwin Erb gave their time freely in the capture and instrumentation of fur seals. John Cox drew the figures.

REFERENCES

Bester, M. N. and Bartlett, P. A., 1990. Attendance behaviour of Antarctic and subantarctic fur seal females at Marion Island. *Antarct. Sci.* **2:** 309–12.

Bonner, W. N., 1968. The fur seal of South Georgia. *Br. Antarct. Surv. Sci. Rep.* **56:** 83 pp.

Boyd, I. L. and Croxall, J. P., 1992 Diving behaviour of lactating Antarctic fur seals. *Can. J. Zool.* **70:** 919–28.

Croxall, J. P., Everson, I. and Kooyman, G. L., 1985. Fur seal diving behaviour in relation to vertical distribution of krill. *J. Anim. Ecol.* **54:** 1–8.

Doidge, D. W. and Croxall, J. P., 1989. Factors affecting weaning weight in antarctic fur seals *Arctocephalus gazella* at South Georgia. *Polar Biol.* **9:** 155–60.

Gordon, A. L. and Molinelli, E. J., 1982. Southern Ocean Atlas. Columbia University Press: Surrey.

Gentry, R. L. and Kooyman, G. L., 1986. Methods of dive analysis. Pp. 28–40 *in* Fur Seals: Maternal Strategies on Land and at Sea ed by R. L. Gentry and G. L. Kooyman. Princeton University Press: Princeton.

Green, K., 1993. Sea surface temperature. Pp. 197–99 *in* Heard Island 1992 ANARE Report ed by K. Green. Antarctic Division: Kingston.

Green, K. and Burton, H. R., 1993. Comparison of the stomach contents of southern elephant seals, *Mirounga leonina*, at Macquarie and Heard Islands. *Mar. Mamm. Sci.* **9:** 10–22.

Green, K., Burton, H. R. and Williams, R., 1989. The diet of Antarctic Fur Seals *Arctocephalus gazella* during the breeding season at Heard Island. *Antarct. Sci.* **1:** 317–24.

Green, K., Williams, R. and Burton, H. R., 1991. The diet of Antarctic Fur Seals during the late autumn and early winter around Heard Island. *Antarct. Sci.* **3:** 359–61.

Green, K., Williams, R. and Burton, H. R., 1997. Foraging ecology of Antarctic Fur Seals *Arctocephalus gazella* around Heard Island. Pp. 105–13 *in* Marine Mammal Research in the Southern Hemisphere Volume 1: Status, Ecology and Medicine ed by M. Hindell and C. Kemper. Surrey Beatty & Sons: Chipping Norton.

Green, K., Williams, R., Handasyde, K., Burton, H. R. and Shaughnessy, P. D., 1990. Interspecific and intraspecific differences in the diets of three species of *Arctocephalus* (Pinnepedia: Otariidae) at Macquarie Island. *Aust. Mamm.* **13:** 193–200.

Gordon, A. L. and Molinelli, E. J., 1982. Southern Ocean Atlas. Columbia University Press: Surrey.

Hindell, M. A., Slip, D. J. and Burton, H. R., 1991. The diving behaviour of adult male and female southern elephant seals, *Mirounga leonina*, (Pinnipedia: Phocidae) *Aust. J. Zool.* **39:** 595–619.

Kooyman, G. L., Davis, R. W. and Croxall, J. P., 1986. Diving behaviour of Antarctic fur seals. Pp. 115–25 *in* Fur Seals: Maternal Strategies on Land and at Sea ed by R. L. Gentry and G. L. Kooyman. Princeton University Press: Princeton.

McCann, T. S. and Croxall, J. P., 1986. Attendance behaviour of Antarctic fur seals. Pp. 102–14 *in* Fur Seals: Maternal Strategies on Land and at Sea ed by R.L. Gentry and G.L. Kooyman. Princeton University Press: Princeton.

North, A. W., Croxall, J. P. and Doidge, D. W., 1983. Fish prey of the Antarctic fur seal *Arctocephalus gazella* at South Georgia. *Br. Antarct. Surv. Sci. Rep.* **61:** 27–37.

Perissinotto, R. and McQuaid, C. D., 1992. Land-based predator impact on vertically migrating zooplankton and micronekton advected to a Southern Ocean archipelago. *Mar. Ecol. Prog. Ser.* **80:** 15–27.

Shaughnessy, P. and Goldsworthy, S., 1990. Population size and breeding season of the Antarctic fur seal *Arctocephalus gazella* at Heard Island. *Mar. Mamm. Sci.* **6:** 292–304.

Shaughnessy, P. and Goldsworthy, S., 1993. Feeding ecology of southern fur seals *(Arctocephalus* spp.) and their management at Heard and Macquarie Islands. *Proc. NIPR Symp. Polar Biol.* **6:** 173–75.

Woehler, E. J. and Green, K., 1992. Consumption of marine resources by seabirds and seals at Heard Island and the McDonald Islands. *Polar Biol.* **12:** 659–65.

Foraging ecology of Antarctic Fur Seals *Arctocephalus gazella* Peters around Heard Island

K. GREEN[1,2], R. WILLIAMS[1] and H. R. BURTON[1]

Antarctic Fur Seals from Heard Island fed mainly on fish, but the prey species changed both seasonally and inter-annually. The majority of prey were pelagic myctophids characteristic of deep, oceanic water, and were generally taken in autumn and winter. The only other fish taken in significant numbers was *Champsocephalus gunnari* which was mostly taken from late winter through early autumn when it was co-dominant in the diet with *Krefftichthys anderssoni*. Males and females foraged in different localities and in different parts of the water column. Males foraged mainly to the south of Heard Island in winter usually diving deep by day, feeding on scattering layers. In summer males also fed on the shelf, presumably to the north and east of the island, on the bentho-pelagic *C. gunnari*. In summer, females fed to the north-east of Heard Island on *K. anderssoni* at shallow depths primarily at night. Although diet studies provided little evidence of feeding on crustaceans, diving data indicate that some males may travel to Antarctic waters in winter to feed on krill.

INTRODUCTION

THE first record this century of pupping by Antarctic Fur Seals *Arctocephalus gazella* from Heard Island was in 1963 (Budd and Downes 1969). Since then the breeding population has increased annually by an average of 23%, with 248 pups born in 1987/88 (Shaughnessy and Goldsworthy 1990). The population throughout the year is, however, dominated by non-breeding adult and sub-adult males, rather than by the breeding population. The numbers of Antarctic Fur Seals at Heard Island are low during winter and spring, dropping to less than 1 000 in December before climbing to a peak of over 21 000 in March when seals are moulting.

Antarctic Fur Seals are major predators on fish in the seas around Heard Island. Data on their diet, however, comes from only two temporally limited studies (Green *et al.* 1989; Green *et al.* 1991). Observations on foraging locations are restricted to records from ships involved in research fishing (Ensor and Shaughnessy 1990; Green *et al.* 1989; Green *et al.* 1991). The present study was designed to study the foraging of fur seals over an entire year. There are few data on the diet of Antarctic Fur Seals over more than one season. The only previous studies were based on the collection of faeces at Heard Island during spring/summer 1987/88 (Green *et al.* 1989) and the South Orkney Islands in summer/autumn 1988 (Daneri and Coria 1992). The latter study confirmed the importance of Antarctic Krill *Euphausia superba* in the diet of fur seals in the South Atlantic sector (see also Bonner 1968 and North *et al.* 1983). The studies at Heard Island, however, suggested that fur seals fed mainly on fish in spring and summer (Green *et al.* 1989) and on fish and squid in autumn/winter (Green *et al.* 1991).

At Heard Island, fish occurred in the diet in similar proportions in all seasons but the relative proportions of prey types changed markedly. Pelagic fish increased in proportion from spring/summer 1987/88 to autumn/winter 1990. At the same time, the representation of the bentho-pelagic *Champsocephalus gunnari* and benthic fish fell. This appeared to be a result of increased foraging by fur seals off the Kerguelen Shelf with the occurrence of squid increasing from 3.4% in spring/summer 1987/88 to 49.3% in autumn/winter 1990. A greater level of erosion of otoliths from a key species (*Gymnoscopelus nicholsi*) also suggested that the seals were foraging further afield in autumn/winter 1990 (Green *et al.* 1991). Although these differences appeared to represent seasonal change, the possibility that the differences were due to sampling in different years could not be discounted.

The aims of the present study were therefore to examine the diet over a full year to determine whether previously observed patterns were inter-seasonal or part of an inter-annual change, and to relate these feeding patterns to the diving behaviour of seals.

METHODS

Food

Apart from one small collection of fur seal faeces made at Red Island on the northwestern tip of Heard Island in February 1992 most sampling was conducted at eastern Heard Island, around the Spit Bay area (Fig. 1). Collections were made on a monthly basis from February 1992 through to March 1993. In winter, scats were only collected on new snow because it was difficult to determine the age of scats exposed by frequent snow thaws.

Faeces were washed through a 0.5 mm sieve and the residue was sorted under a dissecting microscope. Squid beaks were identified by

[1]Antarctic Division, Channel Highway, Kingston, Tasmania, Australia 7050.
[2]Present address: NPWS, Snowy Mountains Region, PMB Cooma, New South Wales, Australia 2630.
Pages 105–13 *in* MARINE MAMMAL RESEARCH IN THE SOUTHERN HEMISPHERE VOLUME 1: STATUS, ECOLOGY AND MEDICINE ed by M. Hindell and C. Kemper. Surrey Beatty & Sons, Chipping Norton. 1997.

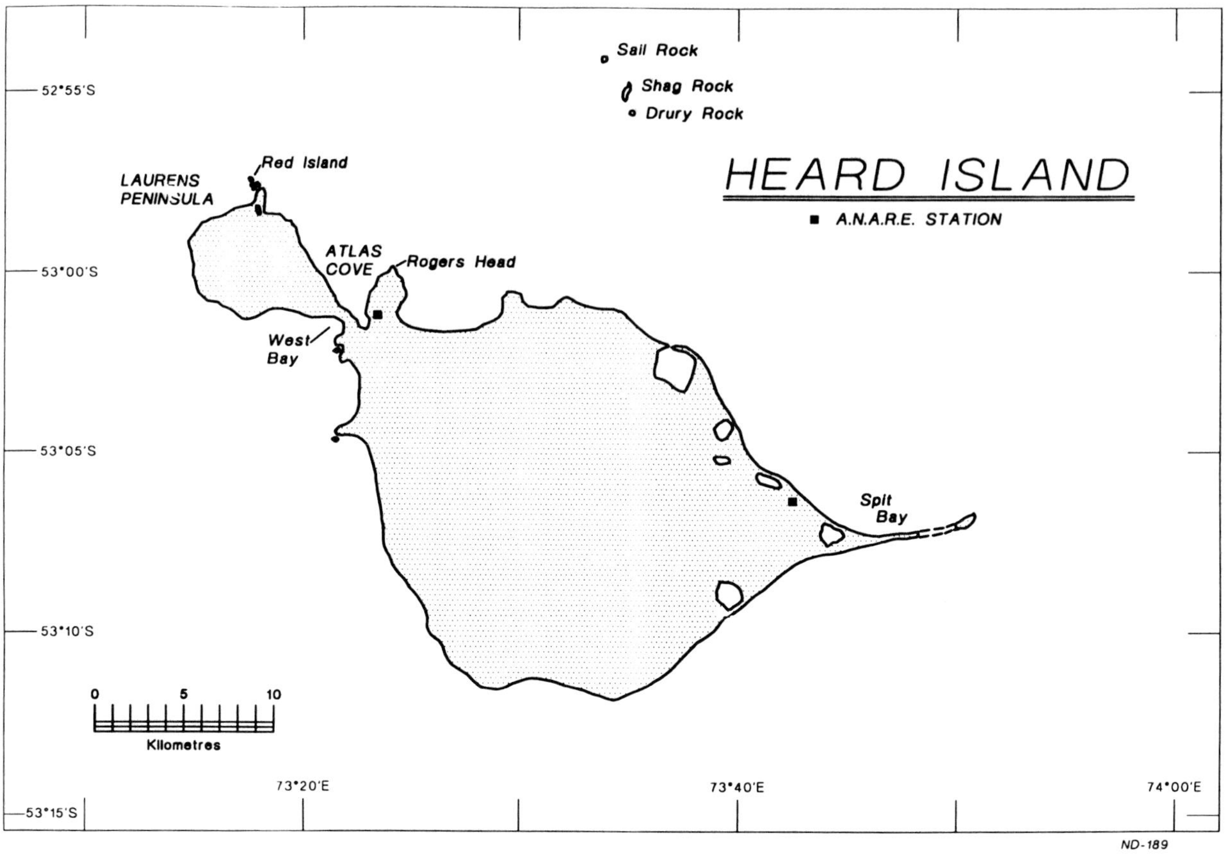

Fig. 1. Heard Island showing sample collection locations.

reference to Clarke (1986) and a reference collection from Southern Elephant Seal stomachs and Antarctic Fur Seal faeces compiled by K. Green. The presence of bony fish and skate was determined by otoliths and/or bones and denticles respectively. Otoliths were identified by reference to Williams and McEldowney (1990). As with previous studies, only otoliths showing little or no signs of erosion were selected for measurement to calculate fish standard lengths (see Green *et al.* 1989). All otoliths measured, both left and right, were used in calculations. All identifiable otoliths (eroded and uneroded) were used to determine the composition of the fish diet.

Diving

Mark 3e Time Depth Recorders (Wildlife Computers, 20 630 NE 150th Street, Woodinville, WA 98072-7641, U.S.A.) were glued to the dorsal midline of the seals with Araldite K268 (Ciba Geigy). For adult males there were three winter deployments consisting of (i) nine units on 23 and 24 April, (ii) five units on 6 July, and (iii) five units on 4 and 5 August, and (iv) one spring/summer deployment of five units on 9 and 10 December. Three adult females were instrumented in March 1992. Nine units were deployed on adult females between parturition and the first foraging trip post-parturition; these were deployed on 30/11/92 (three units) and 8/12/1992 (six units).

On recapture the units were retrieved and returned to the laboratory for down-loading. All dives deeper than 10 m or dives deeper than 4 m but shallower than 10 m and of greater than 30 sec duration were classified as foraging dives (see Green 1997, for details of analysis).

RESULTS

Food

A total of 560 faecal samples containing identifiable food remains was collected between February 1992 and March 1993. Because males outnumbered females by a considerable number during the breeding season, and no females were seen outside the breeding season, most faecal samples were probably from males. No samples could be definitely linked to females and very few samples were found in breeding colonies. The dietary results presented in this study, must therefore relate mainly to males.

Gross composition of the diet

Fish remains predominated, occurring in 82% to 100% of samples (Table 1). Bony fish occurred in 73% to100% of samples, and skate (Rajidae) in 1.8% to 78%. Squid remains occurred in between 0% and 50% of samples.

Of 11 280 otoliths, only 51 were unidentifiable, the remainder being allocated to at least family, and most to species (Table 2). Twenty-seven

Table 1. Percentage frequency of occurence of food items in faecal samples of Antartic Fur Seals.

Months	Numbers of scats	All fish	Bony fish	Skate	Squid	Octopus	Unidentified Cephalopod	Other Mollusc	Crustacean	Polychaete
Feb. (Atlas)	26	96.2	80.8	26.9	15.4			3.8	0.0	
Feb. (Spit)	14	92.9	85.7	21.4	50.0			21.4		
Mar.	41	82.9	73.2	22.0	19.5				9.8	
Apr.	36	97.2	97.2	13.9	25.0			5.6	27.8	
May	49	100.0	91.8	38.8	40.8			2.1	18.4	2.1
June	39	100.0	79.5	59.0	35.9	2.6			17.9	
July	46	100.0	76.1	67.4	8.7				13.0	2.2
Aug.	45	100.0	80.0	62.2	6.7				13.3	8.9
Sep.	26	100.0	76.9	57.7	0.0			11.6	0.0	3.8
Oct.	41	100.0	78.0	78.0*	2.4	2.4	4.9	7.3	2.4	7.3
Nov.	21	100.0	85.7	66.7	19.0		9.5	4.8	19.0	4.8
Dec.	35	100.0	100.0	22.9*	14.3		8.6	8.6	0.0	
Jan.	49	100.0	87.8	26.5	2.0				8.2	
Feb.	56	100.0	98.2	1.8	14.3		3.6	5.4	1.8	1.8
Mar.	36	100.0	88.9	27.8	2.8			2.8		
Total	560									

* Skate egg also found

Table 2. Species of fishes identified from otoliths in faecal samples from Antarctic Fur Seals from Heard Island expressed as a percentage of the total number of otoliths recovered each month.

	Red Is.	Spit Feb.	Mar.	Apr.	May	June	July	Aug.	Sept.	Oct.	Nov.	Dec.	Jan.	Feb.	Mar.
Krefftichthys anderssoni	3.3	5.6	4.5	0.0	1.9	0.0	0.4	1.2	50.1	0.6	84.5	68.8	42.3	48.0	0.3
Paradiplospinus gracilis	0.8	0.2	0.4	0.0	0.5	0.0	0.0	0.0	0.1	0.6	0.0	0.0	0.0	0.0	0.0
Protomyctophum bolini	0.5	0.0	0.0	0.2	0.0	0.0	0.0	0.2	0.0	0.0	0.0	0.2	0.3	0.0	0.1
P. normani	13.0	32.7	19.2	2.0	0.4	0.5	0.0	0.0	0.0	0.3	0.0	0.0	0.0	3.3	6.4
P. tenisoni	0.5	0.4	0.0	0.5	0.5	1.7	0.0	3.1	14.2	19.1	0.6	0.9	0.0	0.0	0.0
Benthalbella elongata	1.1	0.0	0.0	0.0	1.9	1.1	0.0	0.7	0.0	0.0	0.0	0.0	0.0	0.0	0.0
Electrona antarctica	5.1	13.3	9.4	26.3	53.4	85.0	1.8	12.3	1.3	4.9	0.8	13.5	2.1	2.0	4.3
E. subaspera	2.2	34.0	52.1	57.1	22.2	5.3	2.1	13.7	3.4	0.3	1.0	0.0	0.4	4.0	51.7
E. subaspera/carlsbergi	0.0	0.0	0.0	3.9	0.0	0.0	0.0	0.0	0.0	0.0	0.0	0.2	0.0	0.0	0.1
E. carlsbergi	0.0	0.0	0.0	0.0	0.3	0.4	0.0	0.0	0.0	0.0	0.0	0.0	0.0	0.2	0.0
Metelectrona ventralis	0.0	0.0	0.0	0.0	0.0	0.2	0.4	0.0	0.0	0.0	0.0	0.0	0.0	0.0	0.0
Notolepis coatsi	0.3	0.0	0.0	1.1	0.9	1.3	0.0	0.0	0.0	0.0	0.0	0.0	0.0	0.0	0.0
Gymnoscopelus fraseri	1.4	3.9	1.0	3.2	6.7	0.3	0.0	0.0	0.3	0.0	4.9	1.6	5.0	0.0	0.8
G. nicholsi	11.7	8.1	8.8	1.8	5.4	1.1	0.0	0.0	0.0	0.0	0.4	1.1	6.2	0.4	9.5
G. opistopterus	0.0	0.0	0.0	0.0	0.0	0.0	0.0	1.2	0.0	0.0	0.0	0.0	0.0	0.0	0.0
indet myctophid	0.0	0.6	0.2	0.2	0.0	0.0	0.0	0.0	0.1	0.0	0.0	0.1	0.0	0.0	0.0
Magnisudis prionosa	0.3	0.0	0.0	0.0	0.0	0.0	0.0	0.0	0.0	0.0	0.0	0.0	0.0	0.0	0.0
Pelagic (sub total)	40.1	98.8	95.7	96.4	94.2	96.9	4.6	32.5	69.6	25.9	92.3	86.7	56.4	57.9	73.3
Pseudoicichthys australis	0.5	0.2	0.0	0.0	0.0	0.0	0.0	0.0	0.0	0.0	0.0	0.2	0.2	0.0	0.0
Melanostigma gelantinosum	1.6	0.0	0.0	0.0	0.0	0.0	0.0	0.0	0.0	0.0	0.0	0.0	0.0	0.0	0.0
Dissostichus eleginoides	0.0	0.0	0.0	0.0	0.2	0.0	0.0	0.0	0.0	0.0	0.2	0.0	0.0	0.0	0.0
Gobionotothen acuta	0.5	0.6	2.0	1.4	0.0	0.2	0.7	2.2	0.6	6.5	0.7	1.6	0.4	0.2	0.0
Lepidonotothen mizops	1.1	0.0	0.8	0.0	0.2	0.0	0.0	1.4	0.0	1.5	0.0	0.0	0.0	0.0	0.0
L. squamifrons	0.0	0.0	0.0	0.0	0.0	0.0	4.9	0.5	0.0	0.0	0.9	0.0	0.0	0.0	0.0
Notothenia rossii	0.0	0.0	0.0	0.0	0.0	0.0	0.0	0.0	0.0	0.0	0.0	0.0	0.0	0.0	0.0
N. cyanobrancha	0.0	0.0	0.2	0.5	0.0	0.0	0.0	0.0	0.0	0.0	0.0	0.0	0.0	0.0	0.0
indet nototheniid	0.0	0.0	0.2	0.2	0.0	0.0	0.0	0.7	0.4	0.3	0.3	0.0	0.0	0.0	0.0
Channichthys rhinoceratus	2.7	0.2	0.0	0.2	0.0	0.0	1.8	1.0	0.0	0.3	1.2	0.6	0.7	1.0	1.1
indet channichthyid	0.0	0.0	0.0	0.0	0.0	0.0	0.0	0.2	0.0	0.0	0.0	0.0	0.0	0.0	0.0
Harpagifer kerguelensis	0.0	0.0	0.0	0.0	0.0	0.0	0.0	0.0	0.0	0.0	0.0	0.0	0.0	0.0	0.0
Zanclorhynchus spinifer	0.0	0.2	0.0	0.0	0.0	0.0	0.0	0.0	0.0	0.0	0.2	0.3	0.0	0.0	0.0
Benthic (sub total)	6.5	1.2	3.3	2.3	0.5	0.4	7.4	6.0	1.0	8.6	3.6	2.8	1.4	1.2	1.1
Champsocephalus gunnari	53.4	0.0	1.0	1.4	5.3	2.6	87.4	53.1	29.2	65.4	4.2	10.5	41.6	40.5	25.2
Unidentified	0.0	0.0	0.0	0.0	0.0	0.0	0.4	8.2	0.1	0.0	0.0	0.0	0.6	0.3	0.4
Number of otoliths	369	483	489	441	1 232	1 131	284	415	698	324	982	1 418	1 141	1 150	719

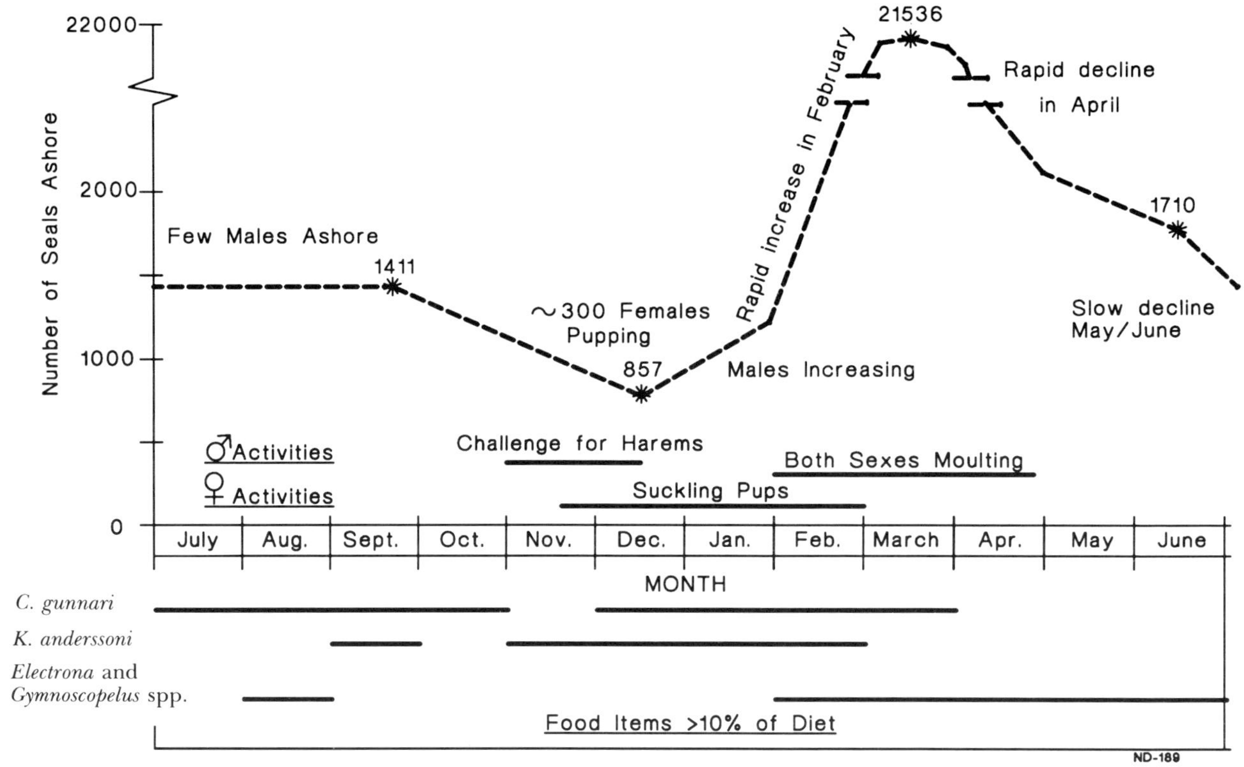

Fig. 2. Numbers of Antarctic Fur Seals ashore on Heard Island in four seasons, their major activities and times of predation on major fish species by number.

species of fish were identified. The main species of squid was identified as *Mastigoteuthis*? (see Green *et al.* 1989), with other species being *Brachioteuthis picta, Gonatus antarcticus, Kondakovia longimana, Moroteuthis ingens, M. knipovitchi, Psychroteuthis glacialis* and *Todarodes filippovae*.

Crustaceans reached a maximum occurrence of 27.8% in April but were generally unimportant with euphausiids *(Euphausia vallentini)* appearing in large amounts (occupying the bulk of the scat) in only four samples.

Size of prey

Size classes of *Electrona antarctica* and *E. subaspera* from scats sampled in late summer/autumn showed a single main peak at 75 mm and 105 mm respectively, with an indication of a smaller size class. *Champsocephalus gunnari* had a peak at approximately 195 and 225 mm standard length for fish taken in July through October and December through March respectively. *Krefftichthys anderssoni* had peaks at approximately 35 mm and 60 mm in November through February but a peak in only the smaller size class in September (Fig. 2).

Inter-seasonal variation

There were major seasonal changes in predation on fish. Otoliths of the myctophids *E. antarctica* and *E. subaspera* were most numerous between February and June. The dominant otoliths from July to February 1993 were from the channicthyid *C. gunnari* and the myctophid *K. anderssoni*. In March 1993 *E. subaspera* returned to prominence but *C. gunnari* replaced *E. antarctica* as the co-dominant species. The myctophids *Gymnoscopelus nicholsi* and *Protomyctophum normani* were important summer prey and *P. tenisoni* was important in September and October. No other fish species exceeded 10% of otoliths at any time (Table 2). There was an apparent decline in predation on *C. gunnari* in November and December (Table 2) which is not totally an artefact caused by 824 otoliths of *K. anderssoni* in a single sample. This one scat would have only affected one month. If only one species, *C. gunnari*, is examined (Fig. 3) then it can be seen that there was still a decline in numbers of this species taken at this time.

Inter-annual variation

The number of fish species identified in the present study (27) was higher than in 1987/88 (17) and 1990 (22). The species composition in the diet was similar between 1990 and 1992 (for May and June) although skate and bony benthic fish were more common in 1992 than 1990 (Table 3). Data over spring/summer 1987/88 bore little resemblance to 1992/93 (Table 3). While occurrence of skate remained high through 1992 (though trailing off in early 1993) the amount of bony benthic fish fell far short of the same period in 1987/88. This was mainly due to the lower predation on *Channichthys rhinoceratus* and *Lepidonotothen squamifrons* (previously *Notothenia* in Green *et al.* 1989). Of the squid species, *B. picta, M. ingens, M. knipovitchi* and *T. filippovae* were not found in previous studies of fur seal diet at Heard Island.

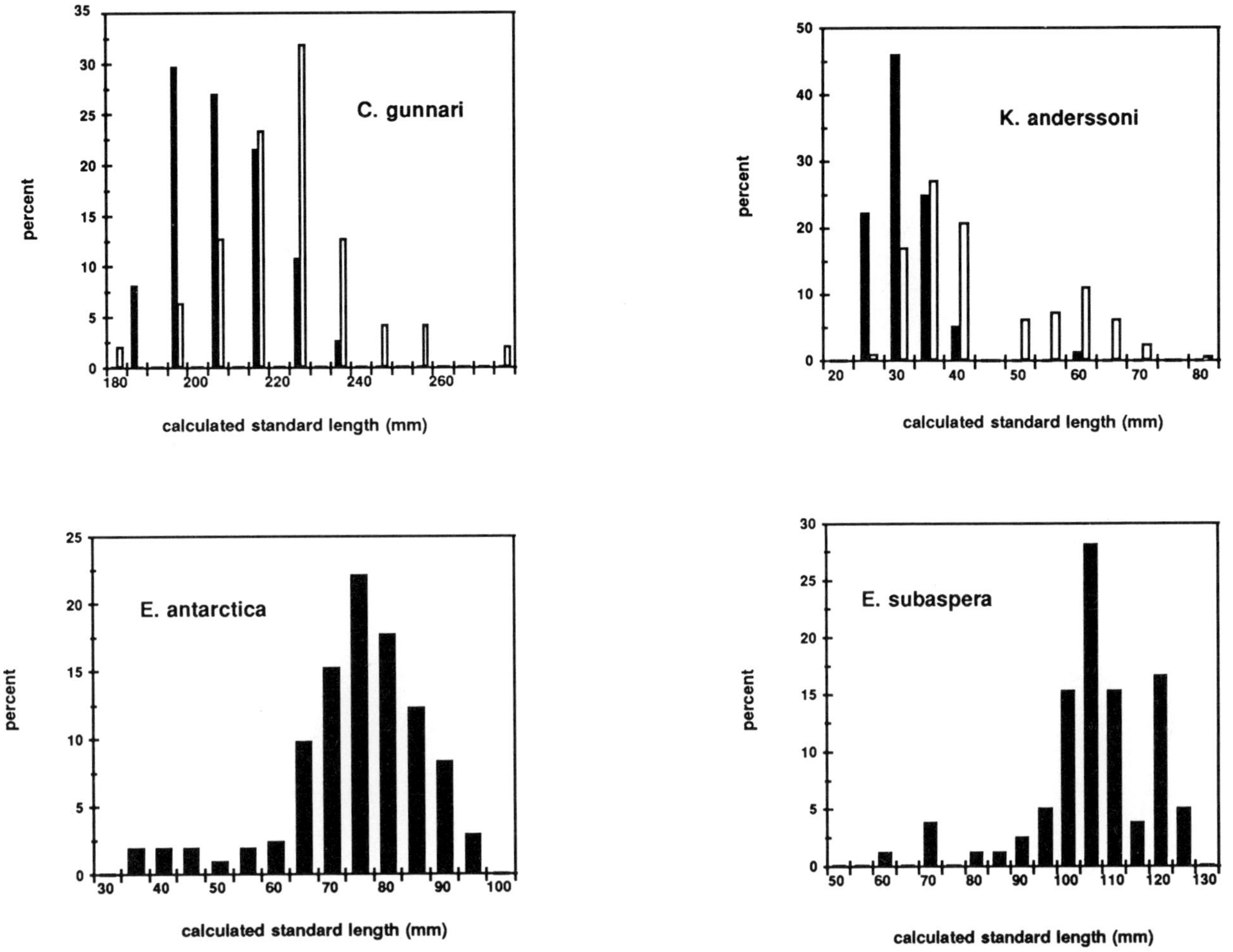

Fig. 3. Length-frequency distribution for *Champsocephalus gunnari, Krefftichthys anderssoni, Electrona antarctica* and *Electrona subaspera*. Closed and open columns are for July–October and December–March for *C. gunari* and September and November–February for *K. anderssoni*.

Table 3. Otoliths of four species of fishes expressed as a paercentage of the total number of otoliths recovered each month compared for this study and 1987/88 (Green *et al.* 1989) and 1990 (Green *et al.*1991). Data from May/June 1990 were from the first and third collections. The results from this study are identified with the letter A, those from May/June 1990 and September 1987–February 1988 with B.

	Study	May	June	Sept.	Oct.	Nov.	Dec.	Jan./Feb. 93
Krefftichthys andersonni	A	1.9	0.0	50.1	0.6	84.5	68.8	45.2
	B	1.2	0.3	0.4	0.0	1.7	0.5	0.9
Electrona antarctica	A	53.4	85.0	1.3	4.9	0.8	13.5	2.0
	B	54.6	71.8	42.4	8.4	0.0	1.0	6.3
E. subaspera	A	22.2	5.3	3.4	0.3	1.0	0.0	2.2
	B	30.0	10.5	3.5	10.9	0.0	1.9	2.6
Champsocephalus gunnari	A	5.3	2.6	29.2	65.4	4.2	10.5	41.0
	B	1.0	1.8	10.5	19.3	50.9	27.9	25.7

Diving location and depth

Few of the TDRs deployed (especially those attached to males) were recovered or contained useable data. Diving behaviour of Antactic Fur Seals around Heard Island is discussed in greater detail elsewhere (Green 1997) and for the purposes of dietary relationships can be summarized as follows:

Males

Male seals foraged mainly to the south and east of Heard Island (see Green 1997). Close to Heard Island in May to October males dived mainly by day and most dives exceeded 200 m. Only 30–45 foraging dives per day were undertaken with total time submerged ranging between 140 to 240 min per day (Table 4). One instrumented male

Table 4. Summary diving statistics for Antarctic Fur Seals in waters around Heard Island and in Antarctic waters.

Tag	Month	Time foraging	Number of dives	Time Submerged (min)	Mean dive duration (min)	Maximum Depth (m)
Males						
(Subantarctic)						
587	May	640 ± 108	34 ± 6	197 ± 41	5.8 ± 0.3	232
572H	May	1 043 ± 330	30 ± 16	142 ± 65	4.9 ± 0.6	234
565	July	591 ± 30	46 ± 5	206 ± 6	4.5 ± 0.4	232
601	Aug.	599*	35	220	6.2 ±	264
597	Oct.	814 ± 105	36 ± 20	237 ± 155	6.5 ± 1.3	444
004	Dec.	979 ± 104	129 ± 60	372 ± 94	3.1 ± 0.8	214
(Antarctic)						
572A	June		206 ± 59	315 ± 106	1.5 ± 0.2	62
608	Oct./Nov.	*c*1 080	73 ± 33**	320 ± 109	4.4 ± 1.1***	130
Females						
(1991/1992)						
564	March	490	110	110	1.0	50
566	March	499	123	117	1.0	49
568	March	813 ± 176	152 ± 87	152 ± 140	1.3 ± 0.3	88
(1992/1993)						
623.1	Early Dec.	476 ± 119	179 ± 112	158 ± 73	1.1 ± 0.5	91
626	Early Dec.	485 ± 138	185 ± 43	250 ± 89	1.4 ± 1.0	106
628.1	Early Dec.	717 ± 185	163 ± 77	228 ± 70	1.6 ± 0.4	120
630.1	Early Dec.	438 ± 50	120 ± 39	227 ± 61	1.9 ± 0.1	162
630.1	Late Dec.	681 ± 121	134 ± 51	226 ± 75	1.8 ± 0.4	192
628.1	Late Dec.	658 ± 300	186 ± 116	224 ± 134	1.2 ± 0.3	120
623.2	Late Dec.	597 ± 197	233 ± 97	273 ± 63	1.3 ± 0.5	138
614.3	Jan.	850 ± 265	279 ± 152	358 ± 162	1.3 ± 0.5	125

*Scaled up by 1.26 from a part of a day sample based on expected close of day.
**Scaled up from seven part days based on an average 18 hours foraging in three full days.
***On the one day when presumed krill-foraging dives predominated 134.8 dives were estimated at an average 2.2 minute duration.

(004), that was challenging for a harem in December, behaved differently, diving by night, and making frequent (129 per night and totalling 372 minutes underwater), and shallow, dives averaging 73.5 ± 47.9 m. Dives of two males which travelled well south of Heard Island into Antarctic waters up to 1 000 km away in June and October/November were shallower and more frequent than animals from subantarctic waters. One male averaged 206 foraging dives per night totalling 315 min submerged and averaged 17.0 ± 10.4 m, compared with 126.4 ± 91.2 m in waters around Heard Island for the same animal.

Females

Female seals foraged mainly to the north-east of Heard Island during the breeding season (Green 1997). They dived mainly by night with most dives being shallow. Three females in March 1992 foraged mainly between 1900 and 0430, although one began foraging close to mid-day on four of six days. The average maximum depth of dives for these females ranged between 42.5 and 45.3 m and they made between 110 and 122 dives per day totalling 110 to 150 min under water. In the second summer (when seals were instrumented earlier in the breeding season), diving bouts also generally began at around 1900 and finished at 0500 (for four females) but started earlier and finished later and became more erratic in late December and January (for four females). Average maximum daily dive depths were deeper in the second summer at between 80 and 130 m and the number of dives per day increased to 120–185 in early December and 135–280 in late December and January. The average time submerged increased from 160–250 min in early December to 225–275 min in late December and 360 min for one animal in January.

DISCUSSION

The results of this, and previous studies (Green *et al.* 1989; Green *et al.* 1991), show that the major component in the diet of Antarctic Fur Seals at Heard Island is fish, in contrast to the predominance of Antarctic Krill in the diet of fur seals around South Georgia (Bonner 1968; North *et al.* 1983) and the South Orkney Islands (Daneri and Coria 1993). Euphausiids *(E. vallentini)* were only found in large amounts in four scats at Heard Island in January 1993, and one sample from November 1992 was full of isopods. The differences in diet between areas in the South Atlantic sector and Heard Island appear to be mainly due to differences in prey availability. Antarctic Krill is abundant in the waters close to South Georgia even occurring in fjords, whereas it is rare around Heard Island (Williams, unpubl. data). However, seals from Heard Island might be consuming quantities of krill when foraging some distance from the island that are not detected in this or previous studies. As the edge of the pack ice is only 480 km south of Heard Island at the time of maximum ice extent in September there is the opportunity to forage there. This latitude (approximately 58°S) is the furthest north that significant quantities of krill have been observed at this longitude (Williams *et al.* 1986). The dive pattern of one animal (572A) that went into the pack ice in winter very much fits the krill feeding pattern (see Croxall *et al.* 1985) of shallow diving mainly at night. Although it is difficult to be sure if an observed dive profile represents krill-feeding there is some similarity between the profile of the "K" dives exhibited by this seal (see Green 1997) and profiles observed in fur seals from South Georgia (Ian Boyd, pers. comm.). This pattern was very different from the foraging of the same animal in Heard Island waters, where over 50% of dives were deep and most were during daylight hours. There were no location data for a second animal but its diving pattern was similar. These data are the first indication that Antarctic Fur Seals from Heard Island may regularly forage in Antarctic waters where they are likely to feed on *E. superba.* Sightings of Antarctic Fur Seals in the pack-ice south of Heard Island are, however, uncommon, but sightings of individual animals have been made in the Vestfold Hills in 1988 (T. Timms, pers. comm.) and the summer of 1991/92 (G. Findlay, pers. comm.) and also further west at Mawson station (Shaughnessy and Burton 1986). The problem of identifying diet away from sampling sites has rarely been addressed (Green and Burton 1993) and the use of devices such as time-depth-location recorders only give an indication of potential prey.

The majority of fish prey consumed around Heard Island were pelagic myctophids (*Electrona, Gymnoscopelus, Protomyctophum* and *Krefftichthys*) which are characteristic of deep, off-shelf water and can constitute the deep scattering layers, usually at 200–500 m (Williams, unpubl. data). The exception is *G. nicholsi* which can also be found near the bottom on the shelf. In the breeding season, female seals had a similar diving pattern to those at South Georgia, despite feeding on fish rather than krill. The majority of dives occurred at night and a high proportion were within the top 20 m of the water column (*cf* Boyd and Croxall 1992). One male that was a challenger at a harem (004) had a dive pattern which was essentially a "female" pattern of night-time feeding, diving to shallow depths and presumably taking *K. anderssoni* that migrate to the near-surface waters at night (Perisinotto and McQuade 1992).

The only other fish taken in significant numbers was *C. gunnari* which was mostly taken from late winter through to early autumn (Fig. 4). *Champsocephalus gunnari* dominated the otoliths in numbers at that time, with *K. anderssoni* as its co-dominant species. The trough in predation on *C. gunnari* in November and December may be due to a reduction in the number of males foraging near the island while they challenged for territories or practiced territorial behaviours, coupled with the arrival of females (Fig. 4). This could result in the predominance in the sampling site of scats with remains of fish collected further out to sea. The predation on *C. gunnari* therefore appears to be moderate during the period when only a few males were on shore from July to October. There was a peak in predation on *C. gunnari* from January to March while males were ashore to moult. Following this peak in the numbers of males on shore, predation on *C. gunnari* was low from March to July. There was a difference in the size of *C. gunnari* between July and October and December to March. This is consistent with fish growth rate and reflects predation on the same age class as it grows.

The dive patterns of male Antarctic Fur Seals may maximize the effectiveness of preying on fish in deep water. This applied both to myctophids in deeper waters and would also be effective in preying on *C. gunnari* near the bottom on the shelf. During the night, shoals of *C. gunnari* are spread through the water column whereas by day the shoals are less dispersed and are closer to the bottom (Williams, unpubl. data). By foraging during the day males could target shoals of *C. gunnari* and thus keep the number of dives required to obtain prey to a minimum.

The co-dominance of *K. anderssoni* with *C. gunnari* is interesting as the dive pattern indicate that the females target pelagic prey such as *K. anderssoni* rising towards the surface at night. Few dives by females even approach the depths that *C. gunnari* are found at night, despite females being able to dive to close to 200 m. *Champsocephalus gunnari* is a fish low in oil content whereas *K. anderssoni* is very oily. It is possible that female fur seals requiring a high fat intake to produce milk for their young, targeted this prey. Similar predation on fish with high oil content can be seen in the shift of Adélie Penguins *Pygoscelis adeliae* to *Pleuragramma antarcticum* once their chicks have hatched (Williams 1989). Off South Georgia, Croxall and Pilcher (1984) showed that female Antarctic Fur Seals select adult female krill high in energy.

There was a change in size classes of *K. anderssoni* taken by seals. In September, when only males were present on shore, only a small size class was preyed upon. From November to February when females were ashore breeding (see Fig. 4), both a small and a large size class were taken. This could be related to a possible

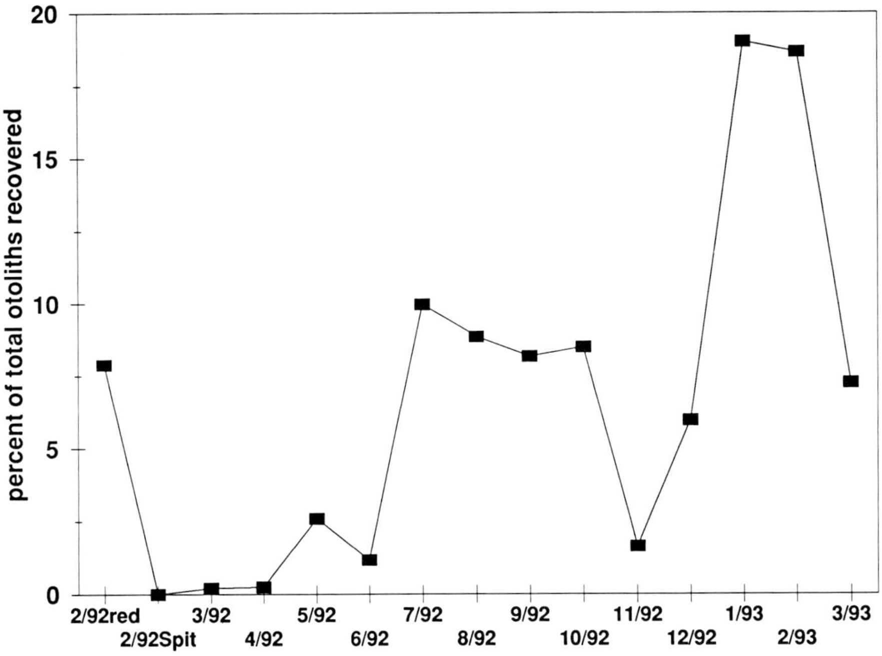

Fig. 4. Numbers of otoliths of *Champsocephalus gunnari* recovered monthly as a percentage of total take over time.

vertical segregation of size classes. Males may only take *K. anderssoni* opportunistically at shallow depths. Females, however, if concentrating their foraging on this species, would feed on both layers. Lengths of *C. gunnari* and *G. nicholsi* were very similar at equivalent times in 1987/88

The change in diet from spring/summer 1987/88 to autumn/winter 1990 was dramatic (Green *et al.* 1991), for example, an increase in the occurrence of squid from a mean of 3.4% in summer 1987/88 to 49.3% in winter 1990. In the present study this trend was similar. Excluding the sample for February 1992 at Spit Bay (when only 14 scats were collected), the occurrence of squid rose between February and June but after June the occurrence of squid dropped dramatically and did not begin to recover until November. Fish occurred in a similar percentage of faeces in spring/summer 1987/88 (95.2%) (Green *et al.* 1989), in autumn/winter 1990 (93.0%) (Green *et al.* 1991) and in the present study (Table 1).

In contrast to the earlier studies there were few benthic/shelf species other than *C. gunnari* taken in spring/summer 1992/93. Their otoliths reached a monthly maximum of 8.3% of all otoliths in October 1992 compared with 52.2% of otoliths in September 1987. Other differences from previous studies were that *Krefftichthys* and *Protomyctophum* were much more abundant in the present study. *Electrona* occurred more in summer in the present study and *G. nicholsi* occurred from February through April rather than December through February. One species, *E. antarctica*, differed with a main peak in size at 75 mm compared to 55 mm with a minor peak at 75 mm in 1987/88 (Green *et al.* 1989).

In the summer of 1987/88, 30.5% of otoliths of *G. nicholsi* were measurable compared to 10.3% in winter. In the present study there were fewer otoliths of this species collected. Of 43 otoliths, 30.2% were measurable in March 1992. This figure is similar to that in 1987/88 and to the 31.1% in the South Orkney Islands where myctophids (primarily *E. antarctica* and *G. nicholsi*) were the main fish prey of Antarctic Fur Seals (Daneri and Coria 1993). In contrast, the figure for March 1993 was one otolith out of 68, indicating that the foraging distance in the 1992/93 breeding season differed greatly from that in 1987/88. Patterns of foraging presented here may not, therefore, be generally applicable across years. This suggests that dietary studies in combination with dive studies are necessary for a better understanding of the data, at least for species which feed close to their haulout areas.

The pattern of foraging documented here is of a large population of non-breeding males, unconstrained by the need to raise young. Feeding in the winter period was on myctophids (or krill), probably far offshore, and while seals were based on land through the summer for the moult they switched on to *C. gunnari* on the shelf (and some myctophids off it). No foraging data exist for females in winter but in summer, breeding females needed to maintain body condition while still delivering milk high in fat to their young. This appears to be achieved by night-time foraging at shallow depths taking fish of high oil content (i.e., myctophids). The choice of prey is simple with both *K. anderssoni* (off the shelf) and *C. gunnari* (on the shelf) close by to the north-east or myctophids (*Electrona* spp.) close by to the south off the shelf. Predictions of the diet of fur seals in non-sampled months were used in a model of consumption of marine prey around Heard Island (Woehler and Green 1992). The present study has shown that short term trends in diet from which Woehler and Green (1992) drew inferences may not continue, that there are inter-annual as well as inter-seasonal differences in diet. Therefore, the full picture of prey consumption cannot be built up piecemeal by examining a part of a season in different years. These smaller studies are extremely useful but are dependent on at least one full year's study (preferably more) as a basis for their interpretation. This study has once again highlighted the need for long-term studies to document natural variation in diet of seals and has further emphasized (see Green and Burton 1993) that there can be sampling location effects. This must be investigated by the use of methods which allow the investigation of the diet at foraging locations remote from land.

ACKNOWLEDGEMENTS

Thanks to Dave Slip who collected some of the fur seal faecal samples, gave much useful advice concerning Time Depth Recorders and commented on the manuscript. Dave Slip, Geoff Moore and Erwin Erb all gave their time freely in the capture and instrumentation of fur seals. Mark Hindell commented on the manuscript.

REFERENCES

Bonner, W. N., 1968. The fur seal of South Georgia. *Br. Antarct. Surv. Sci. Rep.* **56:** 83 pp.

Boyd I. L. and Croxall, J. P., 1992. Diving behaviour of lactating Antarctic fur seals. *Can. J. Zool.* **70:** 919–28.

Budd, G. M. and Downes, M. C., 1969. Population increase and breeding in the Kerguelen fur seal, *Arctocephalus tropicalis gazella,* at Heard Island. *Mammalia* **33:** 58–67.

Clarke, M. R., 1986. A handbook for the identification of cephalopod beaks. Clarendon Press: Oxford.

Croxall, J. P., Everson, I. and Kooyman, G. L., 1985. Fur seal diving behaviour in relation to vertical distribution of krill. *J. Anim. Ecol.* **54:** 1–8.

Croxall, J. P. and Pilcher, M. N., 1984. Characteristics of krill, *Euphausia superba,* eaten by Antarctic Fur Seals, *Arctocephalus gazella,* at South Georgia. *Br. Antarct. Surv. Bull.* **63:** 117–25.

Daneri, G. A. and Coria, N. R., 1992. The diet of Antarctic fur seals, *Arctocephalus gazella,* during the summer-autumn period at Mossman Peninsula, Laurie Island (South Orkneys). *Polar Biol.* **11:** 565–66.

Daneri, G. A. and Coria, N. R., 1993. Fish prey of Antarctic fur seals, *Arctocephalus gazella,* during the summer-autumn period at Laurie Island, South Orkney Islands. *Polar Biol.* **13:** 287–89.

Ensor, P. H. and Shaughnessy, P. D., 1990. Fur seals over the Kerguelen Plateau and elsewhere in the Southern Ocean. *Polar Biol.* **10:** 481–83.

Green, K., 1997. Diving behaviour of Antarctic Fur Seals *Arctocephalus gazella* Peters around Heard Island. Pp. 97–104 *in* Marine Mammal Research in the Southern Hemispshere Vol. 1: Status, Ecology and Medicine ed by M. Hindell and C. Kemper. Surrey Beatty & Sons: Chipping Norton.

Green, K. and Burton, H. R., 1993. Comparison of the stomach contents of southern elephant seals, *Mirounga leonina,* from two breeding poulations. *Mar. Mamm. Sci.* **9:** 10–22

Green, K., Burton, H. R. and Williams, R., 1989. The diet of Antarctic Fur Seals *Arctocephalus gazella* during the breeding season at Heard Island. *Antarct. Sci.* **1:** 317–24.

Green, K., Williams, R. and Burton, H. R., 1991. The diet of Antarctic Fur Seals during the late autumn and early winter around Heard Island. *Antarct. Sci.* **3:** 359–61.

North, A. W., Croxall, J. P. and Doidge, D. W., 1983. Fish prey of the Antarctic fur seal *Arctocephalus gazella* at South Georgia. *Br. Antarct. Surv. Bull.* **61:** 27–37.

Perissinotto, R. and McQuaid, C. D., 1992. Land-based predator impact on vertically migrating zooplankton and micronekton advected to a Southern Ocean archipelago. *Mar. Ecol. Prog. Ser.* **80:** 15–27.

Shaughnessy, P. and Burton, H. R., 1986. Fur seals *Arctocephalus* spp. at Mawson station, Antarctica, and in the Southern Ocean. *Polar Rec.* **23:** 79–81.

Shaughnessy, P. and Goldsworthy, S., 1990. Population size and breeding season of the Antarctic Fur Seal *Arctocephalus gazella* at Heard Island. *Mar. Mamm. Sci.* **6:** 292–304.

Williams, R., 1989. Australian research on Antarctic bird and seal diets. CCAMLR selected scientific papers SC-CAMLR-SSP/S. Pp. 231–49.

Williams, R., Kirkwood, J. M. and O'Sullivan, D. B., 1986. ADBEX 1 Cruize zooplankton data. *A.N.A.R.E. Res. Note* **31.**

Williams, R. and McEldowney, A., 1990. A guide to the fish otoliths from waters off the Australian Antarctic Territory, Heard and Macquarie Islands. *A.N.A.R.E. Res. Note* **75.**

Woehler, E. J. and Green, K., 1992. Consumption of marine resources by seabirds and seals at Heard Island and the McDonald Islands. *Polar Biol.* **12:** 659–65.

Diving and foraging behaviour of juvenile Southern Elephant Seals from Heard Island

D. SLIP[1]

Diving behaviour of two juvenile (9 to 11 months of age) Southern Elephant Seals during their second trip to sea was examined using time depth recorders. Dive behaviour was recorded for 76 days for the juvenile male and 66 days for the juvenile female with 6 652 and 4 566 dives recorded respectively, which represented approximately 65% of the total trip to sea. Mean dive duration was 15.5 ± 5.1 minutes with a maximum of 39 minutes for the juvenile male, and mean dive depth was 323 ± 157 m with a maximum of 834 m, while for the juvenile female, mean dive duration was 21.1 ± 6.4 min with a maximum of 58 min, and mean dive depth was 416 ± 147 m with a maximum of 1270 m. Diving was continuous and over 90% of the time at sea was spent underwater, with only brief periods at the surface. Mean surface intervals were 1.67 ± 2.54 and 1.63 ± 0.94 min, for the male and female respectively, and extended surface intervals greater than three minutes represented less than 2% of all surface intervals for either seal. Both animals remained at sea for the entire period of August to November and showed patterns of continuous, deep, long duration diving with short surface intervals. Dive profiles consistent with pelagic foraging were common. Locations determined from sea temperature profiles suggest that these animals were foraging in mesopelagic waters in mid ocean, both north and south of the Antarctic Polar Front, and at times over the edge of the Kerguelen Plateau.

INTRODUCTION

RECENT applications of time-depth recorder technology have increased our knowledge of the diving and foraging behaviour of free ranging Southern Elephant Seals (e.g., Boyd and Arnbom 1991; Hindell *et al.* 1991a, 1991b; Slip *et al.* 1994). To date, these studies have concentrated on adult animals, however, little is known of the diving or foraging behaviour of juvenile Southern Elephant Seals, and their role in the Southern Ocean ecosystem is poorly understood. Southern Elephant Seal populations in the Indian and Pacific sectors of the Southern Ocean have declined substantially over the past several decades (Laws 1994), and a major factor in these declines appears to be an increase in mortality of seals during their first year at sea (Hindell *et al.* 1994). Therefore, knowledge of the foraging behaviour of seals in their first year at sea is important if we are to understand the processes by which elephant seal populations are regulated, as changes in the distribution and abundance of prey could be a proximate cause of reduced survival in young seals (Hindell 1991).

Although recent studies have identified some of the major prey species taken by Southern Elephant Seals around their subantarctic breeding grounds and haulout sites (e.g., Rodhouse 1992; Green and Burton 1993; Slip 1995), our understanding of the foraging ecology of Southern Elephant Seals is limited by a lack of information on their diet while they are away from these areas. While some of the major foraging areas of adult Southern Elephant Seals have been identified (e.g., Hindell *et al.* 1991a; McConnell *et al.* 1992), little is known of juveniles. Juvenile seals have two distinct periods at sea during their first year (Hindell and Burton 1988). The first begins following the post-weaning fast and the animals have left their natal islands by the end of December. First year animals reappear at these islands from about May until mid-July. The length of this haulout varies among individuals, but most animals have returned to sea by late August, and remain at sea until their first moult which begins in November to December.

The aims of this study were to examine the diving behaviour of juvenile Southern Elephant Seals during their second trip to sea at age 9 to 14 months and integrate this with geographic displacement and dietary information in order to understand more fully the role of juveniles in the Southern Ocean ecosystem.

METHODS

Depth and water temperature data

I collected data on water temperature and depth using microprocessor controlled MK 3 time depth recorders (TDRs) from Wildlife Computers (Woodinville, WA, USA). These units were equipped with a pressure transducer, a temperature probe, an internal clock and 512 kilobytes of memory, all housed in a titanium cylinder (25 × 152 mm). Depth was recorded on two channels; the first operated between 0 and 450 m with an accuracy of ± 2 m, and the second from 450 to 1 500 m with an accuracy of ± 6 m. Water temperature was recorded on the

[1]Department of Zoology, University of Tasmania, GPO Box 252C, Hobart, Tasmania, 7001 and Australian Antarctic Division, Channel Highway, Tasmania, Australia 7050.
Pages 114–24 *in* MARINE MAMMAL RESEARCH IN THE SOUTHERN HEMISPHERE VOLUME 1: STATUS, ECOLOGY AND MEDICINE ed by M. Hindell and C. Kemper. Surrey Beatty & Sons, Chipping Norton. 1997.

third channel and was accurate to ± 0.2°C. I programmed TDRs to record depth every 30 seconds and temperature every 300 seconds, which allowed continuous sampling for up to 80 days.

Deployment of TDRs

In August 1992, I deployed TDRs on three juvenile animals at Heard Island (53°05′S, 73°45′E). The age of these animals was estimated to be between 9 and 11 months old by comparing their snout-tail length and degree of tooth eruption with known age animals. I immobilized the seals using ketamine and diazepam (Slip and Woods 1996), and glued the TDRs on to their backs, midway between the shoulders using Araldite K268 road marker adhesive (Ciba-Geigy). The package of TDR and epoxy weighed about 350 g. All animals went to sea within three hours of recovery from anaesthetic.

Calculating locations

I determined the daily location of each seal using temperature/depth profiles (see Hindell *et al.* 1991a). I compiled temperature/depth profiles for each seal for each day at sea, then compared these profiles to oceanographic data contained in Gordon and Molinelli (1982). These data were integrated over more than a decade. To determine the approximate geographic location of a seal on a particular day, I established the range of temperatures for depths of 0, 100, 200 and 500 m from that day's profile. I traced the region of the Southern Ocean where surface temperatures corresponded to the profile on to a chart, then placed the tracing over the chart of 100 m isotherms and traced the area which corresponded to the surface boundary and the 100 m profile. I then repeated this procedure for the 200 m and 500 m isotherms. This usually produced one to three discrete areas of the Southern Ocean. I assumed a maximum rate of travel of 180 kilometres per day, which is the maximum recorded for adult male Southern Elephant Seals (Slip *et al.* 1994), and eliminated areas outside this limit. As daily temperature/depth profiles were usually similar, I plotted only every tenth day.

Classifying dives

I classified dives according to the shape of the dive profile using a modification of the technique described by Wynen (1993). This technique involved two stages of analyses: a clustering technique was used to produce mathematically derived groups of dives (Belbin 1993), then discriminant function analysis to check the strength of the groupings and produce predictive equations for the groups.

In order to perform the clustering analysis I cut each dive profile into 20 depths relative to the maximum depth of the dive. I randomly selected approximately 1 000 dives from each seal and calculated association matrices using the Bray Curtis dissimilarity index. I then grouped the dives using an hierarchical agglomerative polythetic clustering technique with a homogeneity fusion strategy. The homogeneity of a group is defined as the average of all pair-wise associations between the relative depth measures making up the dive (Belbin 1993). I performed discriminant function analysis using SYSTAT (Wilkinson 1988) to produce predictive equations for each group. I checked the validity of the grouping by reclassifying the original random sample of dives with the derived equations. Every dive was then classified using the predictive equations and the groups plotted to examine the profiles of each of the groups. Groups of similar profiles were combined into a dive type. I classified the groups as foraging type dives or non-foraging type dives by the proportion of bottom time which was defined as the time spent at greater than 85% of the maximum depth of the dive.

I calculated the estimated aerobic dive limit (ADL) of each seal from the equation:

$$ADL = (\text{lean mass} \times TO_2) / RMR,$$

where TO_2 (total available oxygen) = 0.079 l O_2 /kg (Kooyman 1989), RMR (resting metabolic rate) = 0.0113 (lean mass$^{0.75}$) l O_2 /minute (Schmidt-Neilsen 1983), and lean body mass = 0.75 × total body mass (Slip *et al.* 1992).

RESULTS

Recovery of instruments

I recovered two TDRs, one from a male (#111) and one from a female (#113), when the seals returned to moult. Both animals were last sighted at Heard Island on August 12. The male returned on December 9, and the female on November 24, representing 119 and 104 days at sea respectively. The TDRs began recording on August 16, and recorded 76 days of continuous dive data for the male, and 66 days for the female made up of 47 days recording, followed by 12 days where no recording was made due to an electronic fault, then 19 days of recording.

The mass of the male increased from 168 to 192 kg in 119 days, and the female increased in mass from 163 to 188 kg in 104 days. This represents a rate of increase of mass of 0.20 and 0.24 kg day^{-1} for the male and female respectively.

Temperature-Depth profiles

I identified even distinct temperature-depth profiles which could be matched to the profiles of specific areas of the Southern Ocean (Gordon and

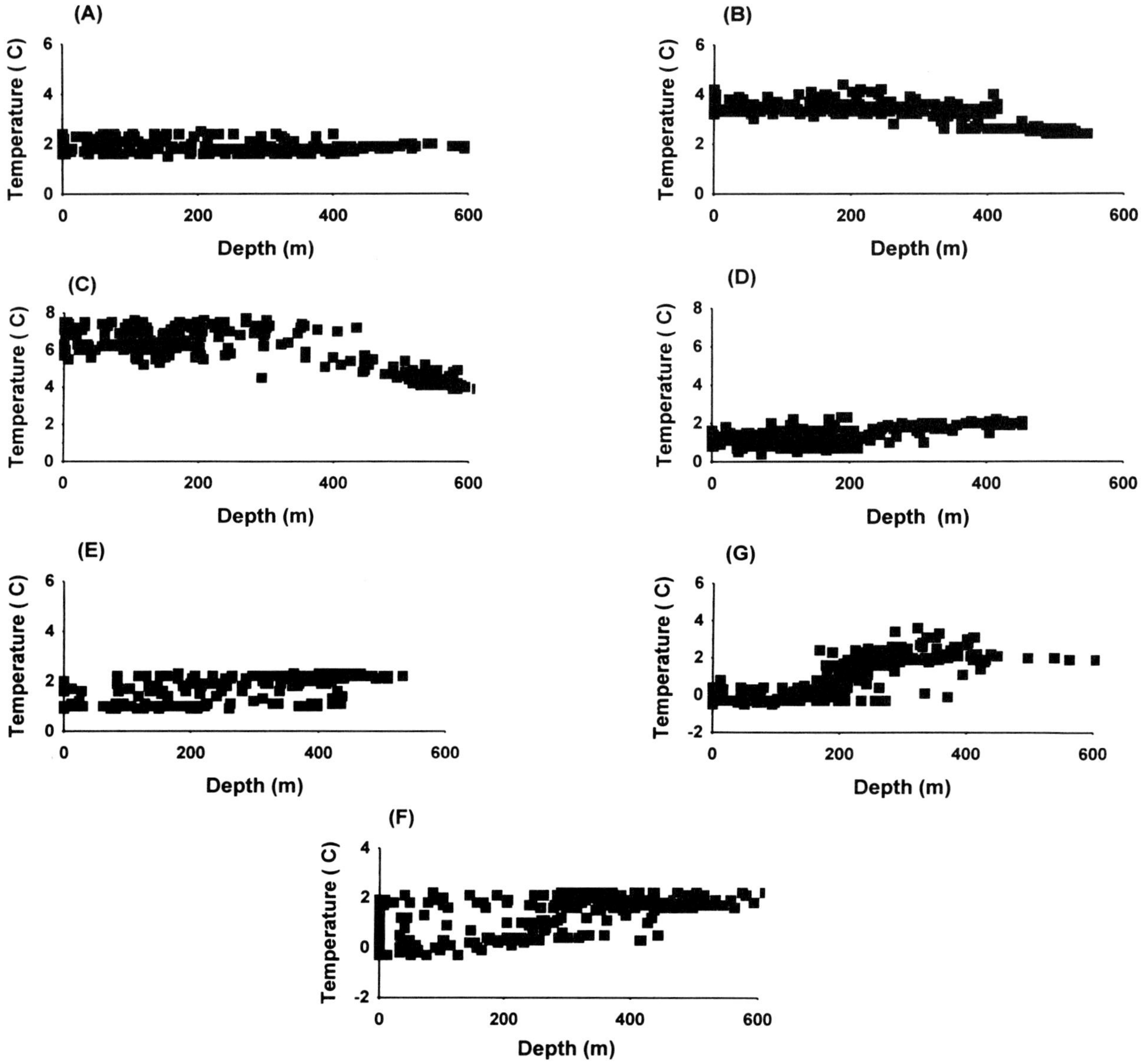

Fig. 1. Temperature/depth profiles from Southern Elephant Seals. (A) Juvenile male 87–111 julien day 230 (B) Juvenile male 87–111 julien day 240 (C) Juvenile male 87–111 julien day 270 (D) Juvenile male 87–111 julien day 300 (E) Juvenile female 87–113 julien day 230 (F) Juvenile female 87–113 julien day 240 (G) Juvenile female 87–113 julien day 290.

Molinelli 1982) for the male and three for the female:

Profile type A (Fig. 1a) was characterized by relatively homogeneous sea temperatures of about 2°C from the surface down to 500 m. This profile was consistent with an area of deep water east north-east of Heard Island (Fig. 2a).

Profile type B (Fig. 1b) was characterized by sea temperatures of between 3 and 4°C from the surface down to 300 m followed by temperatures of about 2.5°C down to 500 m. This profile was consistent with an area in deep water close to the Antarctic Polar Front (APF), north-east of Heard Island (Fig. 2b).

Profile type C (Fig. 1c) had sea temperatures of between 6 and 7.8°C from the surface to 300 m, followed by a gradual decline to about 4.5°C at 500 m. This profile was consistent with an area in deep water north of the APF (Fig. 2c).

Profile type D (Fig. 1d) had sea temperatures of between about 0.5 and 1.8°C from the surface down to 200 m after which temperatures rose slightly to about 2°C at 500 m. This was consistent with an area to the south-east of Heard Island at the edge of the Kerguelen Plateau (Fig. 2d).

Profile type E (Fig. 1e) had surface temperatures of between 1 and 2°C, followed by temperatures of between 1 and 2.2°C from 100 to 200 m, warming to about 2.2°C at 500 m. This profile was consistent with an area east of Heard Island at the edge of the Kerguelen Plateau (Fig. 3e).

Profile type F (Fig. 1f) had sea temperatures of between −0.8 and 2°C from the surface to 100 m, then temperatures of between 0.5 and 2°C at 200 m, rising to between 1.5 and 2°C at 500 m. This was consistent with an area east of Heard Island over seamounts near the edge of the Kerguelen Plateau (Fig. 3f).

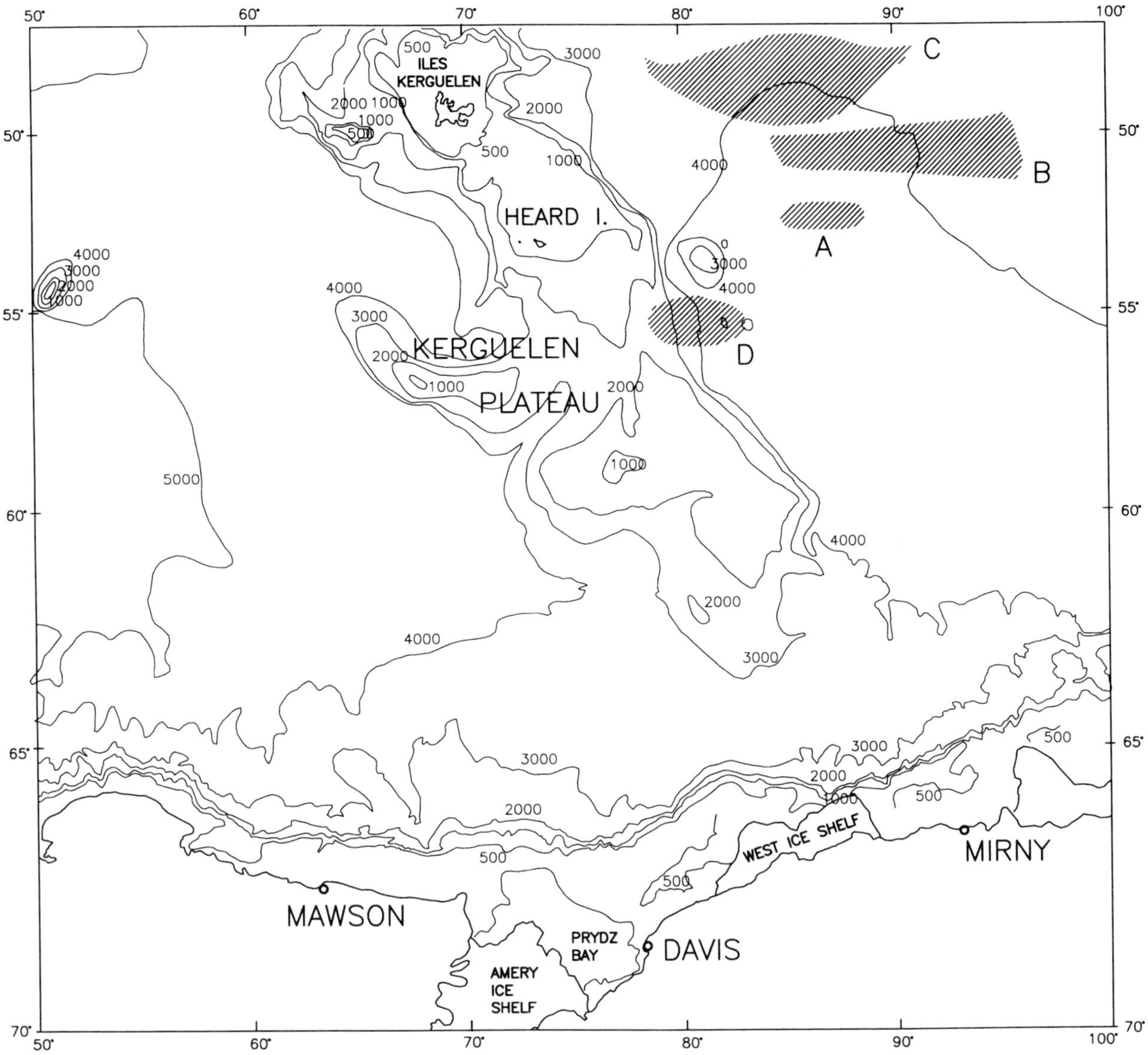

Fig. 2. Estimated location of juvenile male Southern Elephant Seal (87–111) for (A) julien day 230 (B) julien days 240 and 280 (C) julien days 250, 260 and 270 (D) julien days 290, 300 and 304.

Profile type G (Fig.1g) had sea temperatures of between −0.8 and 0.5°C from the surface down to 100 m, between −0.8 and 2°C at 200 m, and about 2°C at 500 m. This was consistent with an area south-west of Heard Island, over the edge of the Kerguelen Plateau (Fig. 3g).

Movements and foraging areas

The above temperature profiles suggest that the juvenile male initially moved east from Heard Island for 10 days then gradually moved north into warmer water for 40 days, before moving south for 10 days to the edge of the Kerguelen Plateau where it remained for 15 days until recording ceased (Fig. 2). The juvenile female moved initially to the east of Heard Island for the first 20 days of recording, after which it moved west to an area off the western edge of the Kerguelen Plateau and remained there for 55 days until recording ceased (Fig. 3).

Diving behaviour

Both animals dived continuously during the time spent at sea, and 6 652 and 4 566 dives were recorded for the male and female respectively, which represented mean diving rates of 3.5 and 2.9 dives per hour (Table 1). Mean dive depths were 323 ± 157 m (median 292, maximum 834 m) for the male and 416 ± 147 m (median 402, maximum 1 270 m) for the female (Table 1).

Classification of dive types

The clustering analysis was performed on dives greater than 10 m and resulted in 29 distinct groups of dives. The number of dives assigned to each of these groups from the discriminant functions was variable and some groups contained less

Table 1. Summary of the dive behaviour of a juvenile male and juvenile female Southern Elephant Seal: the mean and maximum dive depth, the mean and maximum dive duration, the number of dives recorded, the number of days for which dives were recorded, the number of dives per hour, the mean and maximum surface intervals (S.I.), and the number of extended surface intervals (ESIs) (surface intervals greater than 3 minutes) and the percentage of time spent underwater. Numbers in parentheses are standard deviations.

Seal	Days of recording	No. of Dives	Mean Depth (m)	Max. Depth (m)	Mean Duration (min)	Max. Duration (min)	Dives Hour^{-1}	Mean S.I. (min)	Max. S.I. (min)	No. of ESIs	% Time under-water
male	76	6 652	323 (157)	834	15.5 (5.1)	39	3.6	1.67 (2.54)	78.5	128 (1.9%)	90.2
female	67	4 566	416 (147)	1 270	21.1 (6.4)	58	2.6	1.63 (0.94)	30.5	73 (1.76%)	92.9

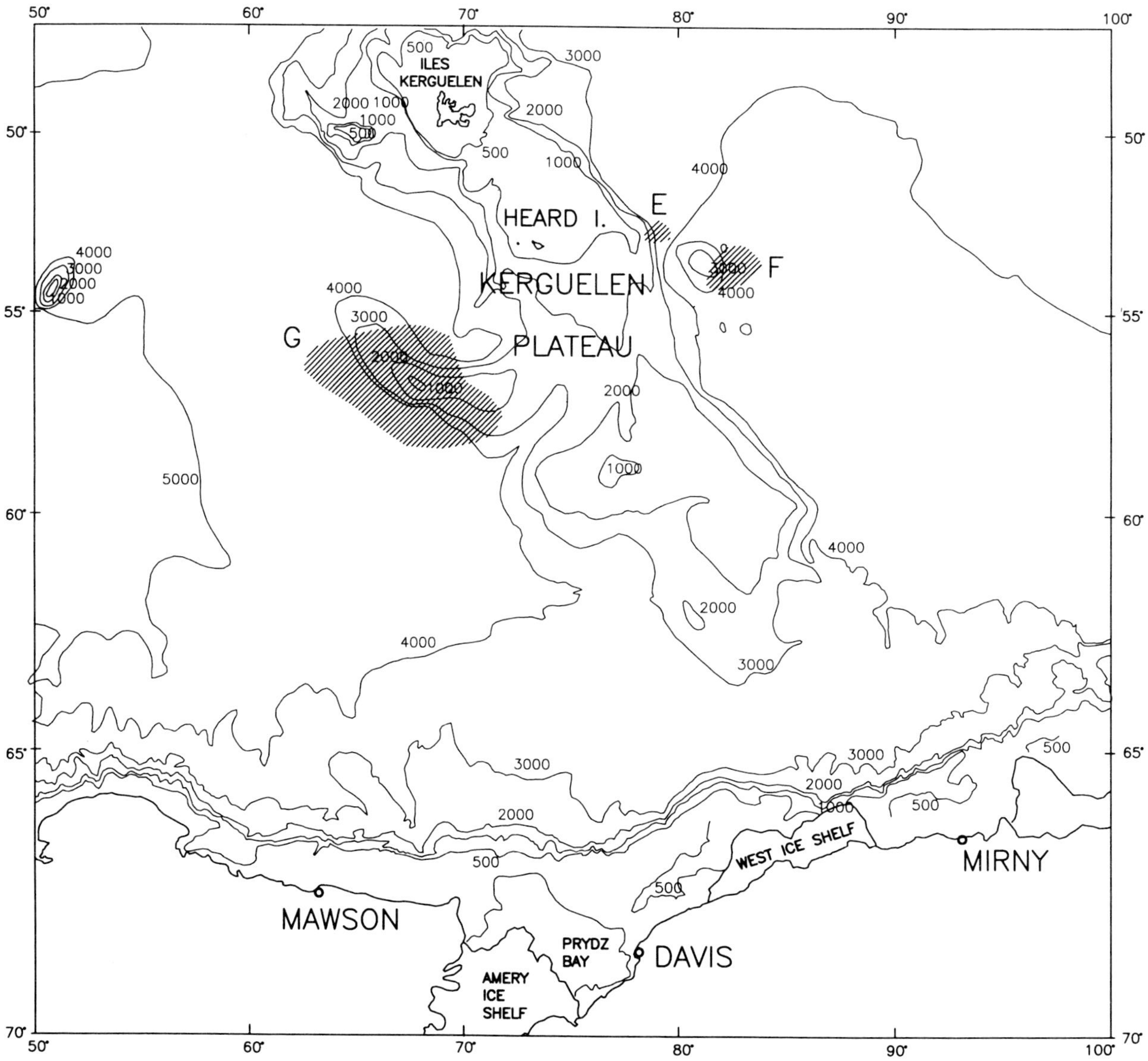

Fig. 3. Estimated location of juvenile female Southern Elephant Seal (87–113) for (E) julien day 230 (F) julien day 240 (G) julien days 250, 260, 270, 286, 290 and 300.

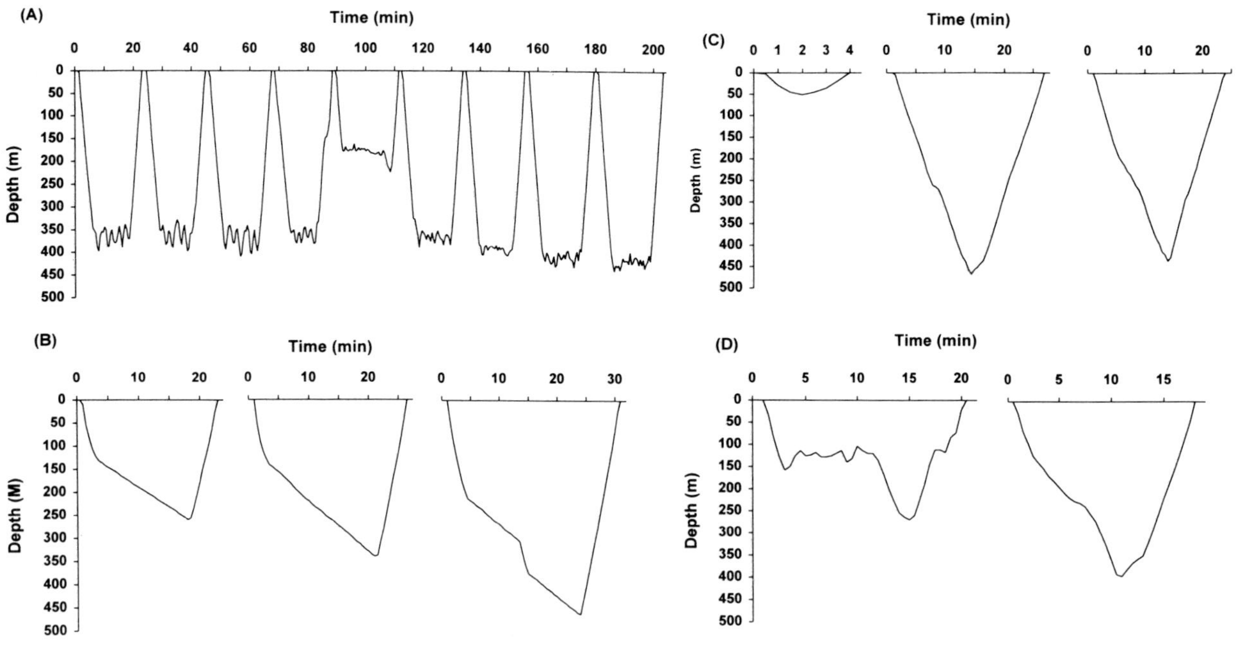

Fig. 4. Examples of dive profiles from two juvenile Southern Elephant Seals (A) Continous series of Type 1 pelagic foraging dives (B) Type 3 processing dives (C) Type 4 transit dives and (D) Type 6 composite dives.

than three dives. Eighteen groups represented pelagic foraging type profiles, analogous to type 1 dives of Hindell *et al.* (1991b) ot type D dives of Le Boeuf *et al.* (1992, 1993). These profiles are characterized by a high proportion of bottom time (>3.8%), and the presence of wiggles at the bottom of the dive (Fig. 4). Five of these groups were small, containing less than six dives, and the remaining 13 groups contained between 20 and 3 064 dives.

Eleven groups had profiles which were non-foraging dives having mean proportions of bottom time of between 5 and 21%. Three of these groups were analogous to the type 3 resting dive (Hindell *et al.* 1991b) or the type C processing dive (Le Boeuf *et al.* 1992; Crocker *et al.* 1994), and one group was analogous to the type 4 travel dive (Hindell *et al.* 1991b) or the type A and B transit dive (Le Boeuf *et al.* 1992; Crocker *et al.* 1994). The other seven non-foraging

Table 2. Percentage of total time at sea spent on each dive type. Also listed are the mean depth, duration and bottom time for each dive type, for each animal.

Seal			Foraging Dives (Type1)		
	N	%Time	Depth (m)	Duration (min)	Bottom time (min)
Male	5 487	76.5	328 ± 161	15.5 ± 5.0	5.9 ± 3.8
Female	3 766	84.7	415 ± 128	21.3 ± 5.4	10.2 ± 3.7
			Non-foraging Dives (Type 3)		
	N	%Time	Depth (m)	Duration (min)	Bottom time (min)
Male	379	6.0	272 ± 95	17.5 ± 5.5	2.0 ± 1.0
Female	103	2.8	343 ± 157	25.4 ± 8.8	5.3 ± 2.3
			Non-foraging Dives (Type 4)		
	N	%Time	Depth (m)	Duration (min)	Bottom time (min)
Male	255	2.8	374 ± 164	12.4 ± 4.8	1.4 ± 1.0
Female	160	2.8	469 ± 326	16.6 ± 10.7	2.9 ± 2.2
			Non-foraging Dives (Type 6)		
	N	%Time	Depth (m)	Duration (min)	Bottom time (min)
Male	370	4.9	280 ± 124	14.8 ± 5.1	1.9 ± 1.6
Female	128	2.7	424 ± 250	20.1 ± 9.8	3.9 ± 2.8

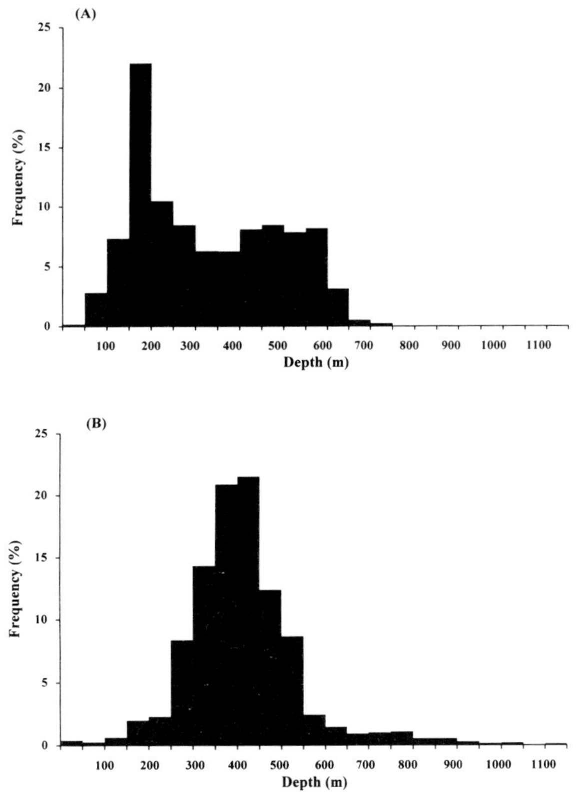

Fig. 5. Frequency distribution (%) of depth of type 1 foraging dives for (A) juvenile male, and (B) juvenile female Southern Elephant Seals.

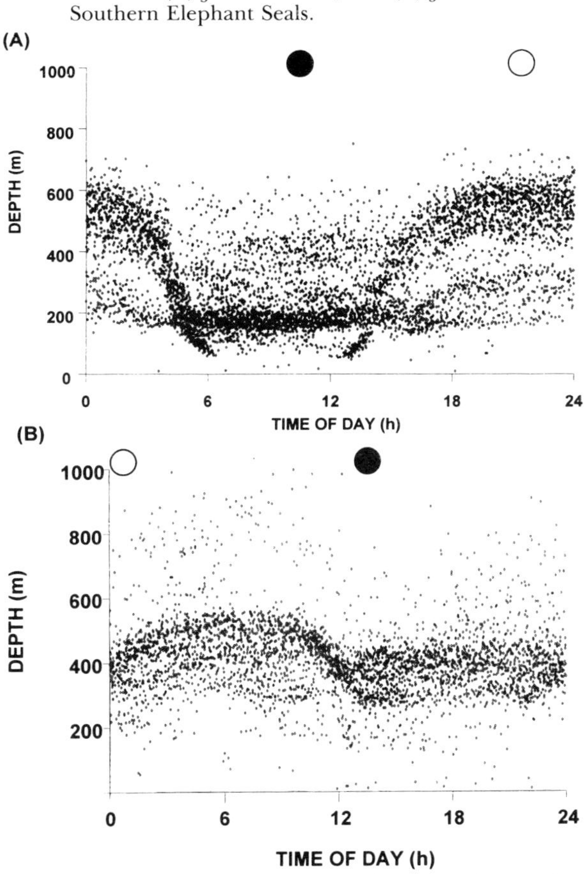

Fig. 6. Maximum depth of foraging dives and time of day for (A) juvenile male and (B) juvenile female Southern Elephant Seals. Time of day is universal time, open circle represents average sunrise, and closed circles represent average sunset over the recording interval.

groups are similar to the type 6 composite dives (Hindell *et al.* 1991b), and four of these groups contained fewer than three dives.

Foraging dives

Type 1 foraging dives accounted for 76.5 and 84.7% of total time at sea for the male and female respectively (Table 2). Mean depth of foraging dives was 328 ± 161 m (median 294) for the male, and 415 ± 128 m (median 402) for the female. More than 90% of foraging dives occurred at depths of between 150 and 600 m (Fig. 5).

Non-foraging dives

Mean depth of type 3 processing dives was 272 ± 95 m for the male and 343 ± 157 m for the female. These dives accounted for 5.7 and 2.8% of total time at sea for the male and female respectively, and over 90% of type 3 dives were between 150 and 500 m in depth. Type 4 travel dives accounted for 2.8% of total time at sea for both the male and female. Mean depth of type 4 dives was 374 ± 164 and 469 ± 326 m for the male and female respectively, and over 90% of these dives occurred between 50 and 900 m. Dives of less than 10 m were classed as type 5 dives and were rare, accounting for 1.8% of the male's and 0.2% of the female's total time at sea. Type 6 dives accounted for 4.8 and 2.7% of total time at sea, with mean depths of 280 ± 124 m and 424 ± 250 m for the male and female respectively. Most type 6 dives were between 100 and 800 m.

Diurnal patterns of dive depth

The female tended to dive deeper in daylight hours and shallower at night or just before sunset (Fig. 6). This pattern is not obvious in the juvenile male, although the deepest mean depths occurred in daylight hours, but the shallowest dives also occured in daylight.

Surface intervals

The mean inter-dive surface intervals were 1.67 ± 2.54 minutes for the male and 1.63 ± 0.94 for the female. Both animals spent relatively little time on the surface with 9.8% and 7.1% of total time at sea spent on the surface for the male and female respectively. Extended surface intervals (ESIs), defined as surface intervals of greater than 3 minutes duration, occurred infrequently and accounted for less than 2% of all surface intervals for both animals (Table 1). The maximum ESI was 78.5 min for the male and 30.5 min for the female. Most ESIs occurred at or just after sunset, although they also occurred at other times. ESIs were not related to duration of the preceeding dive.

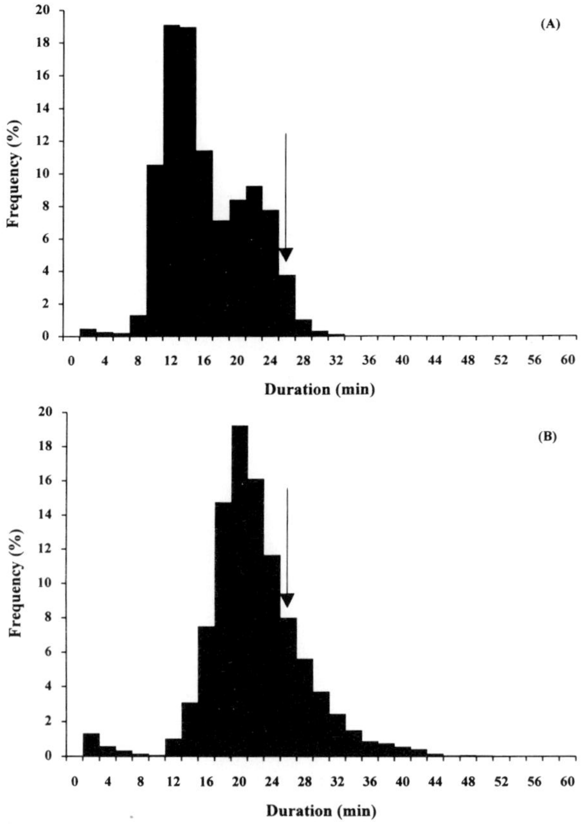

Fig. 7. Per cent frequency distribution of dive duration for (A) juvenile male, and (B) juvenile female Southern Elephant Seals. The arrows mark the aerobic dive limit for each animal.

Dive duration

Mean dive durations were 15.5 ± 5.1 min (maximum 39 min) and 21.1 ± 6.4 min (maximum 58 min) for the male and female respectively. The estimated aerobic dive limit (ADL) was approximately 24 minutes for both animals. The male exceeded the estimated ADL on 5% of dives while the female exceeded it on 22.6% of dives (Fig. 7).

DISCUSSION

Movements and foraging areas

The accuracy of the estimated locations of the foraging areas derived by this method is dependent on the precision of the oceanographic data upon which it is based. These data were compiled over several years during the 1970s (Gordon and Molinelli 1982). However, limited comparisons of recent sea surface temperatures collected by satellite, and temperature-depth profiles collected from adult female Southern Elephant Seals (unpubl. data), whose locations were determined from light intensity curves (Hill 1994), showed little variation from the derived areas. In addition, sea surface temperatures at Heard Island in 1992 (Green 1993) were consistent with Gordon and Molinelli (1982).

By the time juvenile Southern Elephant Seals go to sea for the second time (i.e., at 9 to 11 months of age), they are capable of travelling long distances. Both animals in this study moved considerable distances away from Heard Island (e.g., over 900 km), although they appeared to concentrate foraging activities on relatively small areas, close to the edge of the Kerguelen Plateau. These movements are similar to those of adult females in the post-breeding period who concentrate their foraging on the shelf area of the Kerguelen Plateau as well as in mid ocean areas just off the Kerguelen Plateau (Slip *et al.* 1995). Their behaviour in this regard appears similar to juvenile Northern Elephant Seals: there are no published data available for Northern Elephant Seals on their second trip to sea, however, on their third trip to sea at 1.4 years of age, they migrate along the same general route to distances comparable to adult females, although travelling at significantly slower rates (Le Boeuf *et al.* 1993).

The way in which juvenile seals discover their foraging areas is open to conjecture. It could be that animals which have survived and fed successfully during their first period at sea return to the same area of ocean. The first migrations of young elephant seals may be initially random, as satellite tracks of three juvenile Southern Elephant Seals from South Georgia on their first trips to sea suggests (Fedak, unpubl. data). Once concentrations of food are located foraging behaviour may be learned with increasing levels of sophistication, and a strategy of returning to an area of known food may be a successful one. Longitudinal studies following the same individuals over several foraging trips are necessary to confirm these conjectures.

Prey of juvenile Elephant Seals

Juvenile Southern Elephant Seals appear to spend little time foraging in the euphotic zone (<200 m), with most foraging occurring in the mesopelagic zone (Fig. 5). Most of the prey identified from the stomachs of juvenile Southern Elephant Seals at Heard Island were muscular mesopelagic squid, and although fish made up only a small proportion of prey, two species of mesopelagic myctophid fish were also identified (Slip 1995). Also, fewer species of Antarctic cephalopods have been found in the diet of juveniles than in the diet of adults (Slip 1995), which is consistent with juveniles foraging to the north of Heard Island, in areas close to the Antarctic Polar Front. Nonetheless, the presence of some Antarctic species in the juvenile diet suggests that waters south of Heard Island are also important foraging areas, consistent with the locations determined from water temperature profiles.

Although dietary studies from elephant seals at their haulout sites have shown a predominance of cephalopods with only a few fish remains, it has not yet been possible to sample diet in the seals' foraging areas, and the relative importance of fish in the diet is largely unknown. Stewart and

DeLong (1993) have suggested that mesopelagic fish may be important prey of the Northen Elephant Seal in areas where mesopelagic fish and squid distributions correlate closely. It seems likely that myctophid fish may be important in the diet of juveniles as they occur in large numbers and in dense concentrations in certain areas (Filin *et al.* 1990), and in spring are found concentrated in the mesopelagic zone at depths of between 200 and 600 m (Kozlov *et al.* 1990; Sabourenkov 1990), which corresponds to the depths where most foraging dives occurred (Fig. 5). In addition, the squid *Moroteuthis knipovitchi*, which represented 7% of the cephalopod biomass in the diet of juvenile Elephant Seals at Heard Island (Slip 1995), feeds on myctophid fish ein the mesopelagic layer (Nemoto *et al.* 1985). Thus, myctophids are likely to be an important potential prey of juvenile Southern Elephant Seals, and a high energy one at that, as their energy density is about twice that of muscular squid species, and about four times that of gelatinous squid species (Boyd *et al.* 1994).

Diving behaviour

At the simplest level, the diving performance of the juvenile Southern Elephant Seals was similar to, and in some cases exceeded, that of adult female Southern Elephant Seals in terms of depth and duration (Slip *et al.* 1994), and exceeded that of most other adult pinnipeds (Kooyman 1989). For example, five adult females from Heard Island during the post-breeding summer period at sea had mean maximum dive depths of between 202 ± 121 and 468 ± 196 m, and mean dive durations of between 14.8 ± 6.7 and 19.3 ± 6.17 minutes (Slip *et al.* 1995). This contrasts with the Northern Elephant Seal where the adult pattern of diving is essentially set by the end of the fourth trip to sea when juveniles are two years old with average mass of 260 ± 26 kg (Le Boeuf 1994). It is difficult to make general comparisons with such small sample size, and these differences may simply reflect the wide range of individual differences.

The dive types of these seals were similar to those described in adult Southern and Northern Elephant Seals (Hindell *et al.* 1991b; Le Boeuf *et al.* 1992). The proportion of time at sea spent on each dive type is also similar to that found in adult female Southen Elephant Seals (Hindell *et al.* 1991b; Slip *et al.* 1994). Profiles consistent with benthic foraging, which are common in adult male Southern Elephant Seals (Hindell *et al.* 1991b) were not found in these juveniles. This was expected given that the estimated locations of these animals did not include areas of water shallow enough for these animals to forage on the ocean floor.

The diurnal pattern of dive depth was obscured by two factors: (i) data were averaged over a recording interval where day length changed from between about 11 to 14 hours (Fig. 6), and (ii) the seals moved with respect to longitude, making it impossible to determine exact sunrise and sunset times for the time they were at sea. Nonetheless, by using best approximations for sunrise and sunset there was a tendency to dive deeper in the daylight hours than at night, in the female but not in the male. In both Northern and Southern Elephant Seals, adult females show a similar diurnal pattern of feeding, which corresponds to what is known of the vertical diel migration of their prey, but males generally do not (Hindell *et al.* 1991b; Le Boeuf *et al.* 1993). In addition, juvenile Northern Elephant Seals on their third trip to sea at 1.4 years of age showed sex differences in diurnal pattern of dive depth which was present in females but not in males, which led Le Boeuf *et al.* (1993) to suggest that the different foraging strategies exhibited by males may be important for attaining as well as maintaining large size. Thus, the differences in diel patterns and mean depth and duration of foraging dives between the two juveniles in this study may be due to the early appearance of sex differences in foraging strategies.

The dive pattern of these seals suggests that they are diving aerobically, as surface intervals are generally short and extended surface intervals are rare and not related to the duration of the previous dive. However, my calculation of ADL, assuming available oxygen to be 7.9 ml O_2 kg^{-1} and dive metabolic rate to be the estimated basal metabolic rate (BMR), suggested 5% of dives of the juvenile male and 22.6% of dives of the juvenile female exceeded ADL (Fig. 7). The two factors which determine ADL are oxygen capacity and metabolic rate (Kooyman 1989), and metabolic rate is strongly correlated with swim velocity (Ponganis *et al.* 1990; Williams *et al.* 1991). Crocker *et al.* (1994) suggested that Northern Elephant Seals reduce swim speed during some dives in order to conserve oxygen. Therefore, if swim velocities during long duration dives are relatively low as has been reported in juvenile and adult Northern Elephant Seals (Le Boeuf *et al.* 1992; Thorson and Le Boeuf 1994), diving metabolic rate might also be relatively low and ADL increased. Thorson and Le Boeuf (1994) suggested that juvenile Northern Elephant Seals during their first trip to sea could reduce diving metabolic rate to 0.8 times BMR, resulting in an increase in ADL such that less than 5% of dives exceeded ADL. If juvenile Southern Elephant Seals also reduce diving metabolic rate by behavioural or physiological means to 0.8 times BMR, then all the dives of the juvenile male and less than 1% of dives of the juvenile female exceeded the ADL.

Although it is impossible to determine a general model of foraging ecology of juvenile seals from such a small number of animals, this study has opened a window into the foraging and dive behavior of juvenile Southern Elephant Seals, and showed that Southern Elephant Seals are competent divers by their second trip to sea, capable of covering large distances, and diving to depths and durations comparable to adult animals which are four times their size. Longitudinal studies which follow the dive behaviour of juvenile seals for several years from their first trip to sea are necessary to determine how Elephant Seals locate their feeding grounds on their first trip to sea, or whether the sex differences seen in the foraging behaviour of adults also occur in juveniles.

ACKNOWLEDGEMENTS

This work was undertaken with the approval of the Antarctic Animal Care and Ionising Radiation Usage Ethics Committee (Department of the Environment, Sport, and Territories). Erwin Erb, Ken Green, Geoff Moore and Attila Vrana assisted in the field. Mark Hindell provided statistical advice.

REFERENCES

Belbin, L., 1993. PATN Pattern analysis package. Division of Wildlife Ecology, C.S.I.R.O.: Australia.

Boyd, I. L. and Arnbom, T., 1991. Diving behaviour in relation to water temperature in the Southern Elephant Seal: foraging implications. *Polar Biol.* **11:** 259–66.

Boyd, I. L., Arnbom, T. and Fedak, M. A., 1994. Biomass and energy consumption of the South Georgia population of Southern Elephant Seals. Pp. 98–117 *in* Elephant seals: population ecology, behavior, and physiology ed by B. J. Le Boeuf and R. M. Laws. University of California Press: Berkeley.

Crocker, D. E., Le Boeuf, B. J., Naito, Y., Asaga, T. and Costa, D., 1994. Swim speed and dive function in a female northern elephant seal. Pp. 328–42 *in* Elephant seals: population ecology, behavior, and physiology ed by B. J. Le Boeuf and R. M. Laws. University of California Press: Berkeley.

Filin, A. A., Gorchinsky, K. V. and Kizeleva, V. M., 1990. Biomass of myctophids in the Atlantic sector of the Southern Ocean as estimated by acoustic surveys. Selected Scientific Papers, 1990 (SC-CAMLR-SSP/7). Hobart Australia, CCAMLR. Pp. 417–31.

Gordon, A. L. and Molinelli, E. J., 1982. Southern Ocean Atlas. Columbia University Press: Surrey.

Green, K., 1993. Sea surface temperature. Pp. 197–99 *in* Heard Island 1992 ANARE Report ed by K. Green. Australian Antarctic Division, Channel Highway Kingston, Tasmania. Unpublished report.

Green, K. and Burton, H. R., 1993. Comparison of the stomach contents of Southern Elephant Seals, *Mirounga leonina*, at Macquarie and Heard Islands. *Mar. Mamm. Sci.* **9:** 10–22.

Hill, R. D., 1994. Theory of geolocation by light levels. Pp. 227–36 *in* Elephant seals: population ecology, behaviour, and physiology ed by B. J. Le Boeuf and R. M. Laws. University of California Press: Berkeley.

Hindell, M. A., 1991. Some life-history parameters of a declining population of Southern Elephant Seals, *Mirounga leonina. J. Animal Ecol.* **60:** 119–34.

Hindell, M. A. and Burton, H. R., 1988. Seasonal haul-out patterns of the Southern Elephant Seal *(Mirounga leonina)* at Macquarie Island. *J. Mamm.* **69:** 81–88.

Hindell, M. A., Burton, H. R. and Slip, D. J., 1991a. Foraging areas of Southern Elephant Seals, *Mirounga leonina,* as inferred from water temperature data. *Aust. J. Mar. Freshwater Res.* **42:** 115–28.

Hindell, M. A., Slip, D. J. and Burton, H. R., 1991b. The diving behaviour of adult male and female Southern Elephant Seals, *Mirounga leonina. Aust. J. Zool.* **39:** 595–619.

Hindell, M. A., Slip, D. J. and Burton, H. R., 1994. Possible causes of the decline of Southern Elephant Seal populations in the southern Pacific and southern Indian Oceans. Pp. 66–84 *in* Elephant seals: population ecology, behavior, and physiology ed by B. J. Le Boeuf and R. M. Laws. University of California Press: Berkeley.

Kooyman, G. L., 1989. Diverse divers — physiology and behavior. Berlin: Springer-Verlag. 200 Pp.

Kozlov, A. N., Shust, K. V. and Zemsky, A. V., 1990. Seasonal and inter-annual variability in the distribution of *Electrona carlsbergi* in the Southern polar front area (the area north of South Georgia is used as an example). Selected Scientific Papers, 1990 (SC-CAMLR-SSP/7). Hobart Australia, CCAMLR. Pp. 337–67.

Laws, R. M., 1994. History and present status of Southern Elephant Seal populations. Pp. 49–65 *in* Elephant seals: population ecology, behavior, and physiology ed by B. J. Le Boeuf and R. M. Laws. University of California Press: Berkeley.

Le Boeuf, B. J., 1994. Variation in the diving pattern of northern elephant seals with age, mass, sex, and reproductive condition. Pp. 237–52 *in* Elephant seals: population ecology, behavior, and physiology ed by B. J. Le Boeuf and R. M. Laws. University of California Press: Berkeley.

Le Boeuf, B. J., Crocker, D. E., Blackwell, S. B., Morris, P. A. and Thorson, P. H., 1993. Sex differences in diving and foraging behaviour of northern elephant seals. *Symp. zool. Soc. Lond.* **66:** 149–78.

Le Boeuf, B. J., Naito, Y., Asaga, T., Crocker, D. E. and Costa, D. P., 1992. Swim speed in a female elephant seal: Metabolic and foraging implications. *Can. J. Zool.* **70:** 786–95.

McConnell, B. J., Chambers, C. and Fedak, M. A., 1992. Foraging ecology of Southern Elephant Seals in relation to bathymetry and productivity of the Southern Ocean. *Antarct. Sci.* **4:** 393–98.

Nemoto, T., Okiyama, M. and Takahashi, M., 1985. Aspects of the roles of squid in food chains of marine Antarctic ecosystems. Pp. 415–20 *in* Antarctic Nutrient Cycles and Food Webs ed by W. R. Siegfried, P. R. Condy and R. M. Laws. Springer-Verlag: Berlin.

Ponganis, P. J., Kooyman, G. L., Zornow, M. H., Castellini, M. A. and Croll, D. A., 1990. Cardiac output and stroke volume in swimming harbour seals. *J. Comp. Physiol.* **160B:** 473–82.

Rodhouse, P. G., Arnbom, T. R., Fedak, M. A., Yeatman, J. and Murray, W. A., 1992. Cephalopod prey of the Southern Elephant Seal, *Mirounga leonina* L. *Can. J. Zool.* **70:** 1007–015

Sabourenkov, E. N., 1990. Mesopelagic fish of the Southern Ocean — summary results of recent Soviet studies. Selected Scientific Papers, 1990 (SC-CAMLR-SSP/7). Hobart Australia, CCAMLR. Pp. 433–57.

Schmidt-Neilsen, K., 1983. Animal physiology: adaptation and environment. Cambridge University Press: Cambridge.

Slip, D. J., 1995. The diet of Southern Elephant Seals *(Mirounga leonina)* from Heard Island. *Can. J. Zool.* **75:** 1519–28

Slip, D. J., Burton, H. R. and Gales, N. J., 1992. Determining blubber mass in the Southern Elephant Seal, *Mirounga leonina,* by ultrasonic and isotopic techniques. *Aust. J. Zool.* **40:** 143–52.

Slip, D. J., Hindell, M. A. and Burton, H. R., 1994. Diving behavior of Southern Elephant Seals from Macquarie Island. Pp. 237–52 *in* Elephant seals: population ecology, behavior and physiology ed by B. J. Le Boeuf and R. M. Laws. University of California Press: Berkeley.

Slip, D. J. and Woods, R., in press. Intravenous ketamine and diazepam for immobilization of Southern Elephant Seals. *J Wildl. Manag.*

Stewart, B. S. and DeLong, R. L., 1993. Seasonal dispersion and habitat use of foraging northern elephant seals. *Symp. zool. Soc. Lond.* **66:** 179–94.

Thorson, P. H. and Le Boeuf, B. J., 1994. Developmental aspects of diving in northen elephant seal pups. Pp. 271–89 *in* Elephant seals: population ecology, behavior and physiology ed by B. J. Le Boeuf and R. M. Laws. University of California Press: Berkeley.

Wilkinson, L., 1988. SYSTAT: the system for statistics. SYSTAT, Inc: Evanstown, IL.

Williams, T. M., Kooyman, G. L. and Croll, D. A., 1991. The effect of submergence on heart rate and oxygen consumption of swimming seals and sea lions. *J. Comp. Physiol.* **160B:** 637–44.

Wynen, L., 1993. A multivariate approach to the classification of dive profiles of the Southern Elephant Seal (*Mirounga leonina* L.). Unpublished Hons. thesis Institute of Antarctic and Sothern Ocean Studies, University of Tasmania.

Analysis of the diet of New Zealand Fur Seals *Arctocephalus forsteri* in Tasmania

S. LAKE[1]

The diet of New Zealand Fur Seals *Arctocephalus forsteri* is described using faecal and regurgitate samples collected from southern Tasmania in 1992 and 1993. Fish was the dominant prey type and occurred in 76 ± 12% of samples. Common fish taxa were Jack Mackerel *Trachurus declivis*, Red Bait *Emmelichthys nitidus*, which are harvested for the local commercial fishery, and Myctophidae, which have no commercial value. Breeding and non-breeding seals consumed different fish species suggesting that they utilize different foraging areas. Feeding trials were conducted to quantify both the proportion of remains excreted by New Zealand Fur Seals and the accuracy of estimating prey size from remains obtained from scats. Between 40 and 100% of fish otoliths and cephalopod beaks were excreted. The proportion of otoliths excreted varied between fish species. Estimates of prey size based on otoliths in scats were within 5% of the actual size.

INTRODUCTION

THE diet of New Zealand Fur Seals has been described for New Zealand and Macquarie Island populations, but not for any Australian populations. In Australia, New Zealand Fur Seals breed in Western Australia, South Australia and Tasmania (Shaughnessy *et al.* 1994). In Tasmania, the sole breeding site is Maatsuyker Island (Brothers and Pemberton 1990) where *ca.* 60 pups are born in December each year (Kirkwood *et al.* 1991). Maatsuyker Island (43°38′S, 146°17′E) is 10 km from the south coast of Tasmania and 30–40 km north of the continental shelf break. On the island New Zealand Fur Seals inhabit beaches of jumbled boulders, while Australian Fur Seals *A. pusillus doriferus* which haulout but do not breed there, inhabit flat rock coastline (Brothers and Pemberton 1990).

The prey of New Zealand Fur Seals elsewhere in their range consists of locally abundant pelagic fish, cephalopods and birds (Street 1964; Tate 1981; Green *et al.* 1989; Carey 1992). The New Zealand Fur Seals consume more birds than many otariid species (Green *et al.* 1989) indicating that their foraging behaviour can be relatively predatory (Reidman 1990).

Assessing diet of seals using faecal and regurgitate samples is quick and un-invasive. Fish otoliths (sagittal ear bones) and cephalopod beaks (chitinous mouthparts) are the principal remains used to identify prey. Fish otoliths are excreted in faeces. Cephalopod beaks, due to large size and constraints of the pyloric sphincter, are excreted by regurgitation (Harvey 1989). However, the results from various feeding trials have cast doubt on the accuracy of these methods. For example, few identifiable prey remains were excreted by Australian Sea Lions *Neophoca cinerea*, and it was concluded that faecal analysis is unreliable for that species (Gales and Cheal 1992).

This study therefore had two aims; (i) to assess the reliability of identifying prey species and size from scats in the Australian Fur Seal and (ii) to use scat analysis to describe the diet for New Zealand Fur Seals from Maatsuyker Island.

METHODS

Feeding trials

Experimental design

The experiment involved four mature New Zealand Fur Seals in apparent good health and well established in captivity (Table 1). The seals were housed individually in enclosures with a pool (20 000–40 000 l) and a dry area (2–3 m^2) and fed three experimental diets and a maintenance diet. Prey items in experimental diets were counted and measured (caudal fork and mantle length). The maintenance diet consisted of three fish species not used in the feeding trials and squid with beaks removed. Experimental diets were (i) Squid *Nototodarus gouldi,* (ii) Jack Mackerel or Red Bait, and (iii) Macrouidae.

After a week of maintenance diet, the seals were fed squid for four days followed by four days of maintenance diet (Table 2). The seals were then fed Jack Mackerel or Red Bait for nine days. On day 19, the seals were given the Macrouidae for a single day to compare individual recovery rates. Only three of the seals would eat this species.

Sample collection

Scats (regurgitates and faeces) were collected by draining the pools and hosing the dry areas. The water outlets in the enclosures were covered with mesh (0.3 mm) to prevent loss of diagnostic remains. Samples were collected at least every second day.

Recovery of excreted matter was tested by seeding enclosures with distinct otoliths (as in Harvey 1989). Twenty otoliths were thrown into

[1]Department of Zoology, University of Tasmania, P.O. Box 252C, Hobart, Tasmania, Australia 7001.
Pages 125–29 *in* MARINE MAMMAL RESEARCH IN THE SOUTHERN HEMISPHERE VOLUME 1: STATUS, ECOLOGY AND MEDICINE ed by M. Hindell and C. Kemper. Surrey Beatty & Sons. 1997.

Table 1. Description of New Zealand Fur Seals in feeding trials, Taronga Park Zoo.

	Seal 1	Seal 2	Seal 3	Seal 4
sex	male	male	male	female
age (years)	19	8	12	8
years in captivity	17	7	6	7
mass (kg)	120	72	100	42
consumption (kg/day)	8	6	8	4

Table 2. Experimental design of feeding trials. Maintenance diet was three fish species and squid with beaks removed. Samples were collected from day 0 to day 28.

Diet	Days fed	No. seals	No. prey
maintenance	−7–0	4	—
cephalopod	1–4	2	265
maintenance	5–8	4	—
Jack Mackerel	9–18	2	434
Red Bait	9–18	2	455
Macrouidae	19	3	85
maintenance	20–28	4	—

each enclosure three times during the experiment. The number of seeded otoliths collected from each enclosure was counted.

Diet at Maatsuyker Island

Sample collection and processing

Seven breeding and two non-breeding colonies on Maatsuyker Island were sampled; these were all areas not inhabited by Australian Fur Seals. At least 15 samples of faeces and all regurgitate samples were collected from as many sites as possible on four visits to the island between November 1992 and March 1993.

Individual faecal and regurgitate samples were stored frozen in plastic bags, then rinsed through a sieve (mesh size 0.3 mm). Otoliths, beaks and feathers were extracted and stored dry. The presence of gelatinous zooplankton and crustaceans was noted. Otoliths were identified using reference collections (Hecht 1987; Gales and Pemberton, unpubl.; Bulman, unpubl.).

Maximum length and width of uneroded otoliths were measured on a video-linked stereo microscope. Only right otoliths were used to prevent estimating the size of individual fish twice. Regressions of otolith length to fish length for Jack Mackerel and Red Bait were available from the Tasmanian Department of Sea Fisheries.

Analysis of results

The diet of the seals was analysed by percentage dominance (i.e., the proportion of samples in which each otolith type is numerically dominant (Pierce and Boyle 1991)) and by frequency of occurrence, total number, and mean number of otoliths per sample.

The Mantel test, a multivariate, non-parametric test, was used to examine differences between seasons. To avoid bias due to different sample sizes, data sets were randomly sub-sampled to give equal sample sizes.

RESULTS

Feeding trials

Recovery of Jack Mackerel and Red Bait otoliths

Over 95% of otoliths seeded in the enclosures were collected, the majority the first time the enclosure was cleaned. Occasionally, fewer than 5% of otoliths were collected in the second cleaning.

Recovery rate of Jack Mackerel otoliths from the faeces was 59 ± 16% (two animals, $n = 434$ fish), and that of Red Bait otoliths was 61 ± 12% (two animals, $n = 455$ fish). However, of these otoliths, only 50 ± 6% of Jack Mackerel otoliths, and 39 ± 8% of Red Bait otoliths, were identifiable to species, indicating a degree of erosion.

Length of Jack Mackerel and Red Bait

The mean length of Red Bait estimated from recovered otoliths was 19 ± 2 cm ($n = 198$), which was a significant underestimate of 5%, or 1.4 cm (independent T test, $T_{555} = 8.92$, $P < 0.001$). However, the estimated length of Jack Mackerel (25 ± 4 cm, $n = 184$) was not significantly different from the size of fish fed to the seals (independent T test $T_{224} = 2.48$, $P = 0.804$).

Recovery of other prey remains

The absolute recovery rate of Macrouidae otoliths was 96 ± 2% (three animals, $n = 85$), significantly higher than the absolute recovery rates of Jack Mackerel and Red Bait otoliths (*ANOVA*, $P = 0.03$ Jack Mackerel; $P = 0.01$ Red Bait).

Recovery rate of cephalopod beaks was 93 ± 10% (two animals, $n = 265$). This was also significantly higher than recovery rates of Jack Mackerel and Red Bait otoliths (*ANOVA*, $P = 0.04$ Jack Mackerel; $P = 0.05$ Red Bait).

Approximately 4% of otoliths in samples were from fish species that seals had not consumed in captivity. These otoliths were relatively small and are presumed to have been contained in the digestive tracts of fish and squid fed to the seals and thus secondarily consumed.

Diet of free-ranging animals

Fish otoliths occurred in 76 ± 12% of samples. Gelatinous zooplankton also occurred frequently in samples, especially in January (54% of samples). Cephalopod, crustacean and bird remains had frequencies of occurrence of less than 7% (Table 3), and are not considered further.

Table 3. Prey types in New Zealand Fur Seal diet (frequency of occurence of prey types as a percentage of total number of samples (n = 279)). Remains considered were: otoliths for fish; beaks for cephalopods; tissue for gelatinous plankton; exoskeleton for crustacean; and feathers for birds.

	November	December	January	March
fish	87	87	65	66
gelatinous plankton	13	4	54	1
cephalopods	0	0	1	0
birds	0	7	7	3

Fish species

Only three fish taxa had percentage dominance of greater than 7% (Table 4). These were Red Bait (22%), Jack Mackerel (21%) and Myctophidae (18%). Two species of myctophids were identified. The most abundant species was *Symbolophorus barnardi* and the other *Gymnoscopelus piablis*. The mean length of Jack Mackerel was 31 ± 5 cm (n = 145). The mean length of Red Bait was 16 ± 4 cm (n = 315).

Comparison of diets

The fish diet of breeding and non-breeding seals was significantly *different* (*Mantel test P* = 0.01). Jack Mackerel and Red Bait otoliths were in samples from both breeding and non-breeding sites (Table 5), but samples from non-breeding seals contained many more myctophid otoliths. Also, a relatively large proportion of samples from breeding sites did not contain otoliths.

DISCUSSION

Feeding trials

Comparison with other seal species

Recovery rates of fish otoliths and cephalopod beaks from New Zealand Fur Seals were higher than recovery rates reported for Australian Sea Lions (Gales and Cheal 1992) but similar to recovery rates reported from other studies on New Zealand Fur Seals (Fea and Harcourt 1997), South American Fur Seals *A. australis* and Californian Sea Lions *Zalophus californianus* (Dellinger and Trillmich 1988). Intensity of digestion seems to differ between and within species. This may be due to retention of stones in the stomach (Needham 1997) in some species or individuals. Differential retention rates highlight the need for caution when trying to infer the number of prey consumed from otoliths found in seal scats, although identification of prey species for frequency of occurrence analysis seems to be adequate. Considering each fish or cephalopod has two otoliths or beaks, recovery rates of 50% or more of remains from New Zealand Fur Seals indicated there was reasonable likelihood that all prey species consumed would be represented in scats.

Otolith measurements and estimating fish length

The accuracy of estimating fish size varied between fish species, and depends on the size and fragility of the otoliths (Fea and Harcourt 1997), However, this study has shown that if sufficient care is taken, the size of the principal prey items of fur seals in southern Tasmania can be estimated quite accurately.

The length of Red Bait was underestimated by 5% when estimated from otoliths recovered in scats, while the estimated length of Jack Mackerel did not differ from the known size of the fish fed to the seals. Red Bait otoliths were also more difficult to identify to species due to greater erosion than Jack Mackerel otoliths. Thus Red Bait otoliths appeared to be relatively susceptible to digestion. Macrouidae, which have large otoliths, had almost 100% representation in samples. Other feeding studies (e.g., Hawes 1983; Da Silva and Neilson 1985) have also found that otolith types vary in susceptibility to digestion.

Nearly 100% of cephalopod beaks were excreted in feeding trials, higher than in other studies. This indicates that the squid component of New Zealand Fur Seal diet can be reliably estimated, but only if the seals regurgitate exclusively on land.

Diet of free-ranging animals

The dominant prey of New Zealand Fur Seals at Maatsuyker Island are the pelagic schooling fish, Jack Mackerel, Red Bait and Myctophids. Jack

Table 4. Fish species in New Zealand Fur Seal diet (percentage dominance and per cent frequency of occurence in samples (n = 279), mean number of otoliths per sample, and per cent total number of otoliths n = 2119)).

Common name	Scientific name	Percentage Dominance	% frequency Occurence	Total no. otoliths %	Otoliths/ sample
no otoliths		25	25	—	—
Red Bait	*Emmelichthys nitidus*	16	22	17	6 ± 6
Lantern Fish	*Myctophidae*	14	18	54	24 ± 49
Jack Mackerel	*Trachurus declivis*	10	21	9	3 ± 3
Jack Mackerel or Red Bait		5	21	8	4 ± 6
Cardinal Fish	*Apogonops anomalus*	2	4	3	7 ± 10
Leather Jacket	*Meuschenia australis*	2	4	3	2 ± 1
Barracouta	*Thyrsites atun*	1	1	>1	2 ± 2
Red Bait juvenille	*emmelychthys nitidus*	1	3	3	6 ± 9
Blue Grenadier juvenille	*Macruronus novaezelandiae*	1	1	>1	3 ± 1
Cucumber Fish	*Chloropthalmus nigripinni*	1	>1	>1	3 ± 1

*Table 5.*Comaprison of breeding and non-breding animals fish diet, January and March, 1993 (frequency of occurence of otoliths in samples (n = 71).

otolith type	breeding	non-breeding	ratio
Jack Mackerel	21	43	2.0
red Bait	23	11	0.5
myctophids	15	28	1.9
no otoliths	46	7	0.2

Mackerel and Red Bait form surface schools to feed on krill *Nyctiphanes australis* and Myctophids form dense schools in areas of up-welling water such as the subtropical convergence (Hulley 1986) and the continental shelf break (May and Blaber 1989). Myctophids constituted 90% of fish biomass at the upper continental slope, east coast of Tasmania, 1984/85, and were 100 times more abundant than other fish species, including Jack Mackerel (May and Blaber 1989). Gelatinous plankton such as salps were another important component of New Zealand Fur Seal diet. These animals also form dense aggregations in suitable areas by rapid growth and asexual reproduction. New Zealand Fur Seals therefore appear to forage pelagically for regionally abundant prey.

Breeding and non-breeding differences in diet

The diets of breeding and non-breeding New Zealand Fur Seals were significantly different. A large proportion of samples from breeding areas did not contain otoliths. Relatively severe digestion in these samples suggested that breeding animals had slower passage rates and probably consumed less food. Further, breeding animals consumed few myctophids, suggesting that the foraging area of breeding animals was restricted to the continental shelf. The foraging areas of breeding female otariids is restricted by their need to regularly return to suckle pups (Gentry and Kooyman 1986; Costa 1991), and female New Zealand Fur Seals alternate between foraging and suckling their young for 10–11 months each year (Goldsworthy and Shaughnessy 1991). This means that New Zealand Fur Seals are restricted to continental shelf regions around Maatsuyker Island for most of the year.

There is a commercial fishery for Jack Mackerel and Red Bait in Tasmania. The fishery has landed up to 40 000 tonnes of fish per annum since 1985 (Jordan *et al.* 1992). Although fishing is mainly on the east coast vessels also operate in south-west coastal waters (N. Brothers, pers. comm.). As the fish caught by the industry are the same size as those consumed by New Zealand Fur Seals there is the potential for competition between fisheries and breeding New Zealand Fur Seals at Maatsuyker Island.

ACKNOWLEDGEMENTS

Feeding trials were conducted at Taronga Park Zoo. Tasmanian Department of Sea Fisheries provided digital image analysis equipment and fish/otolith regressions. Cathy Bulman, Rosemary Gales and David Pemberton provided otolith reference collections. Irynej Skira collected November samples. Ken Green gave advise for statistics. Assistance in the field was from Andrew Irvine, Mary-anne Lea, Amanda Lorken and Sam Thalmann. Australian Maritime Safety Authority, Helicopter Resources and fishing boats provided transport. The project was funded by Southwest World Heritage Area Flora and Fauna Research Grants, Parks and Wildlife Service.

REFERENCES

Brothers, N. and Pemberton, D., 1990. Status of Australian and New Zealand Fur Seals on Maatsuyker Island, southwestern Tasmania. *Wildl. Res.* **17:** 563–69.

Carey, P., 1992. Fish prey species of the New Zealand Fur Seal (*Arctocephalus forsteri*, Lesson). *NZ. J. of Ecol* **16:** 41–46.

Costa, D. P., 1991. Reproductive and foraging energetics. Pp. 300–44 *in* Behaviour of Pinnipeds ed by D. Renouf. Chapman and Hall: London.

Croxall, J. P., McCann, T. S., Prince, P. A. and Rothery, P., 1988. Reproductive performance of seabirds and seals at South Georgia and Signey Island, South Orkney Islands 1976–1987: Implications for Southern Ocean monitoring studies. Pp. 261–85 *in* Antarctic Ocean and Resources Variability ed by Sahrhage. Springer-Verlag: Berlin Heidelberg.

Da Silva, J. and Neilson, J. D., 1985. Limitations of using otoliths recovered in scats to estimate prey consumption in seals. *Can. J. Fish. Aquat. Sci.* **42:** 1439–442.

Dellinger, T. and Trillmich, F., 1988. Estimating diet composition from scat analysis in otariid seals (Otariidae): is it reliable? *Can. J. Zool.* **66:** 1865–870.

Fea, N. and Harcourt, R., 1997. Assessing the use of faecal and regurgitate analysis as a means of determining the diet of New Zealand Fur Seals. Pp. 143–50 *in* Marine Mammal Research in the Southern Hemisphere Volume 1: Status, Ecology and Medicine ed by M. Hindell and C. Kemper. Surrey Beatty & Sons: Chipping Norton.

Gales, N. J. and Cheal, A. J., 1992. Estimating diet composition of the Australian sea-lion *(Neophoca cinerea)* from scat analysis: an unreliable technique. *Wildl. Res.* **19:** 447–56.

Gales, R. and Pemberton, D., 1994. Diet of the Australian fur seal in Tasmania. *Aust. J. Mar. Freshwater Res.* **45:** 653–64.

Gentry, R. L. and Kooyman, G. L., 1986. Introduction. Pp. 3–27 *in* Fur Seals: Maternal Strategies on Land and at Sea ed by R. L. Gentry and G. L. Kooyman. Princeton University Press: New Jersey.

Goldsworthy, S. D. and Shaughnessy, P. D., 1991. Status report on eared seals (Otariidae) in Australia and its territories. Australian National Parks and Wildlife Service.

Green, K., Williams, R., Handasyde, A., Burton, H. R. and Shaughnessy, P. D., 1989. Interspecific and intraspecific differences in the diets of fur seals, *Arctocephalus* species (Pinnipedia: Otariidae), at Macquarie Island. *Aust. Mammal.* **190:** 193–200.

Harvey, J. T., 1989. Assessment of errors associated with harbour seal *(Phoca vitulina)* faecal sampling. *J. Zool., Lond.* **219:** 101–11.

Hawes, S. D., 1983. An evaluation of Californian sea lion scat samples as indicators of prey importance. MA thesis, San Francisco State University, California. 49 pp.

Hecht, T., 1987. A guide to the otoliths of Southern Ocean fishes. *S. Afr. J. Antarct. Res.* **17:** 2–85.

Hulley, P. A., 1986. Myctophiformes in Smiths Sea Fishes (6th Edition) ed by M. M. Smith and P. C. Heemstra, Springer-Verlag: Berlin.

Jordan, A., Pullen, G. and Williams, H., 1992. Jack Mackerel Resource Assessment in South Eastern Australian Waters. Final report to the fishing industry research and development council. Government Printer: Tasmania.

Kirkwood, R., Pemberton, D. and Copson, G., 1991. The conservation and management of seals in Tasmania. A report to Tasmanian Department of Parks, Wildlife and Heritage.

May, J. L. and Blaber, S. J. M., 1989. Benthic and pelagic fish biomass of the upper continental slope off eastern Tasmania. *Marine Biol.* **10:** 11–25.

Needham, D. J., 1997. The role of stones in the sea lion stomach — investigations using contrast radiography and fluoroscopy. Pp. 164–69 *in* Marine Mammal Research in the Southern Hemisphere Volume 1: Status, Ecology and Medicine ed by M. Hindell and C. Kemper. Surrey Beatty & Sons: Chipping Norton.

Pierce, G. J. and Boyle, P. R., 1991. A review of methods for diet analysis in piscivorous marine mammals. *Oceanogr. Mar. Biol. Annu. Rev.* **29:** 409–86.

Reidman, M., 1990. The pinnipeds: seals, sea lions and walruses. University of California Press: Berkley.

Shaughnessy, P. D., Gales, N. J. and Dennis, T. E., 1994. Distribution and abundance of New Zealand Fur Seals *Arctocephalus forsteri* in South Australia and Western Australia. *Wildl. Res.* **21:** 667–95.

Street, R. J., 1964. Feeding habits of the New Zealand Fur Seal, *Arctocephalus forsteri. New Zealand Marine Dept. Fisheries Technical Report. No.* **9:** 1–20.

Tate, M. L., 1981. The autumn-winter diet of the New Zealand Fur Seal with special reference to its cephalopod prey. Thesis: Wildlife Management: Otago.

Top left: Bottlenose Dolphins *Tursiops truncatus,* 8th November 1985. Photo: P. Corkeron.

Top right: Australian Seal Lion *Neophoca cinerea* female, Dangerous Reef, South Australia, 3rd February 1995. Photo: C. Kemper.

Left: Antarctic Fur Seals *Arctocephalus gazella* at Heard Island, illustrating an adult male, and adult female and her pup. Photo: Simon Goldsworthy.

Bottom left: Female Australian Fur Seals *Arctocephalus pusillus doriferus* at Cape Gantheaume, Kangaroo Island, South Australia. Photo: Simon Goldsworthy.

Bottom centre: Two fur seal pups at Heard Island. The left pup is a Subantarctic Fur Seal *Arctocephalus tropicalis* exhibiting the uniform glossy black coat. The pup on the right is an Antarctic Fur Seal *Arctocephalus gazella* with a more grizzled appearance, pale muzzle and belly. Photo: Simon Goldsworthy.

Bottom right: New Zealand Fur Seal *Arctocephalus forsteri* pup. Photo: Simon Goldsworthy.

Prey of Hooker's Sea Lions *Phocarctos hookeri* based at Otago Peninsula New Zealand

C. LALAS[1]

Species, numbers and sizes of individual prey were determined from diagnostic remains found in scats and regurgitates of Hooker's Sea Lions *Phocarctos hookeri* collected fortnightly from June 1991 to May 1992. Prey were diverse in taxa (crustaceans, cephalopods and fish) and habitats (benthic, demersal and pelagic). Foraging was apparently restricted to the continental shelf and changed seasonally, with inshore prey eaten more during winter and spring. Their disappearance from the diet indicated that sea lions foraged further from shore during summer and autumn. Only one species, Barracouta *Thyrsites atun* was eaten throughout the year.

INTRODUCTION

HOOKER'S Sea Lion *Phocarctos hookeri* is restricted to the subantarctic islands and southern mainland of the New Zealand region (Crawley 1990; Woodley and Lavigne 1993). The species was eliminated from the New Zealand mainland (North Island and South Island) following human settlement (Smith 1989), but now occurs along the southeastern coastline of South Island (Hawke 1986). Small numbers, mainly subadult males, are resident in this area (Beentjes 1989), with the greatest concentration at Papanui Beach (45°52′S, 170°44′E) on Otago Peninsula (Hawke 1993). The year-round presence of male Hooker's Sea Lions makes Papanui Beach an ideal site for a long-term investigation of their diet at the northern limit of their distribution.

There are no published quantitative assessments of the diet of Hooker's Sea Lions. Data at present are restricted to anecdotal accounts that require substantiation (Cawthorn *et al.* 1985; Crawley 1990; Riedman 1990). Described as opportunistic feeders (Cawthorn *et al.* 1985), they eat a broad variety of food including algae, crustaceans (krill and crabs), molluscs (cephalopods, bivalves and gastropods), fish (teleosts and elasmobranchs), seabirds and seals. Arrow Squid *Nototodarus sloanii* is seasonally important and is considered to be the main prey for breeding Hooker's Sea Lions around Auckland Islands, the species population base (Anon. 1991).

The diet of Hooker's Sea Lions in this investigation was deduced from diagnostic, indigestible remains in scats (faeces) and regurgitations. All methods for the analysis of diet in marine mammals are subject to biases (Jobling and Breiby 1986; Dellinger and Trillmich 1988; Harvey 1989; Pierce and Boyle 1991). Biases in analyses of scats and regurgitates were highlighted in diet validation experiments on Australian Sea Lions *Neophoca cinerea* by Gales and Cheal (1992).

This investigation of the prey of Hooker's Sea Lion at Otago had four aims:(i) To identify prey taxa (genera) and rank them in order of frequency of occurrence and proportion of wet weight, (ii) to record any seasonal changes in the occurrence of prey, (iii)To deduce foraging behaviour from the distributions and habits of prey, and (iv) to compare these with sympatric New Zealand Fur Seals *Arctocephalus forsteri*.

METHODS

Papanui Beach (45°52′S, 170°44′E) was visited at fortnightly intervals from June 1991 to May 1992. Visits lasted for two hours, usually between 1200 h and 1600 h. The 350 m long sandy beach and abutting 150 m foreshore dune (mapped by Beentjes 1989) were closely checked. All scats and regurgitates seen were collected, and the number and sex of sea lions ashore was noted.

Scats were grouped according to their degree of dryness and their location on the beach. Each group contained 1–12 scats. This system avoided biases associated with point sampling. Scat samples were washed through a 0.7 mm square mesh sieve and diagnostic remains retained and stored dry. Regurgitates were bagged individually and sorted later. Any pebbles in samples were retained and stored dry but are not considered in this paper.

Prey remains in scats and regurgitates were identified to genus or species by comparison with a reference collection of diagnostic remains from known length and mass (wet weight) animals. This collection, held by the author, included otoliths, spines, mouthparts, teeth, exoskeletons and vertebrae but not fish scales. With the exception of two species, the reconstituted lengths and wet weights of prey items were calculated by interpolation from diagnostic remains in the reference collection. This method compensated for erosion or breakage of diagnostic remains (especially otoliths) in samples. For calculations

[1]Box 31, Portobello, Dunedin, New Zealand.
Pages 130–36 *in* MARINE MAMMAL RESEARCH IN THE SOUTHERN HEMISPHERE VOLUME 1: STATUS, ECOLOGY AND MEDICINE ed by M. Hindell and C. Kemper. Surrey Beatty & Sons. 1997.

of wet weight, otoliths were considered too eroded if their edges were delaminated to expose growth increments, and so were disregarded.

The reconstituted lengths and wet weights of the two most numerous prey items, Paddle Crabs *Ovalipes catharus* and Ahuru *Auchenoceros punctatus* were calculated from equations (see below). Paddle Crabs typically appeared in regurgitates as compacted but fully articulated, decalcified exoskeletons that unfurl when suspended in water. Carapace width (CW, mm) was measured directly or, if ripped or disintegrated, could be calculated if a pleopod was intact as

$$CW = 43.9 \times PA^{0.43} \ (n = 14, r^2 = 0.98)$$

where PA (cm^2) is the area of one pleopod. Reconstituted wet weight (g) was calculated as

$$Wt = 1.75 \times 10^{-4} CW^{3.02} \ (n = 14, r^2 = 0.98)$$

Otoliths of Ahuru (illustrated in Paulin 1983) have a fine posterior tip prone to breakage and digestive erosion. Otolith depth (OD mm, the maximum vertical measure) was relatively unaffected by digestion and could be measured under a binocular microscope to the nearest 0.01mm. Reconstituted fish total length (TL mm) was calculated as

$$TL = 36.8 \times OD^{1.31} \ (n = 25, r^2 = 0.94)$$

Reconstituted wet weight (g) was calculated as

$$Wt = 5.5. \times 10^{-6} TL^{2.98} (n = 43, r^2 = 0.94)$$

Often diagnostic remains were either too few or too eroded to produce an exact indication of prey size. The average size for that taxon at that time of year was then used. Common names, scientific names and taxonomic listing for fish follow Paulin *et al.* (1989).

RESULTS

Numbers of sea lions, scats and regurgitates

The number of sea lions ashore at fortnightly intervals varied through the year as did monthly totals for numbers of scats (n = 490) and regurgitates (n = 29) collected (Table 1). All the sea lions seen ashore were males except one tagged female who was resident throughout the year. Sea lion numbers were lowest between November and January when relatively few scats were collected. Most scats were collected between February and May, when there was no obvious peak in the number of sea lions ashore. The occurrence of regurgitates was sporadic and did not show any seasonal pattern.

Diversity of prey

Forty-one taxa (genus or species) were recorded as prey from remains in scats or regurgitates (Table 2): one polychaete, four crustaceans, one gastropod, two cephalopods (one squid and one octopus), and 34 fish (one lamprey, four elasmobranchs and 29 teleosts). There were 770 items with a calculated total reconstituted wet weight of 379 kg, equivalent to approximately 0.5 kg per item.

Table 1. Numbers of Hooker's Sea Lions ashore on Papanui Beach at fortnightly intervals, June 1991–May 1992, and monthly totals for numbers of scats and vomits analysed. *Included a tagged female (see text).

Month	Sea Lions	Scats	Vomits
June	6 & 3	32	2
July	4 & 10	27	0
August	9* & 5*	35	5
September	10* & 2	31	5
October	7 & 5	37	3
November	6 & 0	16	0
December	0 & 1	13	1
January	3 & 6	32	2
February	8* & 8*	51	1
March	5 & 5*	90	2
April	6* & 6	73	1
May	10 & 3*	53	7

A more realistic impression of the number of major prey items is gained with the elimination of taxa that contributed only a trace of the total wet weight, i.e., total weight equal to 0.0 kg in Table 2. The diversity of prey from Hooker's Sea Lions foraging off Otago Peninsula then stands at 33 genera: one crustacean, two cephalopods and 30 fish. With one exception, these genera averaged at least 0.10 kg per prey item. The exception was Ahuru, a small species that averaged 9 g per individual.

Other diagnostic remains in scats and reguritates

Remains of parasitic isopods and parasitic copepods were a regular occurrence and were assumed to be associated with the larger prey of sea lions. Gastropod and bivalve shells regularly appeared in association with pebbles. In all cases, shells were old and worn, usually broken and often encrusted inside and out. No matching pairs of bivalve shells were found. The combination of these features indicated that these mollusc shells did not represent prey items. However, scats included remains of one hermit crab and two small, oval chitinous opercula for which no corresponding gastropods were found (Table 2). The only records of algae were occasional small fragments of seaweed.

Comparison between scats and regurgitates

Ten genera of fish were represented in regurgitates compared with 32 genera of fish in scats (Table 3). Although regurgitates yielded two species not recorded from scats, overall the analysis of regurgitates did not alter results derived from scats for the fish component in the sea lion diet. In contrast, most Paddle Crabs and Arrow Squid, and more than one-third of Octopus *Octopus maorum* were derived from regurgitates.

Composition of diet

Composition of the sea lion diet was calculated using number and reconstituted wet weight of

Table 2. Prey of Hooker's Sea Lions of Otago Peninsula. Annotated listing of species with diagnostic remains in scats or vomits collected at Papanui Beach. Identification (ID) method: the main diagnostic remains used to identify, and quantify size of, prey items where E = exoskeleton; M = mouthparts; O = otoliths; Op = operculum; S = spines; T = teeth; Ta = tail. "Total weight" represents the calculated total reconstituted green weight per prey taxon, where 0.0 kg indicates less than 50 g and a question mark indicates an estimate.

			Number of Prey Items Per Month June 1991–May 1992															
Polychaeta		ID Method	J	J	A	S	O	N	D	J	F	M	A	M	Total No. of Items	Proportion Quantified No.	%	Total Weight (kg)
Bristle Worm	cf. *Platynereis* sp.	M	—	—	—	—	—	—	—	1	—	—	—	—	1	0	0	0.0
Galatheidae																		
Lobster Krill	*Munida gregaria*	E	—	—	—	—	—	—	—	17+	2+	—	15+	1+	35+	0	0	0.0
Paguroidea																		
Hermit Crab	unidentified sp.	E	—	1	—	—	—	—	—	—	—	—	—	—	1	0	0	0.0
Hymenosomatidae																		
Penny Crab	unidentified sp.	E	1	—	—	—	—	—	—	—	—	—	—	—	1	0	0	0.0
Portunidae																		
Paddle Crab	*Ovalipes catharus*	E	1	—	74	56	25	—	—	—	—	—	1	—	157	118	75	16.5
Gastropoda																		
Snail	unidentified sp.	Op	—	—	—	—	—	—	—	—	—	2	—	—	2	0	0	0.0
Ommastrephidae																		
Arrow Squid	*Nototodarus sloanii*	M	—	—	—	—	—	—	—	—	—	4	—	6	10	10	100	7.1
Octopodidae																		
Octopus	*Octopus maorum*	M	2	—	2	5	—	—	—	—	—	9	1	18	37	34	92	53.4
Geotriidae																		
Lamprey	*Geotria australis*	M	—	—	1	1	—	—	—	—	—	—	—	—	3	0	0	?0.6
Squalidae																		
Spiny Dogfish	*Squalus acanthias*	TS	—	—	—	1	1	—	—	—	—	—	—	—	2	2	100	2.4
Triakidae																		
School Shark	*Galaeorhinus galeus*	T	—	—	—	—	—	—	—	—	1	—	—	—	1	0	0	?2.0
Rajidae																		
Skate	*Raja* sp.	ST	1	—	—	2	2	—	—	2	2	—	—	4	13	13	100	36.9
Callorhinchidae																		
Elephant Fish	*Callorhinchus milii*	T	1	—	—	—	—	—	—	—	—	—	—	—	1	0	0	?2.0
Moridae																		
Ahuru	*Auchenoceros punctatus*	O	40	5	1	117	4	—	—	—	—	—	—	4	171	166	97	1.6
Rock Cod	*Lotella rhacinus*	O	2	—	1	—	1	—	—	—	—	—	—	—	4	4	100	0.8
Red Cod	*Pseudophycis* sp.	OM	5	—	3	1	3	—	—	2	1	4	2	2	23	20	87	14.1
Macrouridae																		
Oblique Banded Rattail	*Coelorinchus aspercephalus*	O	—	—	—	—	—	—	—	—	—	—	—	1	1	1	100	0.0
Ophidiidae																		
Ling	*Genypterus blacodes*	O	—	—	—	—	—	—	—	1	—	—	2	2	5	5	100	16.5
Trachichthyidae																		
Common Roughy	*Paratrachichthys trailli*	O	1	—	1	7	—	—	—	—	—	—	—	2	11	10	91	1.1
Scorpaenidae																		
Jock Stewart	*Helicolenus* sp.	O	—	—	—	—	—	—	—	—	—	3	1	1	5	5	100	2.0
Congiopodidae																		
Pigfish	*Congiopodus* sp.	O	—	—	—	1	1	—	1	—	—	—	—	—	3	3	100	0.7
Carangidae																		
Jack Mackerel	*Trachurus* sp.	O	2	—	—	—	—	—	—	1	11	28	24	3	69	47	68	48.9
Arripidae																		
Kahawai	*Arripis trutta*	O	—	—	—	—	—	—	—	—	1	—	—	—	1	1	100	1.6
Latrididae																		
Moki	*Latridopsis* sp.	O	—	1	—	—	—	—	—	—	—	—	—	—	1	1	100	0.3
Mugilidae																		
Yelloweyed Mullet	*Aldrichetta forsteri*	O	—	—	—	—	—	1	—	—	—	—	—	—	1	1	100	0.2
Labridae																		
Wrasse	*Notolabrus* sp.	MO	4	1	2	5	1	2	—	—	1	1	—	8	25	22	88	6.8
Odacidae																		
Greenbone	*Odax pullus*	MO	10	5	3	7	2	—	—	—	—	1	—	11	39	36	92	16.0
Nototheniidae																		
Black Cod	*Paranotothenia* sp.	O	—	1	—	—	—	—	—	—	—	—	—	—	1	1	100	0.2
Uranoscopidae																		
Spotted Stargazer	*Genyagnus monopterygius*	O	—	1	—	—	—	—	—	—	—	—	—	—	1	1	100	0.2
Giant Stargazer	*Kathetostoma giganteum*	O	—	—	1	1	—	—	—	—	—	—	—	—	2	2	100	1.9
Leptoscopidae																		
Estuary Stargazer	*Leptoscopus macropygus*	O	—	—	—	—	1	—	—	—	—	—	—	—	1	1	100	0.4
Pinguipedidae																		
Blue Cod	*Parapercis colias*	OT	—	—	—	—	—	—	—	—	2	3	—	3	8	8	100	2.5
Tripterygiidae																		
Triplefin	cf *Triptergyion* sp.	O	—	—	1	—	—	—	—	—	—	—	—	—	1	1	100	0.0
Clinidae																		
Crested Weedfish	*Cristiceps aurantiacus*	Ta	—	2	—	—	—	—	—	—	—	—	—	—	—	2	0	0.0
Gempylidae																		
Barracouta	*Thyrsites atun*	OT	4	3	1	—	3	3	4	4	2	3	5	6	38	15	39	94.7
Centrolophidae																		
Blue Warehou	*Seriolella brama*	O	—	—	1	—	—	—	—	4	—	—	—	—	5	5	100	0.6
Bothidae																		
Witch	*Arnoglossus scapha*	O	—	—	—	—	—	—	—	—	—	1	2	1	4	4	100	1.3
Pleuronectidae																		
Brill	*Colistium guntheri*	O	—	—	—	—	4	—	—	—	—	—	—	—	4	4	100	8.2
Lemon Sole	*Pelotretis flavilatus*	O	—	—	1	—	—	—	—	—	—	—	—	—	1	1	100	0.0
Sole	*Peltorhamphus* sp.	O	11	2	6	1	1	—	1	—	—	—	—	1	23	23	100	2.2
Flounder	*Rhombosolea* sp.	O	1	10	24	5	2	2	6	—	—	—	—	5	55	54	98	34.5
Monacanthidae																		
Smooth Leatherjacket	*Parika scaber*	OT	—	—	—	1	1	—	—	—	—	—	—	—	2	2	100	0.3

Table 3. Prey with diagnostic remains in vomits. The same prey could be found in both scats and vomits, hence Total Number may be less than Number of Scats plus Number in Vomits. Taxonomic listings, where numbers represent numbers of prey items.

Prey	Number in scats	Number in Vomits	Total Number	Proportion only in vomits
Single occurence in vomits:				
Spiny Dogfish	1	1	2	50%
Elephant Fish	0	1	1	100%
Oblique Banded Rattail	0	1	1	100%
Ling	4	1	5	20%
Common Roughly	10	1	11	9%
Greenbone	38	1	39	3%
Witch	3	1	4	25%
Multiple occurences in vomits:				
Paddle Crab	15	142	157	90%
Arrow Squid	2	8	10	80%
Octopus	23	16	37	38%
Red Cod	21	3	23	9%
Wrasse	21	4	25	16%
Barracouta	37	4	38	3%

Table 4. Composition of the diet by number and by reconstituted wet weight of prey in samples from data in Table 2.

Proportion by Number of Items			Proportion by reconstituted Greenweight		
1	Ahuru	23%	1	Barracouta	25%
2	Paddle Crab	21%	2	Octopus	14%
3	Jack Mackerel	9%	3	Jack Mackerel	13%
4	Flounder	7%	4	Skate	10%
5	Greenbone	5%	5	Flounder	9%
6	Barracouta	5%	6	Paddle Crab	4%
7	Octopus	5%	7	Ling	4%
8	Wrasse	3%	8	Greenbone	4%
9	Red Cod	3%	9	Red Cod	4%
10	Sole	3%	10	Brill	2%
Top 10	taxa summed	85%	Top 10	taxa summed	90%

prey items and compared for the 10 most important taxa (Table 4). Although seven taxa appear in both categories, rankings are different between the two analyses. In particular, the combination of ahuru and paddle crab accounted for 44% of the total number but only 4.5% of the total wet weight of prey items.

Seasonally adjusted results

The number of scats and regurgitates analysed per month fluctuated through the year (Table 1). This translated into a ten-fold difference between the minimum and maximum estimates for monthly samples of prey reconstituted wet weight. Seasonally adjusted results were calculated in order to compensate for these differences in monthly sample size. Proportions by wet weight were calculated for each prey taxon each month and each month was given equal ranking. Each month then contributed one-twelfth towards the annual diet (Table 5). This final assessment of composition of annual diet bore nine genera in common with the top 10 ranking from the unadjusted results in Table 4 although there were minor rearrangements in order.

Five of the 10 main prey taxa occurred sporadically throughout the year without any discernible seasonal trends. Of the other five, only Barracouta *Thyrsites atun* was recorded as prey continually throughout the year. For the other four taxa, Jack Mackerel *Trachurus* sp. was the greatest proportion of weight between December and May, and Flounder *Rhombosolea* sp., Paddle Crab and Greenbone *Odax pullus* between May and November.

The seasonal occurrence of Arrow Squid as prey was similar to Octopus (Fig. 1). Recorded in only two months, the total reconstituted wet weight of Arrow Squid as prey was 7.1 kg. In contrast, the respective wet weight of Octopus from the same two months was 32.7 kg.

Prey sizes

The size range of prey items is shown in Table 6. The heaviest prey items recorded from this investigation was an Octopus 1.5 m total length and a Ling *Genypterus blacodes* weighing 11 kg.

Observations relevant to feeding

Hooker's Sea Lions are rarely seen with prey. All known records from Otago (Table 7)

Table 5. Prey ranked in order of importance from seasonally adjusted results for reconstituted wet weight of prey in samples.

		Proportion of Monthly Total Reconstituted Wet Weight of Prey												Annual
	Prey	June	July	Aug	Sep	Oct	Nov	Dec	Jan	Feb	Mar	Apr	May	Diet
1	Barracouta	36%	45%	5%	—	25%	67%	76%	44%	28%	21%	28%	20%	33%
2	Jack Mackerel	5%	—	—	—	—	—	—	4%	35%	46%	44%	1%	11%
3	Flounder	1%	26%	46%	12%	4%	16%	21%	—	—	—	—	3%	11%
4	Octopus	10%	—	12%	35%	—	—	—	—	—	13%	—	36%	9%
5	Skate	8%	—	—	9%	22%	—	—	25%	13%	—	—	21%	8%
6	Greenbone	14%	17%	4%	7%	5%	—	—	—	—	1%	—	4%	4%
7	Red Cod	15%	—	5%	2%	—	4%	—	8%	6%	6%	2%	0%	4%
8	Paddle Crab	0%	—	20%	16%	8%	—	—	—	—	—	0%	—	4%
9	Ling	—	—	—	—	—	—	—	15%	—	—	24%	3%	4%
10	Wrasse	3%	1%	2%	8%	1%	11%	—	—	1%	1%	—	1%	2%
Monthly wet weight														
All prey (kg)		31.4	15.2	39.8	37.3	30.3	8.1	13.0	21.4	22.4	38.8%	45.3	76.2	—
Top 10 taxa summed		91%	89%	93%	88%	65%	98%	97%	98%	83%	87%	98%	90%	90%

Table 6. Size ranges of prey. Results for the ten most important genera in the annual diet (Table 5) and for Ahuru, the most numerous prey. Length measure: FL = fork length; TL = total length; CW = carapace width. Sample size: excludes atypically small fish from three species — a Barracouta 49 cm long (0.7 kg); an Octopus weighing 10 g; and a Wrasse 9 cm long (8 g).

Prey		Length Measure	Sample Size	Length (cm) Range (average)	Wet Weight (kg) Range (average)
1	Barracouta	FL	14	68– 86 (77)	1.9 – 3.2 (2.5)
2	Jack Macherel	FL	47	27– 45 (38)	0.3 – 1.2 (0.7)
3	Flounder	TL	54	16– 50 (35)	0.1 – 1.7 (0.6)
4	Octopus	~TL	33	30–150 (120)	0.1 – 4.3 (1.5)
5	Skate	~TL	13	53– 93 (68)	1.0 – 5.7 (2.8)
6	Greenbone	TL	36	22– 49 (31)	0.1 – 1.9 (0.4)
7	Red Cod	TL	20	20– 50 (36)	0.1 – 1.3 (0.6)
8	Paddle Crab	CW	118	4– 11 (7)	0.02– 0.3 (0.1)
9	Ling	TL	4	68–124 (n.a.)	1.6 –10.9 (n.a.)
10	Wrasse	TL	21	15– 40 (23)	0.1 – 1.3 (0.3)
Ahuru		TL	166	7– 13 (10)	2 g – 12g (9 g)

Table 7. Observations from Otago of Hooker's Sea Lions with prey. Includes records more recent than the diet analysis.

Date	Place	Distance from Papanui Beach	Time (NZST)	Prey	Water Depth	
4 September 1991	Sandfly Bay	8 km SW	1045–1115 h	3 Spiny Dogfish	5–10 m	G. Loh, D. Nelson and
			1600 h	1 Skate	5–10 m	G. Legge (pers. comm.)
21 October 1991	Otago Harbour	7 km NW	afternoon	1 Salmonid	1–2 m	D. McLeod (pers. comm.)
7 November 1982	Campbell Point	95 km SW	1200 h	1 Skate	1–2 m	Lalas (pers. obs.)
3 December 1992	Campbell Point	95 km SW	0830 h	1 Skate	1–2 m	Lalas (pers. obs.)
25 July 1993	Otago Harbour	13 km W	afternoon	1 Flounder	1–3 m	*Otago Daily Times*

are of males taking fish during daylight in shallow water close to shore, including within Otago Harbour. Of the eight observations of fish being eaten, three were Skate *Raja* sp. and three were Spiny Dogfish *Squalus acanthias*. Elasmobranchs were, therefore, greatly over-represented in observations of feeding compared to analyses of scats and regurgitates. Of the four types of fish recorded, one was not documented from scats or regurgitates at Papanui Beach although the observation was contemporary with the diet analysis and the location was relatively nearby. This was a large salmonid, either Quinnat Salmon *Oncorhynchus tshawytscha* or Brown Trout *Salmo trutta* an estimated 60 cm long (D. McLeod, pers. comm.). All but the flounder were swallowed whole.

DISCUSSION

This investigation has qualified the annual diet of Hooker's Sea Lion at the northern tip of their distribution. I have used seasonally adjusted results for reconstituted wet weights of prey. An interpretation derived solely from numbers of prey items would have produced a very different, perhaps deceptive, result. In particular, the combination of Ahuru and Paddle Crab accounted for almost half the total number but less than one-twelfth of the total wet weight of prey items.

Biases in analyses of prey remains in scats and regurgitates have been documented in several studies (Jobling and Breiby 1986; Dellinger and Trillmich 1988; Harvey 1989; Pierce and Boyle 1991; Gales and Cheal 1992; Pierce *et al.* 1993).

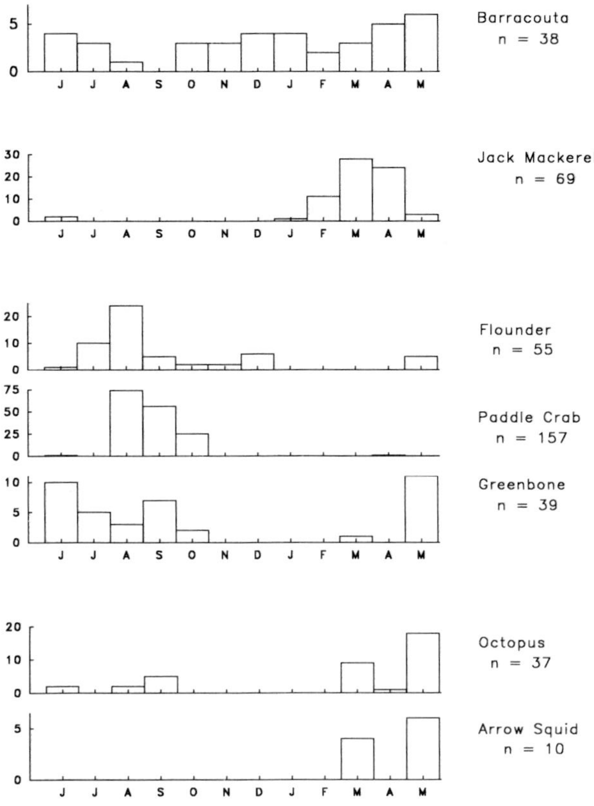

Fig. 1. Seasonal occurrence of important prey from data in Table 2.

Attempts were made in this study to either minimize or account for these biases. Inaccuracies inherent in the back-calculation of prey sizes from eroded diagnostic remains were minimized by matching diagnostic remains in samples against those held in a reference collection. Although relatively small items may have originated from the gut contents of larger prey, this problem was left unaddressed in this investigation because small items contributed only a trivial amount towards the total wet weight of prey. Limited observations of sea lions feeding indicated that prey were swallowed whole and so the diagnostic remains were not lost.

Typically, cephalopod remains are over-represented in regurgitates and under-represented in scats, while the situation is reversed for fish remains (Pierce and Boyle 1991; Gales and Cheal 1992). This was also the case in the present investigation, but there was also an over-representation of elasmobranchs and crabs in regurgitates. Elasmobranchs were also over-represented in observations of sea lions with captured fish. After accounting for duplication, prey remains from scats and regurgitates were combined for analysis. The validity of this simple summation is questionable but this shortfall is regarded as an inevitable outcome of an unobtrusive sampling technique. Given the likelihood of inaccuracies in the quantification of diet, the presented results appear over-precise. These results should only be considered as indicators of the relative importance of the various prey taxa.

The diet of Hooker's Sea Lions at Otago was dominated by a variety of teleost fish, although elasmobranch fish (mainly Skate) and cephalopods (mainly Octopus) were also important. This is in contrast to the variety of marine organisms listed by Cawthorn *et al.* (1985). Algae, krill, bivalves and gastropods were only minor components of the diet at Otago and probably were either taken incidentally or present in the gut of larger prey items. Also, neither seabirds nor marine mammals were recorded as prey in this one-year investigation at Papanui Beach.

Barracouta was the only important prey taken throughout the year. Five of the other important taxa showed distinct seasonal occurrences. Jack Mackerel were recorded only from January to June and can be considered to be summer-autumn prey. No Flounder, Greenbone, or Paddle Crab were recorded from January or February with relatively few recorded from November to April. These three taxa are considered as winter-spring prey.

The habits of important prey species of Hooker's Sea Lions at Otago can be categorized from descriptions in Graham (1974), Paul (1986) and Annala (1993). Barracouta and Jack Mackerel are schooling fish throughout the water column. The remainder are demersal, dwelling at or near the bottom. In particular Flounder, Skate and Paddle Crabs are benthic, while Octopus, Greenbone and Wrasse *Notolabrus* sp. are reef dwellers. Paddle Crabs and Greenbone are restricted to bottom depths less than 10 m and 20 m, respectively. The other species spread across the continental shelf with Red Cod *Pseudophycis bachus* and Ling extending beyond to depths of 700 m.

These results indicate a seasonal shift in foraging zones by Hooker's Sea Lions at Otago. Benthic or reef-dwelling species, including two restricted to shallow water, are prominent among winter-spring prey. Consequently, foraging during winter and spring must have been concentrated in shallow waters typically less than 20 m deep. The absence of these prey in the sea lion diet during the other six months of the year is not attributable to unavailability. For example Flounder, an exclusively winter-spring prey, is taken throughout the year by local commercial fisheries (Ritchie *et al.* 1975). Therefore, sea lions foraged further from shore during summer and autumn. The seasonal restriction of Jack Mackerel as the main summer-autumn prey probably reflects a seasonally restricted availability. Their population dynamics is in a state of flux off southern New Zealand and their seasonality remains undocumented (Annala 1993).

Deepwater species were notably absent among prey items. The only record was for one Oblique Banded Rattail *Coelorinchus aspercephalus*. Locally this demersal species occurs at a minimum depth of 150 m (Lalas, unpubl. data). Consequently foraging by Hooker's Sea Lions off Otago is probably restricted to the continental shelf. In contrast, sympatric New Zealand Fur Seals forage over the continental shelf edge (Hawke 1991; Harcourt *et al.* 1995) or beyond (Harcourt and Davis 1997) and deepwater species are prominent in the fish component of their diet (Tate 1981; Carey 1992). Also, the combination of Arrow Squid and Octopus forms the bulk of the diet of fur seals off Otago (Street 1964; Tate 1981), of which only Octopus is targeted by Hooker's Sea Lions.

The seasonal importance of Arrow Squid as prey of Hooker's Sea Lions at Auckland Islands (Anon.(?) 1991) was not reflected at Otago. Instead, Arrow Squid formed only a small portion of the diet throughout the year of this investigation. Any detailed comparison between the two locations must await a quantitative assessment of the prey of Hooker's Sea Lions at Auckland Islands.

ACKNOWLEDGEMENTS

The preparation of this paper for publication was funded by the Otago Conservancy, Department of Conservation (DOC). Papanui Beach has restricted public access, therefore I am indebted to the adjacent landowners, Rod and Heather McKay and David and Sarah McKay, for access, hospitality and encouragement. Thanks to Peter Krygsman and Carol Sutherland for field assistance; John Jillett for access to facilities at Portobello Marine Laboratory, University of Otago; Leeane Stephens (DOC) for typing tables, Hiltrun Ratz for typing text; Robert Ardern for preparing graphs; David McLeod, Graeme Loh (DOC), Dean Nelson (DOC) and Gordon Legge for incidental observations. I thank Nick Gales (DOC), Rob Harcourt (University of Otago) and James Harvey and another anonymous reviewer for critical comments on the manuscript.

REFERENCES

Annala, J. H., 1993. Report from the Fishery Assessment Plenary, May 1993: stock assessments and yield estimates. Unpublished report held in MAF Fisheries Greta Point library, Wellington.

Anon., 1991. Hooker's sea lion Public Discussion Paper June 1991. Ministry of Agriculture and Fisheries and Department of Conservation, Wellington.

Beentjes, M. P., 1989. Haul-out patterns, site fidelity and activity budgets of male Hooker's Sea Lions *(Phocarctos hookeri)* on the New Zealand mainland. *Mar. Mamm. Sci.* **5:** 281–97.

Carey, P. W., 1992. Fish prey species of the New Zealand fur seal *(Arctocephalus forsteri,* Lesson). *N.Z. J. Ecol.* **16:** 41–46.

Cawthorn, M. W., Crawley, M. C., Mattlin, R. H. and Wilson, G. J., 1985. Research on pinnipeds in New Zealand. *Wildlife Research Review* **7**. Wildlife Research Liason Group: Wellington.

Crawley, M. C., 1990. New Zealand sea lion. Pp. 256–62 *in* The Handbook of New Zealand Mammals, ed by C. M. King. Oxford University Press: Auckland.

Dellinger, T. and Trillmich, F., 1988. Estimating diet composition from scat analysis in otariid seals (Otariidae): is it reliable? *Can. J. Zool.* **66:** 1865–870.

Gales, N. J. and Cheal, A. J., 1992. Estimating diet composition of the Australian sea-lion *(Neophoca cinerea)* from scat analysis: an unreliable technique. *Wildl. Res.* **19:** 447–56.

Graham, D. H., 1974. A Treasury of New Zealand Fishes. Reed Reprint Library Edition. A. H. and A. W. Reed: Wellington.

Harcourt, R. and Davis, L. S., 1997. The use of satellite telemetry to determine fur seal foraging areas. Pp. 137–42 *in* Marine Mammal Research in the Southern Hemisphere Volume 1: Status, Ecology and Medicine ed by M. Hindell and C. Kemper. Surrey Beatty & Sons: Chipping Norton.

Harcourt, R. G., Schulman, A., Davis, L. S. and Trillmich, F., 1995. Summer foraging by lactating female New Zealand fur seals *(Arctocephalus forsteri)* off Otago Peninsula, New Zealand. *Can. J. Zool.* **73:** 678–90.

Harvey, J. T., 1989. Assessment of errors associated with Harbour seals *(Phoca vitulina)* faecal sampling. *J. Zool. (Lond.)* **219:** 101–11.

Hawke, D. J., 1986. Observations of Hooker's sea lion, *Phocarctos hookeri,* at a hauling ground on Otago Peninsula, New Zealand. *N.Z. J. Mar. fresh. Res.* **20:** 333–37.

Hawke, D. J., 1991. Ship-borne observations of New Zealand fur seals across the continental shelf near Otago Peninsula, New Zealand. *N.Z. Nat. Sci.* **18:** 45–49.

Hawke, D. J., 1993. The presence of female Hooker's Sea Lions on the south-east coast of New Zealand. *N.Z. Nat. Sci.* **20:** 75–77.

Jobling, M. and Breiby, A., 1986. The use and abuse of fish otoliths in studies of feeding habits of marine piscivores. *Sarsia* **72:** 255–60.

Paul, L., 1986. New Zealand fishes: an identification guide. Reed Methuen: Auckland.

Paulin, C. D., 1983. A revision of the family Moridae (Pisces: Anacanthini) within the New Zealand region. *Rec. Nat. Mus. N.Z.* **2:** 81–126.

Paulin, C., Stewart, A., Roberts, C. and McMillan, P., 1989. *Nat. Mus N.Z. Misc. Ser.,* **19.**

Pierce, G. J. and Boyle, P. R., 1991. A review of methods for diet analysis in piscivorous marine mammals. *Ocean. Mar. Biol.* **29:** 409–86.

Pierce, G. J., Boyle, P. R., Watt, J. and Grisley, M., 1993. Recent advances in diet analysis of marine mammals. *Symp. Zool. Soc. Lond.* **66:** 241–61.

Riedman, M., 1990. The pinnipeds: seals, sea lions, and walruses. University of California Press: Berkeley.

Ritchie, L., Saul, P. and O'Sullivan, K., 1975. The wetfish report 1941–1979. *N.Z. Min. Agric. Fish., Fisheries Tech. Rep.* **137.**

Smith, I. W. G., 1989. Maori impact on the marine megafauna: pre-European distributions of New Zealand sea mammals. Pp. 76–108 *in* Saying so does not make it so: Papers in Honour of B. Foss Leach, ed by D. G. Sutton. N. Z. Archaeological Association, Monograph **17**.

Street, R. J., 1964. Feeding habits of the New Zealand fur seal. *N. Z. Mar. Dept. Fisheries Tech. Rep.* **9**.

Tate, M. L., 1981. The autumn-winter diet of the New Zealand fur seal Arctocephalus forsteri (Lesson) with special reference to its cephalopod prey. Unpublished Dip. Sci. Thesis, University of Otago.

Woodley, T. H. and Lavigne, D. M., 1993. Potential effects of incidental mortalities on the Hooker's sea lion (*Phocarctus hookeri*) population. *Aquat. Cons. Mar. Freshwater Ecosys.* **3:** 139–48.

The use of satellite telemetry to determine fur seal foraging areas

R. HARCOURT[1] and L. DAVIS[1]

In recent years much has been learnt about diving and foraging behaviour of pinnipeds, but there is still a paucity of information on where the animals undertake these activities. We report the first attempt to locate foraging New Zealand Fur Seals using satellite telemetry. We attached Platform Terminal Transmitters (PTTs) to the heads of two females during the summer months of 1993/94 and to the backs of three females in late autumn 1994 at the Fuchsia Gully rookery (45°52′S, 170°44′E) on Otago Peninsula, South Island New Zealand. In summer the females were located over the continental shelf in water less than 200 m, while in autumn they were predominantly in deep water (<500 m) beyond the continental slope. Back-mounting of PTTs proved more successful in the proportion of successful uplinks received. For small marine mammals such as female fur seals, head-mounting of PTTs may be inappropriate due to poor retention of the device and irritation to the animal.

INTRODUCTION

In recent years there has been a growing interest in determining how and where fur seals forage at sea. Many populations of fur seals are expanding, both in size and range (e.g., Antarctic Fur Seal: Payne 1977; Croxall *et al.* 1988; Shaughnessy and Goldsworthy 1990. New Zealand Fur Seal: Taylor 1982, 1992; Crawley 1990; Lalas and Harcourt 1995; Shaughnessy *et al.* 1995). However, other populations are declining (e.g., Northern Fur Seal: Fowler 1987; Trites 1992). With expansions or declines in numbers, problems may arise for management. Where populations are increasing, they may be perceived to compete for the same stock as commercial fisheries (Harwood and Croxall 1988; Lipinski and David 1990). When populations are declining they may be adversely affected by fisheries either through this same competition, or by being caught in fishing gear (Shaughnessy and Payne 1979; Mattlin 1989, 1991). These interactions have lead to an increased interest in determining where fur seals forage and what they prey on.

Lactating female fur seals alternate brief periods ashore nursing their pups with longer periods at sea (Gentry and Kooyman 1986). The use of Time Depth Recorders (TDRs) has allowed inferences to be made about how far seals forage from shore, based on broad assumptions of swim speed and assuming that animals swim in approximately a straight line to foraging grounds, combined with knowledge of local bathymetry (e.g., Gentry and Kooyman 1986). Recently TDRs have incorporated a Geographic Location device which gives a prediction of the animal's location based on a combination of measurements of sea surface temperature and day length (DeLong *et al.* 1992). However, whilst these programmes are useful for detailing movements of animals which undergo long migrations (e.g., Northern Elephant Seals), they do not have sufficient accuracy to determine foraging movements of animals over relatively small distances (Stewart and DeLong 1993).

VHF radio tags have also been used to track fur seals at sea, with animals located from aeroplanes flying transects or by following individual seals by ship (Antonelis *et al.* 1987; Loughlin *et al.* 1987; Goebel *et al.* 1991). The high cost and logistic difficulties preclude the use of this method for most studies. VHF transmitters have also been used with ground-based receiving stations, but the limited range of the VHF radios (5–20 km at sea level, Thompson 1993), means that for many species, foraging occurs beyond the range of the transmitters.

In this paper we describe the use of an alternative method to determine the location of seals at sea. UHF satellite linked transmitters have been successfully used to track phocid seals and penguins at sea (e.g., Ancel *et al.* 1992; Davis and Miller 1992; Heide-Jorgensen *et al.* 1992; McConnell *et al.* 1992a; Hammond *et al.* 1993). We describe the first application of this technology on fur seals. The aims of the study were two-fold. First to identify the best attachment site for transmitters to maximize the number of locations at sea. Second, to use satellite telemetry to determine if female New Zealand Fur Seals show seasonal variation in foraging location.

[1]Department of Zoology, University of Otago, P.O. Box 56, Dunedin, New Zealand.

Pages 137–42 *in* MARINE MAMMAL RESEARCH IN THE SOUTHERN HEMISPHERE VOLUME 1: STATUS, ECOLOGY AND MEDICINE ed by M. Hindell and C. Kemper. Surrey Beatty & Sons, Chipping Norton. 1997.

METHODS

Study site and animals

Summer

In November and December 1993 we captured and physically restrained 18 female fur seals at Fuchsia Gully (45°52′S, 170°44′E), Otago Peninsula, South Island, New Zealand. All of the females had recently given birth. Females were weighed to the nearest 0.5 kg using a Salter 100 kg balance and tagged in each fore flipper using numbered Jumbo Rototags (Dalton Supplies, Australia). We also marked 14 of the females by clipping the guard hairs on the dorsal surface and bleaching the underfur (Schwarzkopf Brilliant Blonde).

We attached TDRs (Wildlife Computers Mk4) to four seals, and a dummy TDR to a fifth. The TDRs weighed 100 g and represented 0.25% of the seal's mass. The TDRs were glued to neoprene pads (4 mm thickness) and the pads were then glued to the females on the dorsal surface between the shoulder blades. Two of the females fitted with TDRs were also fitted with Satellite Transmitters (Platform Transmitter Terminals or PTTs, Telonics ST-6) which uplink to the ARGOS system (ARGOS 1989). The PTTs had a mass of 262 g which represents 0.65% of the average female's body mass. The PTTs were glued to neoprene pads 4 mm thick, and then glued to the rear of the head (Fedak *et al.* 1983).

Autumn

In April 1994, six more female fur seals were caught and tagged. Outside of the breeding season the animals were much more wary, and it was not possible to observe whether females were with pups immediately before capture. We determined if these females were still lactating by examining their teats. If distended, we assumed they were still lactating and nursing pups.

We deployed five TDRs as above. PTTs were deployed on three of these animals, but the deployments differed from the summer, in that the PTTs were affixed to the high curve of the arch of the back. The PTTs were retrieved at the same time as the TDRs.

The PTTs were programmed to transmit for 18 hours in 24 to extend battery life. The repetition rate of the PTTs was 50 sec. The ARGOS system provides location data of variable quality, depending on the number of uplinks received within a single satellite pass of approximately 15 minutes. At the time of this study, ARGOS provided a four Class Location Quality (LQ), ranging from Class 0, when only two uplinks were received, through to Class 3. ARGOS estimate that error in LQ has a standard deviation accuracy of 1 km for a Class 1, but does not present a quality control for Class 0. To determine the appropriate location, the locations were run through a forward/backward averaging filter which eliminated successive locations which would require the animal to swim at greater than $2 ms^{-1}$ (see McConnell *et al.* 1992b).

Effect of instrumentation

It is impossible to determine whether the instruments affect the animal's behaviour at sea, because there are no alternative means of collecting the information. However, as female attendance patterns may in part reflect how hard they are working at sea (Trillmich 1990), we compared the summer attendance patterns of females with TDRs (including the dummy TDR) to those which were only tagged and bleach marked. Attendance patterns for the females were obtained by scanning the rookery with binoculars between 0750 h and 1720 h and estimated to the nearest day. For all females with TDRs (including those with PTTs), the wet/dry periods recorded by the devices were compared with observations at the rookery.

RESULTS

Effect of instruments

Dates of deployment of the instruments in the two periods are shown in Table 1. The instruments did not affect the foraging patterns of female fur seals during the summer. Days at sea, days on land, and proportion of time at sea did not differ significantly between females with and females without instruments (Nested ANOVA, all ns, Table 2). Therefore, we assumed that the locations of female fur seals with attached instruments were not the result of aberrant responses to the devices. The attendance records from the observations at the rookery also corresponded with the wet/dry records of the TDRs (Harcourt *et al.* 1995).

The PTTs also did not appear to unduly affect the females behaviour at sea. In summer, females with PTTs and TDRs were intermediate between the other two TDR only females in maximum and mean dive depth, and maximum and mean dive duration (Table 3). In autumn, the PTT/TDR females also overlapped the TDR only females for mean and maximum dive depths and dive duration (Table 3).

Table 1. Weights of female New Zealand Fur Seals and dates and periods of deployment of satellite transmitters.

Female ID	Mass (kg)	PTT	Date deployed	Date retrieved	Days deployed
T0036	39	20887	29/12/93	13/01/94	16
T0045	37.5	20888	29/12/93	25/01/94	28
T0091	43.5	20889	06/04/94	28/04/94	22
T0093	42	20888	26/04/94	08/05/94	12
T0094	45.5	20887	26/04/94	24/05/94	28

Table 2. Summer attendance of New Zealand Fur Seal females: with and without instruments (TDRs and PTTs or TDRs only). All means are shown with ± 1 standard deviation.

Group	Sample size	Mean foraging trip length (days)	Mean shore attendance (days)	Mean % time spent at sea
With instruments	5	3.70 ± 0.78	3.72 ± 3.07	49.33 ± 10.96
Without instruments	13	2.98 ± 1.15	2.69 ± 0.78	53.80 ± 9.58

Table 3. Maximum and Mean Dive Durations and Depths for female New Zealand Fur Seals foraging in summer and autumn off the Otago Peninsula, New Zealand.

Female ID	Max duration (min)	Mean duration (min)	Max depth recorded (m)	Max depth extrapolated*	Mean depth
SUMMER					
T0034	6.17	1.18	163	163	18.1
T0036	4.75	0.71	115	115	14.8
T0038	3.17	0.67	108	108	12.1
T0045	4.67	15.7	107	107	15.7
AUTUMN					
T0089	5.67	0.85	198	198	18.7
T0091	6.17	1.28	127	348	32.2
T0092	6.67	1.21	225	382	39.9
T0093	4.33	1.14	177	177	33.5
T0094	7.17	1.02	106	314	21.5

*During certain individual dives, the maximum depth of the TDR was occasionally exceeded. In these cases, an estimate of the maximum depth was made based on the assumptions that the dive was V-shaped, and that the mean rate of descent and the mean rate of ascent did not change below the maximum TDR depth. The extrapolated depth was then calculated by taking the mid-point of the dive, and multiplying this time period by the descent rate, and separately the ascent rate, and taking the lesser of the two resulting depths. For 60 randomly selected, 150 m + V-shaped dives of known depth, this method resulted in estimates within ± 3.98%.

Location of foraging areas

Summer

The two PTTs produced eight separate locations. Female T0036 was located twice (both LQ = 0) during a seven day trip to sea, at 1815 h on January 11 (day 6), and 7.5 hours later (day 7) (see Fig. 1). Female T0045 was located on two separate trips. On her first trip (2 days) she was located four times, once on the night of December 30 at 1913 h (30/12/93, LQ = 0), and three times on the morning of December 31, at 0230 h (LQ = 1), 0410 h and 0500 h (31/12/93, LQ = 0) (Fig. 1). She was also located during a second trip of eight days: on January 12, at 1542 h (Day 7, LQ = 0), and at 1720 h (Day 7, LQ = 1) (Fig. 1). The furthest location from the rookery was 78 km for T0036, and 27 km for T0045, both within the estimated range based on transit times from the TDRs (Harcourt *et al.* 1995). None of the locations in the summer months were beyond the 200 m depth contour (Fig. 1).

Autumn

The tracks obtained from the deployment of the three PTTs in autumn are shown in Figure 2. Only four locations were obtained for female T0094, despite the transmitter being deployed for 28 days. The first two (both LQ = 0) were on the second and fourth days of a five-day foraging trip and placed her beyond the continental slope, approximately 180 km east of Fuchsia Gully. The third location indicated that she hauled out at another fur seal rookery, Nugget Point, about 110 km south-south-west of Otago Peninsula, before returning to Fuchsia Gully (Fig. 2), and the fourth location at sea occurred after the TDR memory was full. The other two deployments provided locations on several days. T0093 was located on the second, sixth, seventh, eighth and ninth days of a nine day trip (LQ = 1 on day 2 and 9, otherwise LQ = 0). She was located up to 145 km east of the rookery, and all but the last location (when she appeared to be in transit from her TDR record) were beyond the continental slope (200 m). T0091 was located each day from the second to the fifth of a six day trip (LQ's all = 0), up to 220 km east of the rookery and always beyond the 200 m contour. She was also located on the second and ninth day (both LQ = 0) of a 10 day trip, once beyond the continental slope, and once over it. It appears that in contrast to the summer deployments, the females were travelling beyond the edge of the continental slope and foraging in deep water (Fig. 2). Maximum dive depths and duration's were consistently greater during the autumn period (Table 3).

Table 4. Success of uplinks from satellite telemeters on two different attachment sites on female New Zealand Fur Seals.

Female ID	Attachment Site	PTT	Location per day at sea	Proportion of days uplinks received:days at sea	Proportion of foraging trips for which uplinks were received
T0036	Head	20887	1/10	10%	1/3
T0045	Head	20888	2/9	22%	2/2
T0091	Back	20889	7/18	39%	2/2
T0093	Back	20888	5/9	56%	1/1
T0094	Back	20887	2/23	9%	2/3

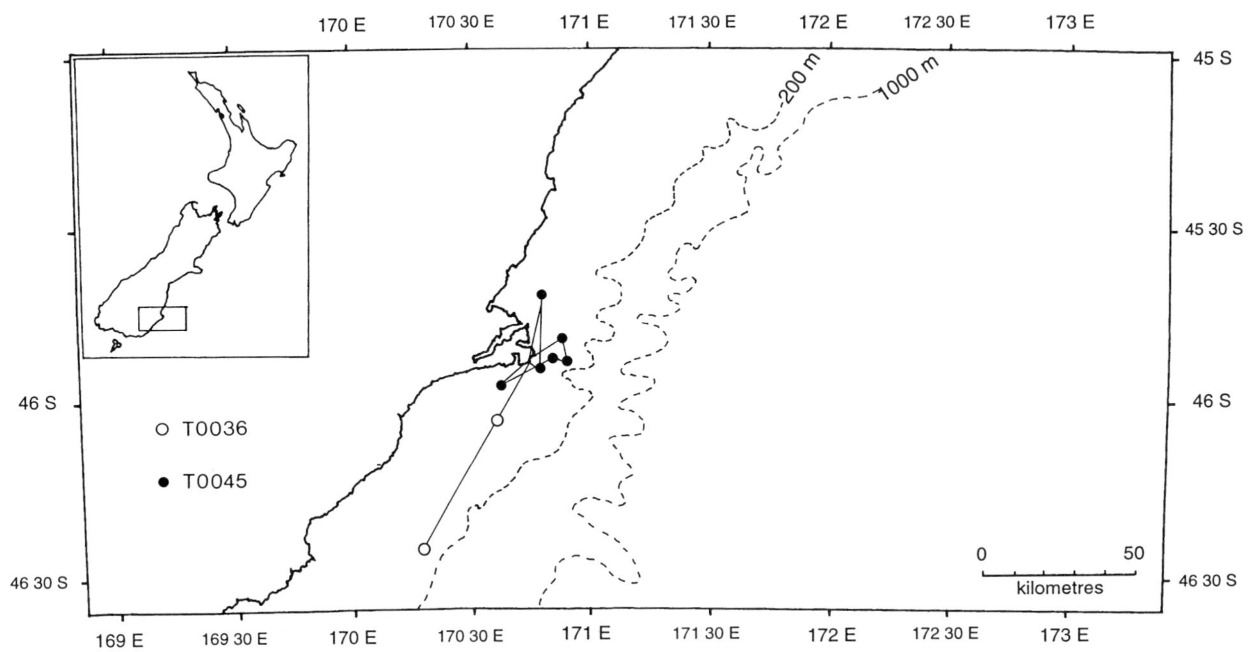

Fig. 1. Map showing the locations (at sea only) received from two female New Zealand Fur Seals foraging off the Otago coastline during summer 1993/1994.

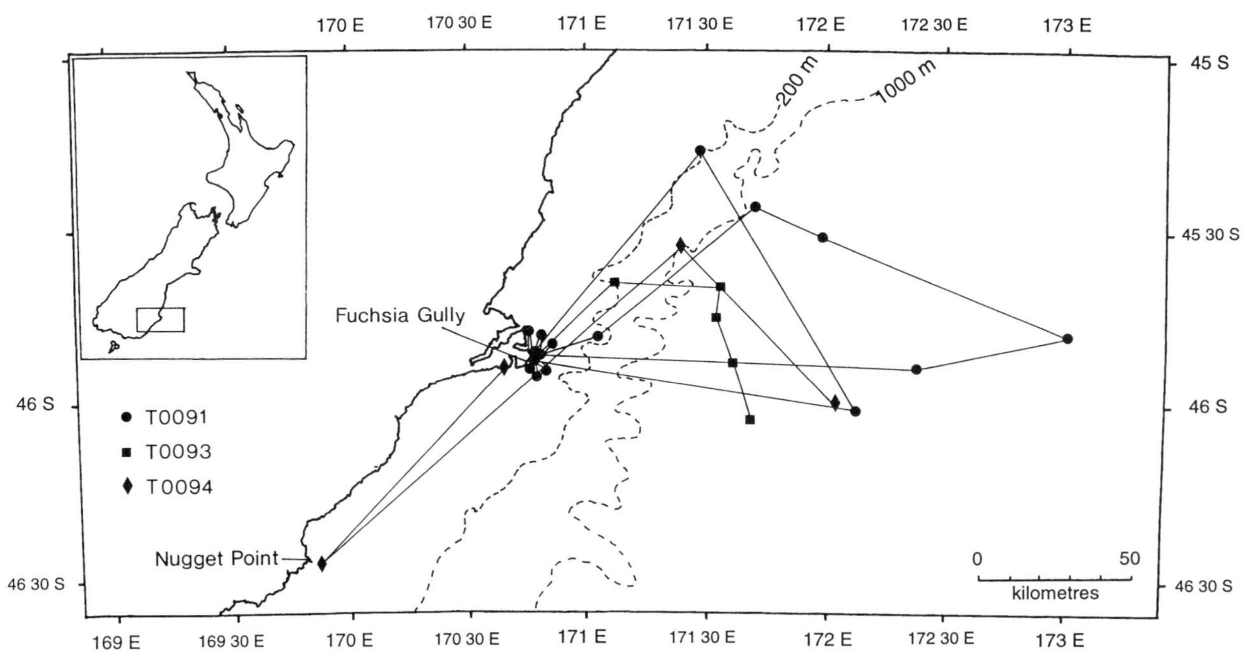

Fig. 2. Movements at sea of three female New Zealand Fur Seals tracked by satellite off the Otago coastline during autumn 1994.

Attachment of transmitter

The proportion of successful uplinks at sea for the two attachment sites, in relation to the number of days at sea per female are shown in Table 4. Attachment on the high arch of the back returned a higher proportion of useable uplinks for two of the transmitters, with successful uplinks on every trip to sea. One of the transmitters (20887) returned a very poor proportion of uplinks during both deployments (10% (summer) and 9% (autumn) of days at sea respectively). It also gave lower quality locations during testing prior to deployment.

Observations of the females at the rookery suggest that the head mounted PTTs irritated the animals. The two females with head mounted PTTs were seen to rub them for extended periods on the rocks. Both units showed wear and had partially lifted from the skin on recovery. The back mounted units did not appear to cause the same level of irritation. They were not subjected to pronounced rubbing at the rookery, abrasion of the units was less apparent and they were firmly attached on recovery.

DISCUSSION

Because of the nature of the marine environment, insights into the foraging habits of marine mammals must normally, come from indirect methods (but see Lyons 1989; Hoelzel 1992; Harcourt 1993). Our use of satellite telemetry has allowed us to determine foraging areas of New Zealand Fur Seals in an economically and technologically feasible way.

As satellite telemetry had previously only been reported from phocid seals we had to determine the most successful methodology with fur seals. We found differences in the proportion of successful uplinks depending on the site at which the PTT was attached. In phocid seals, uplinks at sea are most likely when the transmitter sits high on the head (McConnell *et al.* 1992a). Early attempts placing the transmitter on the back of grey seals, resulted in locations only when the animals were hauled out (McConnell *et al.* 1992a). In contrast, the much smaller female fur seals used in this study were more likely to be located at sea when the transmitter was on the back. Different behavioural patterns whilst at sea may account for this difference. Phocid species in which satellite telemetry has proved successful, tend to dive for long periods, and then spend a short but constant time at the surface with their nose exposed (Fedak and Thompson 1993). In contrast, fur seals dive for short periods, and tend to rest with their head below the surface, and their backs exposed (Harcourt, pers. obs.). The mass of the transmitter when placed on the relatively small female fur seal head may also have influenced the animal's behaviour at sea in a manner which reduced the time the antenna was exposed. This effect would be less likely with the much larger grey and elephant seals. There were sufficient uplinks to provide useable data when the transmitter was placed on the back, where it also did not appear to be irritating to the animal and was held more securely. For longer term deployment the back might be the only possible site of attachment, although further advances in miniaturization of transmitters may alleviate this problem.

The differences between the successful locations in the summer and the autumn are consistent with a seasonal shift in prey. From analysis of scats at the rookery, it appears in summer the seals fed primarily on species which inhabit the continental shelf and/or slope such as Red Cod, Ahuru and Lanternfish. In the autumn, Arrow Squid was the dominant item (Fea and Harcourt, unpubl. data). The duration of foraging trips by females also increased from a mean of 3.4 days in summer, to 7.6 days in autumn. That would allow them time to forage further from the rookery. Corresponding with the above, dive patterns also changed considerably between the two periods, from relatively shallow dives with maximum depths at dusk and dawn during the summer (Harcourt *et al.* 1995), to much deeper dives, which continued deep throughout the night in autumn. The maximum depth attained by any instrumented female in summer was 163 m, in autumn, at least three of the females frequently dove deeper than 225 m, and the maximum depth of the other two was also deeper than summer. These depths are only possible over, or beyond the edge of the continental slope, consistent with the foraging locations.

In conclusion, the use of satellite telemetry in combination with other methods has identified possible seasonal changes in foraging by female New Zealand Fur Seals. Further work is required to determine if this is an artefact of the small sample size. Future improvements in satellite technology, particularly further miniaturization of instruments, may improve our understanding of the ecology of these marine mammals.

ACKNOWLEDGEMENTS

This research was funded by the New Zealand Lottery Board and the University of Otago, and conducted under permit by the Department of Conservation, New Zealand. RGH was supported by a University of Otago Post-Doctoral Fellowship. John Croxall and Mike Fedak were extremely helpful in discussions on how to interpret dive records and satellite tracks. We thank Alan and Kay Dickson for access to Fuchsia Gully, and A. Schulman, S. Davenport,

N. Whitmore, C. Arbuckle, S. Hellyer and students from the Department of Zoology, University of Otago for assistance in the field. Bernie McConnell very kindly prepared the location maps, and Fritz Trillmich loaned us the TDRs. Thanks to Callan Duck and the two reviewers, Tony Preen and Brent Stewart for greatly improving the manuscript.

REFERENCES

Ancel, A., Kooyman, G. L., Ponganis, P. J., Gendner, J.-P., Lignon, J., Mestre, X., Huin, N., Thorson, P. H., Robisson, P. and Le Maho, Y., 1992. Foraging behaviour of Emperor penguins as a resource detector in winter and summer. *Nature* **360:** 336–39.

Antonelis, G. A., Stewart, B. S. and Perryman, W. F., 1990. Foraging characteristics of female northern fur seals *(Callorhinus ursinus)* and California sea lions *(Zalophus californianus). Can. J. Zool.* **68:** 150–58.

Argos, 1989. Guide to the Argos system. Argos CLS: Toulouse.

Crawley, M. C., 1990. New Zealand Fur Seal. Pp. 246–55 *in* The Handbook of New Zealand Mammals ed by C. M. King. Oxford University Press: Auckland.

Croxall, J. P., McCann, T. S., Prince, P. A. and Rothery, P., 1988. Reproductive performance of seabirds and seals at South Georgia and Signy Island, South Orkney Islands, 1976–1987: Implications for southern ocean monitoring studies. Pp. 261–85 *in* Antarctic Ocean and Resources Variability ed by D. Sarhage. Springer-Verlag: Berlin.

Davis, L. S. and Miller, G. D., 1992. Satellite tracking of Adelie penguins. *Polar Biol.* **12:** 503–06.

DeLong, R. L., Stewart, B. S. and Hill, R. D., 1992. Documenting migrations of Northern elephant seals using daylength. *Mar. Mamm. Sci.* **8:** 155–59.

Fedak, M. A., Anderson, S. S. and Curry, M. G., 1983. Attachment of a radio tag to the fur of seals. *J. Zool., Lond.* **200:** 298–300.

Fedak, M. A. and Thompson, D., 1993. Behavioural and physiological options in diving seals. Pp. 333–48 *in* Marine Mammals: Advances in behavioural and population biology ed by I. L. Boyd. Symposia of the Zoological Society of London No. 66. Oxford University Press: Oxford.

Fowler, C. W., 1987. Marine debris and northern fur seals: a case study. *Mar. Pollut. Bull.* **18:** 326–35.

Gentry, R. L. and Kooyman, G. L., 1986. Fur seals: maternal strategies on land and at sea. Princeton University Press: Princeton, N.J.

Goebel, M. E., Bengtson, J. L., DeLong, R. L., Gentry, R. L. and Loughlin, T. R., 1991. Diving patterns and foraging locations of female northern fur seals. *Fish. Bull.* **89:** 171–79.

Hammond, P. S., McConnell, B. J. and Fedak, M. A., 1993. Grey seals off the east coast of Britain: distribution and movements at sea. Pp. 211–24 *in* Marine Mammals: Advances in behavioural and population biology ed by I. L. Boyd. Symposia of the Zoological Society of London No. 66. Oxford University Press: Oxford.

Harcourt, R., 1993. Individual variation in predation on fur seals by Southern sea lions *(Otaria byronia). Can. J. Zool.* **71:** 1908–911.

Harcourt, R. G., Schulman, A., Davis, L. S. and Trillmich, F., 1995. Summer foraging by lactating female New Zealand Fur Seals *(Arctocephalus forsteri)* off Otago Peninsula, New Zealand. *Can. J. Zool.***73:** 678–90.

Harwood, J. and Croxall, J. P., 1988. The assessment of competition between seals and commercial fisheries in the North Sea and the Antarctic. *Mar. Mamm. Sci.* **4:** 13–33.

Heide-Jorgensen, M. -P., Stewart, B. S. and Leatherwood, S., 1992. Satellite tracking of ringed seals, *Phoca hispida*, off Northwest Greenland. *Ecography* **15:** 56–61.

Hoelzel, A. R., 1991. Killer whale predation on marine mammals at Punta Norte, Argentina; food sharing, provisioning and foraging strategy. *Behav. Ecol. Sociobiol.* **29:** 197–204.

Lalas, C. and Harcourt, R., 1995. Pup production of the New Zealand Fur Seal on Otago Peninsula, New Zealand. *J. Roy. Soc. N.Z.* **25:** 81–88.

Lipinski, M. R. and David, J. H. M., 1990. Cephalopods in the diet of the South African fur seal *(Arctocephalus pusillus pusillus). J. Zool., Lond.* **221:** 359–74.

Loughlin, T. R., Bengtson, J. L. and Merrick, R. L., 1987. Characteristics of feeding trips of female northern fur seals. *Can. J. Zool.* **65:** 2079–84.

Lyons, K. J., 1989. Individual dietary preferences in the female Californian sea otter. *Amer. Zool.* **29**(4)**:** 137A.

Mattlin, R. H., 1989. New Zealand Fur Seal — fisheries interaction: a growing concern. Proceedings of the 8th Biennial conference on the biology of Marine Mammals, Pacific Grove, California, Dec. 7–11, 1989.

Mattlin, R. H., 1991. Incidental catch of fur seals in a New Zealand trawl fishery. Proceedings of the 9th Biennial conference on the biology of Marine Mammals, Chicago, Illinois, Dec. 5–9, 1991.

McConnell, B. J., Chambers, C., Nicholas, K. S. and Fedak, M. A., 1992a. Satellite tracking of grey seals *(Halichoerus grypus). J. Zool., Lond.* **226:** 271–82.

McConnell, B. J., Chambers, C. and Fedak, M. A., 1992b. Foraging ecology of southern elephant seals in relation to the bathymetry and productivity of the Southern Ocean. *Ant. Sci.* **4:** 393–98.

Payne, M. R., 1977. Growth of a fur seal population. *Trans. Roy. Soc. Lond. B.* **279:** 67–79.

Shaughnessy, P. D. and Payne, A. I. L., 1979. Incidental mortality of Cape fur seals during trawl fishing activities in South African waters. *Fish. Bull. S. Afr.* **12:** 20–25.

Shaughnessy, P. D. and Goldsworthy, S. D., 1990. Population size and breeding season of the Antarctic fur seal *Arctocephalus gazella* at Heard Island — 1987/88. *Mar. Mamm. Sci.* **6:** 292–304.

Shaughnessy, P. D., Goldsworthy, S. D. and Libke, J. A., 1995. Changes in the abundance of New Zealand Fur Seals, *Arctocephalus forsteri,* on Kangaroo Island, South Australia. *Wildl. Res.* **22:** 201–15.

Stewart, B. S. and DeLong, R. L., 1993. Seasonal dispersion and habitat use of foraging northern elephant seals. Pp. 171–94 *in* Marine Mammals: Advances in behavioural and population biology ed by I. L. Boyd. Symposia of the Zoological Society of London No. 66. Oxford University Press: Oxford.

Taylor, R. H., 1982. New Zealand Fur Seals at the Bounty Islands. *N.Z. J. Mar. Fresh. Res.* **16:** 1–9.

Taylor, R. H., 1992. New Zealand Fur Seals at the Antipodes Islands. *J. Roy. Soc. N.Z.* **22:** 107–22.

Thompson, P. M., 1993. Harbour seal movement patterns. Pp. 225–40 *in* Marine Mammals: Advances in behavioural and population biology ed by I. L. Boyd. Symposia of the Zoological Society of London No. 66. Oxford University Press: Oxford.

Trillmich, F., 1990. The behavioural ecology of maternal effort in fur seals and sea lions. *Behaviour* **114:** 3–20.

Trites, A. W., 1992. Northern fur seals: why have they declined? *Aquat. Mamm.* **18:** 3–18.

Assessing the use of faecal and regurgitate analysis as a means of determining the diet of New Zealand Fur Seals

N. FEA[1] and R. HARCOURT[1]

Feeding trials were carried out on three captive New Zealand Fur Seals *Arctocephalus forsteri*. These seals were fed up to five different prey species (four teleosts and one cephalopod) of known size in order to measure percentage recovery and erosion of otoliths and cephalopod beaks in faeces and regurgitates. Glass beads were fed inside the experimental meals to provide an indication of defecation time and a comparative measure of recovery. Digestion coefficients (DC's) to account for the proportion of erosion during digestion were constructed for four teleost species. The DC's accurately predict the size range of ingested prey. The mean passage half-time for fish otoliths was 51.3 hours, and for squid beaks 62.3 hours. 64.8% of otoliths consumed were recovered, as were 58.0% of squid beaks. Most species showed comparable recovery of otoliths, with the exception of English mackerel where few of their fragile otoliths were recovered. These findings will have application in estimations of ingested prey number and size from otoliths and cephalopod beaks collected from wild seals.

INTRODUCTION

SEALS and other marine mammals are commonly perceived by the fishing industry to be in direct competition with commercial and recreational fisheries (Harwood and Croxall 1988; Harwood 1992). In order to assess the impact of seals on fisheries, accurate knowledge of their prey is needed. However, although several techniques have been used to determine the prey items of seals, no single method has proved adequate. Recently, there have been considerable advances in the technology used to determine where seals forage (e.g., McConnell *et al.* 1992, 1993; Harcourt and Davis 1997). However, a fundamental understanding of prey types and the proportions consumed by seals is lacking for most species.

Two principal methods have been widely used in the past to try and determine the prey species of seals in the wild. These are; (i) the collection of faecal and regurgitate material and (ii) the removal of stomach contents. Large numbers of scat and regurgitate samples can be collected with comparatively little effort and it is less intrusive than the alternative method. Stomach contents can be studied by removal of digestive tracts from dead animals, or by stomach lavage. Whilst the analysis of stomach contents enables the age and sex of the seal to be identified, shooting seals for stomach content analysis is not only wasteful due to loss, e.g., two-thirds of the sea-lions shot at sea by Fiscus and Baines (1966) sank and could not be retrieved, but stomachs are often empty when the animals are sampled, and so the animals are needlessly sacrificed (Street 1964; Murie and Lavigne 1986; Lipinski and David 1990; Beck *et al.* 1993). Lavage has been successfully used with pinnipeds (e.g., Antonelis *et al.* 1987), but in order to achieve large sample sizes, many animals must be lavaged, which is extremely time-consuming, can cause major disturbance to colonially breeding animals, and also entails a high level of risk for large pinnipeds which must be anaesthetized for the procedure. In contrast, faecal and regurgitate analysis is non-invasive. For studies of protected species, it is often the only possible method.

The hard remains of prey that resist total digestion found in faecal and regurgitate samples, primarily teleost sagittal otoliths and cephalopod beaks, have been used to identify prey species for many species of seal (e.g., Treacy and Crawford 1981; Tate 1981; North *et al.* 1983; Aurioles *et al.* 1984; Green *et al.* 1989; Carey 1992). However, certain biases have been shown to be associated with such methods. Some prey species may not appear in faecal samples due to lack of indigestible hard parts, and the proportional recovery of some identifiable remains may differ, leading to underestimations of particular species (Harvey 1989; Gales and Cheal 1992).

To gauge the overlap in prey of pinniped species with local fishing industries, studies have focused on the prey biomass consumption of seals (Prime and Hammond 1990; Frost and Lowry 1981; Beck *et al.* 1993). The length and

[1]Department of Zoology, University of Otago, P.O. Box 56, New Zealand.
Pages 143–50 *in* MARINE MAMMAL RESEARCH IN THE SOUTHERN HEMISPHERE VOLUME 1: STATUS, ECOLOGY AND MEDICINE ed by M. Hindell and C. Kemper. Surrey Beatty & Sons, Chipping Norton. 1997.

mass of fish and cephalopods can be positively correlated with measures of their otoliths and beaks (Jobling and Breiby 1986; Harvey 1989). This relationship has been used to calculate the original size of the prey consumed from the otoliths and beaks found in the scats and regurgitates. However, such calculations/ estimations of ingested prey size can significantly underestimate the size and/or presence of the original prey, as the effects of digestion may reduce the size of excreted hard parts, or completely dissolve fragile hard parts (Jobling and Breiby 1986; Dellinger and Trillmich 1988).

Previous diet studies on New Zealand Fur Seals have attempted to minimize any biases caused by digestion in different ways. Street (1964) obtained dietary information from examinations of gut contents and regurgitated food remains, as well as from observations of seals feeding at sea. Prey identified from gut contents were expressed as original total weight of the sample if there was sufficient prey flesh present with the associated hard part. Items that were retrieved from regurgitates were discussed in terms of prevalence by occurrence. Tate (1981) focused primarily on the cephalopod prey of the New Zealand Fur Seal, but did include some analysis of faecal remains. Weights of identified fish and cephalopod items recovered from scats and regurgitates were determined by reference to growth relationships compiled from fresh specimens of each species. No corrections were made for digestion. Green *et al.* (1990) compensated for such potential sources of error by only using otoliths that showed little or no signs of erosion for measurements used in prey size estimates. Carey (1992) did not have access to otolith size/fish size relationships compiled from items unaffected by digestion and so prey importance was described in terms of numerical abundance and frequency of occurrence in scats.

In order to account for any biases introduced through the effects of digestion, it is necessary to examine the effects of digestion on otolith and cephalopod beak recovery and the changes they may undergo during passage. Results from captive feeding experiments on different seal species have varied widely (e.g., Harbour Seals *Phoca vitulina*, da Silva and Neilson 1985; Harvey 1989; Markussen 1993. Grey Seals *Halichoerus grypus* Prime and Hammond 1987. California Sea Lions *Zalophus californianus*, South American Fur Seals *Arctocephalus australis*, Dellinger and Trillmich 1988 and Australian Sea Lions *Neophoca cinerea*, Gales and Cheal 1992). In a study on grey seals, Prime and Hammond (1990) calculated digestive efficiency from faecal remains, and concluded that, after correcting for the effects of digestion, no major component of the diet was missed using this method. From work on South American Fur Seals and California Sea Lions, Dellinger and Trillmich (1988) concluded that whilst estimates of biomass from faecal analysis are probably unreliable, it could be used to correctly estimate the numerical proportion of different prey species, because smaller otoliths showed similar passage times to larger ones. In contrast, Gales and Cheal (1992) reported that for the Australian Sea Lion retention times were highly variable, and they suggested that at least for this species, scat analysis was completely unreliable. They did however suggest, given the large degree of variability between studies, that for any given pinniped species, feeding experiments may help determine if faecal analysis is a valid technique for that species.

This study investigates the utility of using otoliths and squid beaks found in faeces and regurgitates to assess the prey consumption of New Zealand Fur Seals *Arctocephalus forsteri*. Passage times show how much time may elapse for the egestion of otoliths and beaks of different prey species. In the wild these findings may then be combined with knowledge of prey distribution and seal movements to estimate the foraging range of the seal colony being studied. The primary objectives were: (i) to estimate how long it took for otoliths and beaks of fish and cephalopods fed to captive New Zealand Fur Seals to appear in faeces and regurgitates; (ii) to compare differences in numbers of otolith and beak recovery between different prey species and individual fur seals; (iii) to determine the size reduction of recovered otoliths and beaks; and (iv) to derive Digestion Coefficients to predict the underestimation of prey size based on measurements from digested otoliths and beaks. Investigating the limits and potential of such approaches in diet studies has relevance for studies on free-living New Zealand Fur Seals, in the interpretation of information gained from faecal and regurgitate analysis.

METHODS

Feeding trials were conducted on three adult New Zealand Fur Seals (one female, two males) at the Marineland complex in Napier, New Zealand during March and April, 1994. The animals ranged from 9 to 19 years of age and 48–152 kg mass with the single female being the youngest and smallest. The two males were introduced to captivity after being injured in the wild, and the female was born at Marineland. The seals were fed maintenance meals of decapitated fish for two weeks before the feeding trials commenced to clear their digestive tracts of otoliths and beaks.

For each trial, the experimental animal was isolated from the other two in an enclosure with a seawater pool 2.3 m deep and 6.1 m diameter,

75 000 l capacity. The other two seals remained in the larger adjacent enclosure with a semi-connecting pool. Each seal was subjected to two separate sets of experimental feeding regimes. These involved a series of experimental meals of prey species that each seal was already known to eat whole, as each seal had a known history of feeding preferences. During the regime, each seal was fed consecutive meals of the different prey species. After each prey species had been fed once, the seals were fed only headless fish for the remainder of the trial, which was at least four days after the last headed prey was consumed, and when at least two consecutive days of sample washing's contained no otoliths or beaks. Species most likely to be accepted whole by the seals were offered to the seals twice daily, and if a meal was refused it was kept until the next feeding time when the seal would accept it. The duration of each feeding trial therefore ranged between five and 10 days depending on what prey species were, and when it was accepted by the seal. Each seal was subjected to two feeding trials. The entire experiment's duration was 43 days.

The males accepted whole prey from a wider size range than the female. Males consumed Red Cod *Pseudophycis bachus*, Jack Mackerel *Trachurus declivis*, English Mackerel *Scomber australisicus*, Koheru *Decapterus koheru*, Yellow-eyed Mullet *Aldrichetta forsteri*, Quinnat Salmon Smolt *Oncorhyncus tschawytcha*, and Arrow Squid *Nototodarus sloanii* but the female only accepted whole mullet, smolt and arrow squid. For all meals, the prey was offered head first to the seal to avoid fish spines being caught in the seals throat. The weights each half-day meal provided was roughly half the estimated daily intake of a seal, approximately 10% of its body weight. The lengths (mm) (squid, dorsal mantle length; fish, fork length) and weights (g) of the entire prey specimens were recorded prior to feeding.

In order to indicate rate of passage for particular meals, and also to give a comparative measure of collecting efficiency in addition to the recovery of otoliths and beaks, gelatine capsules containing indigestible glass marker beads were inserted into the opercula cavity of the fish, and in the mantle cavity of the squid. The possibility that prey items may be retained for long periods (Bigg and Fawcett 1985) is not addressed in this study due to the limited time frames of the feeding trials.

Scats and regurgitates were collected each morning at 8 a.m. Those found on dry ground in the enclosure were scooped up into a plastic bag using a plastic spatula. The floor of the enclosure was then hosed down, which washed any further material into the pool. A bucket end with 0.5 mm mesh netting connected to one end was wedged into the outlet pipe of the pools drainage well. The outlet was then opened to allow the pool to begin draining. Material on the floor of the pool was swept towards the drain using a long handled broom from the edge of the pool. After the pool had been thoroughly swept, the outlet pipe was closed, and the netting containing scats, regurgitates and other debris, was removed. Distinction between the recovery of items in scats and regurgitates was not possible due to the mixing of samples in the pool. The seals were observed egesting items via both forms in the pool.

Samples collected were washed through two sieve sizes, 2.0 mm, and 0.5 mm. Hard parts and other identifiable remains were removed, and stored in plastic vials containing fresh water. The colour and number of the plastic marker beads was recorded and the otoliths and beaks were then isolated and stored dry in air-tight plastic bags.

The measurements of the otoliths correspond to the orientation of the otoliths *in situ*, with length, width and thickness measures to ± 0.01 mm using digital callipers, which were also used to measure upper and lower rostral lengths of the squid beaks. Weights of both otoliths and beaks were measured to ± 0.0001 gm using a H51 Mettler balance.

A test of otolith recovery efficiency was conducted once, between feeding trials, after the pool had been drained and cleared of debris. Eighty beads, 24 Jack Mackerel, three Herring and 20 Smolt otoliths plus nine squid beaks were scattered in to the pool, which was subsequently drained and material collected in the outlet catch-net. One-hundred per cent of otoliths and beaks and 95.1% of glass beads were recovered.

In order to determine the relationship between undigested fish and otolith size for each prey species, regressions were constructed from a reference collection loaned by Dr C. Lalas (Marine Science, University of Otago, PO Box 56, Dunedin, New Zealand). Otoliths and beaks removed from prey specimens not fed to the seals at Marineland were also measured and added to this data.

RESULTS

Mean recovery of otoliths for the three seals in all trials was 63.2% (SD ± 24.1), ranging from 6.3% to 100%. There were no significant differences between seals for recovery of prey items or beads for each species and so the data for individual species recovery was pooled for all three seals (Fig. 1, Table 1). The glass marker beads had an overall recovery of 81.3% (range 30% to 100%, Fig. 2). There were no differences in otolith and beak recovery between the individual seals (ANOVA $F_{2,1}$ = 2.49, n.s., Fig. 3). For the prey species fed in more than one feeding trial

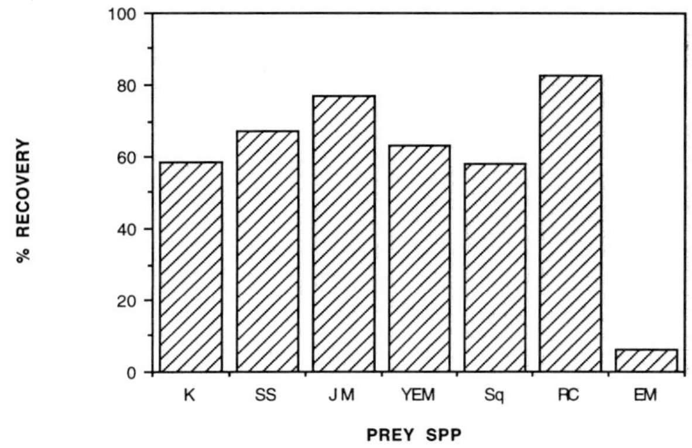

Fig. 1. Per cent otolith and beak recovery for all fish and squid species fed to seals (K = Koheru, SS = Salmon Smolt, JM = Jack Mackerel, YEM = Yellow-eyed Mullet, Sq = Arrow Squid, RC = Red Cod, EM = English Mackerel).

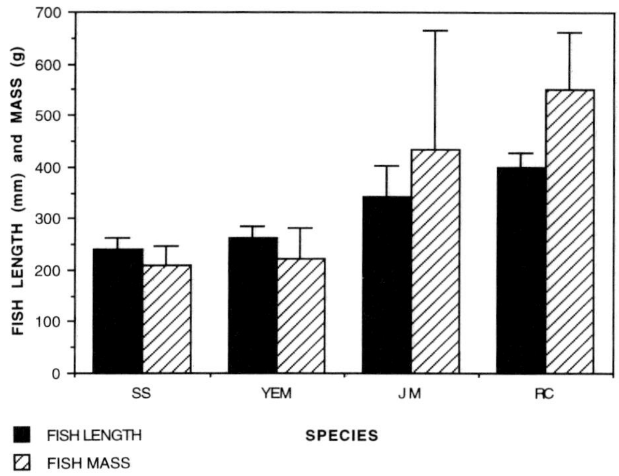

Fig. 4. Mean weights (gm) and lengths (mm) of fish species fed to seals (SS = Salmon Smolt, YEM = Yellow-eyed Mullet, JM = Jack Mackerel, RC = Red Cod).

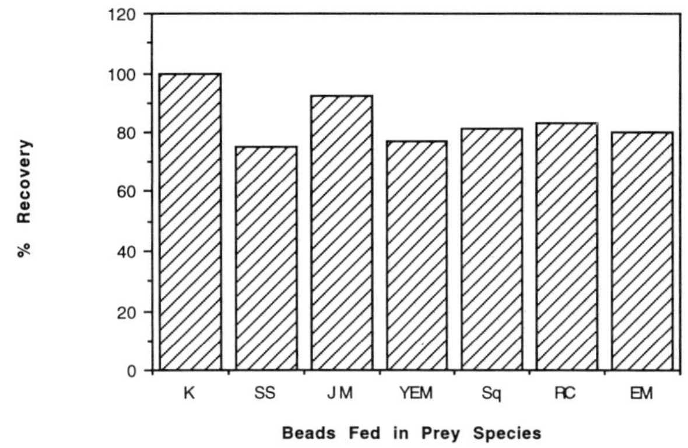

Fig. 2. Per cent recovery of beads fed with prey species to seals (K = Koheru, SS = Salmon Smolt, JM = Jack Mackerel, YEM = Yellow-eyed Mullet, Sq = Arrow Squid, RC = Red Cod, EM = English Mackerel).

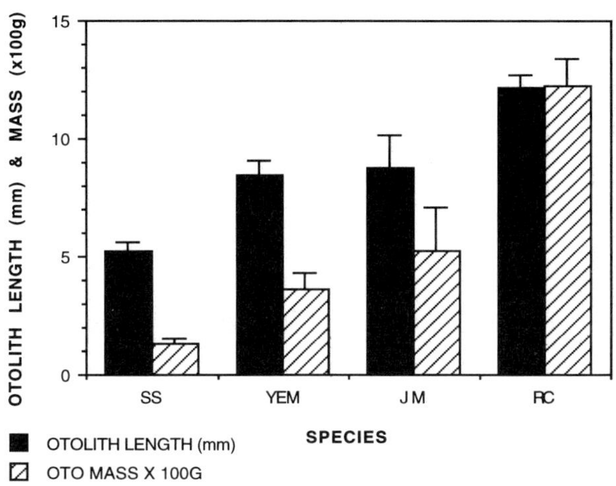

Fig. 5. Mean weight (gm) and lengths (mm) of undigested otoliths (using regression equations constructed from reference collection of undigested otoliths).

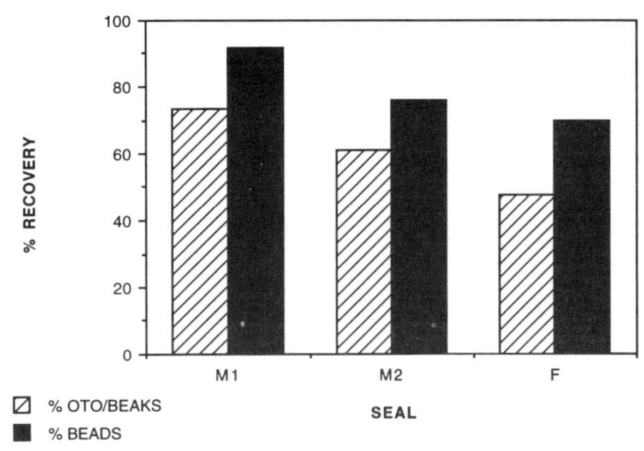

Fig. 3. Per cent recovery of otoliths and beaks combined, and beads for the three seals (M1 = Older Male Seal, M2 = Younger Male Seal, F = Female Seal).

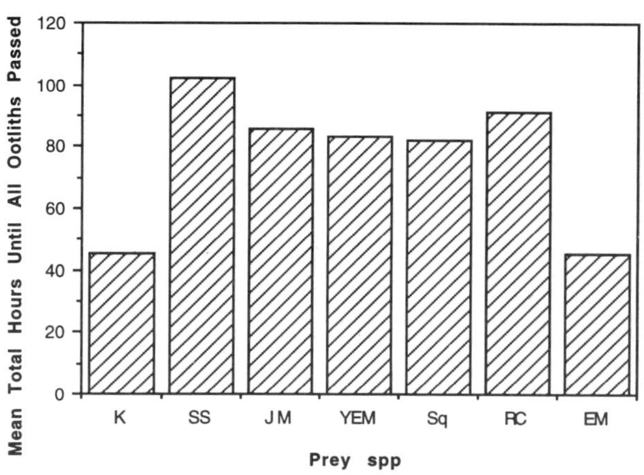

Fig. 6. Total passage time (hrs) of prey species fed to seals (K = Koheru, SS = Salmon Smolt, JM = Jack Mackerel, YEM = Yellow-eyed Mullet, Sq = Arrow Squid, RC = Red Cod, EM = English Mackerel).

Table 1. Proportion recovered (otoliths, beaks or beads), proportion of actual feed recovered, passage half-times and Digestion Coefficients for six species of fish and one squid fed to three New Zealand Fur Seals.

Species	% Otolith/ beak recovered	% Bead recovered	% Actual feed	Pt/2 (hours)	Digestion co-efficients	No. prey in total	Mean length (range)	Mean weight (range)	No. trials fed in (n seals)
Salmon Smolt	67.1	75.0	72.0	46.3	2.50 ± 1.94	94	240.39 (120–278)	208.78 (75–370)	6 (3)
Yellow-eyed Mullet	63.2	76.78	65.4	47.0	2.71 ± 3.35	98	262.73 (218–320)	223.29 (105–360)	6 (3)
Jack Mackerel	77.0	92.08	82.3	50.0	1.46 ± 0.51	32	340.5 (262–482)	433.59 (195–1 025)	4 (2)
Red Cod	82.6	83.33	85.7	55.0	1.95 ± 0.65	15	397.93 (356–456)	550.33 (430–790)	2 (1)
Arrow Squid	58.0	81.18	71.1	62.3	N/A	58	254.57 (190–400)	320.91 (140–940)	6 (3)
English Mackerel	6.3	80.0	12.5	45	N/A	8	319.8 (306–337)		1 (1)
Koheru	58.3	100.0	83.3	45	N/A	6	366.7 (350–391)	686.7 (630–745)	1 (1)

(i.e., English Mackerel and Koheru were only available for one trial each), mean recovery values were high when the results from the seals were grouped (Red Cod 82.6%, Jack Mackerel 77.0%, Yellow-eyed Mullet 63.2%, Salmon Smolt 63.1%, Table 1). Red Cod and Jack Mackerel were larger than Yellow-eyed Mullet and Salmon Smolt (Fish Mass: ANOVA, $F_{3,235}$ = 88.11, $p < .001$; Fish Length: ANOVA, $F_{3,235}$ = 170.79, $p < .001$; Fig. 4). Their otoliths appeared more robust, were significantly longer and weighed more (Otolith Length: ANOVA, $F_{3,235}$ = 648.99, $p < .001$; Otolith Mass: ANOVA, $F_{3,235}$ = 712.00, $p < .001$; Fig. 5). In contrast, English Mackerel have thin and fragile otoliths in vivo and only 6.3% of these otoliths were recovered.

To back-calculate the actual numbers of each fish species consumed by the seals, left and right otoliths of the fish species were counted, as were the upper and lower beaks of the arrow squid. The half with the highest number was then used to estimate the amount eaten. For the prey species used in more than one trial, the mean estimate of the number of fish ingested was 73.0% (SD ± 23.9) of the actual number. This did not vary greatly between these species (ANOVA $F_{1,4}$ = 0.412, n.s.) with means ranging from 65.4% to 85.7% (Table 1). The single feeding trial using English Mackerel showed recovered otoliths underestimated the amount of ingested prey by 87.5%.

The time before all recovered otoliths and beaks were recorded (Total Pt) had a wide variance between prey species (excluding Koheru and English Mackerel which had small sample sizes). This ranged from a minimum of 17 hours recorded for an Arrow Squid trial to 185 hours for smolt (Fig. 6). The passage half-time (Pt/2) is the number of hours it takes for half the total recovered otoliths and beaks to reappear (Dellinger and Trillmich 1988). This measure was similar amongst prey species used in more than one trial (ANOVA $F_{1,4}$ = 0.36, n.s., Table 1).

Using intact fish specimens for those species in Table 1, otolith size predicted fish length better than mass, except for Jack Mackerel where the reverse was true (Table 2). We applied these relationships to the digested otoliths, and then compared the predicted outcome with the actual fish fed to the seals. Without correction for digestion, the size of the fish of all species was greatly underestimated (Salmon Smolt by 21.5 %, Yellow-eyed Mullet by 19.7%, Jack Mackerel by 28.4% and Red Cod by 35.4%). Back calculations from these overall values did accurately predict the mass of ingested prey, but there was a high degree of variation between individual seals (Fig. 7). The proportion of loss of otolith material varied by species. We constructed digestion coefficients (DC) using the formula:

DC = Undigested otolith mass/digested otolith mass

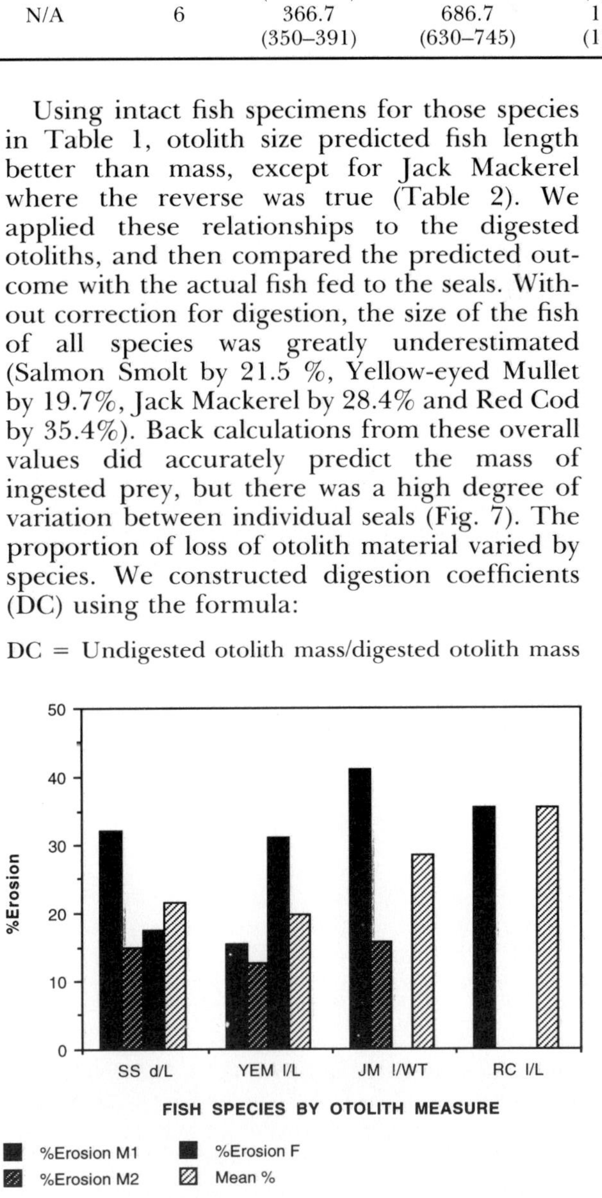

Fig. 7. Mean per cent size underestimation of fish fed to seals from digested otoliths (SS d/L = Salmon Smolt otolith depth/Fish length, YEM l/L = Yellow-eyed Mulley otolith length/Fish length, JM l/WT = Jack Mackerel otolith length/Fish weight, RC l/L = Red Cod otolith length/Fish length, M1 = Older Male Seal, M2 = Younger Male Seal, F = Female Seal).

Table 2. Regression equations derived from the otoliths of undigested fish, used to predict fish length and mass. All predict fish length except Jack Mackerel which predicts fish mass.

Species	Equation	R^2	'n'	Mean (range) Fish length (mm) Fish weight (g)
Salmon Smoult	$y = -17.689 + 43.905x + 1.0057x^2$ (x = o length)	0.927	55	237.73 (41–410)
(y = fish length)	$y = 54.185 + 1.6165e + 4x - 1.6787e + 5x^2$ (x = mass)	0.936		120.26 (0.79–292)
Yellow-Eyed Mullet	$y = -29.231 + 41.101x - 0.7508x^2$ (x = length)	0.965	90	197.55 (46–375)
(y = fish length)	$y = 64.647 + 7503.4x + -5.2822e + 4x^2$ (x = mass)	0.976		150.91 (0.74–622)
Jack Mackerel	$y = 5.02.39 - 220.39x + 24.511x^2$ (x = length)	0.842	34	307.97 (136–505)
(y = fish mass)	$y = -39.452 + 8338.6x + 3.7149e + 4x^2$ (x = mass)	0.832		635.74 (25–1 640)
Red Cod	$y = -26.273 + 13.219x - 1.8346x^2$ (x = length)	0.972	51	259.29 (51–635)
(y = fish length)	$y = 45.159 + 3511.0x - 4566.2x^2$ (x = mass)	0.985		258.05 (1.1–1 850)

based on predicted mass of undigested otoliths constructed from the fish length: otolith mass (for Jack Mackerel, fish mass: otolith mass) regressions outlined above, and a random selection of digested otoliths.

The digestion coefficients for each species are shown in Table 1. There was no significant difference in DC between the two smaller species (Salmon Smolt and Yellow-eyed Mullet, *t*-test, d.f. = 190, t = .278, n.s.), but there was between the two larger species (Red Cod vs Jack Mackerel, *t*-test, t = 7.996, $p < 0.01$). The two larger fish species showed proportionally lower loss of otolith material than the two smaller fish species (Small vs Large, *t*-test, d.f. = 45, t = 6.031, $p < 0.05$), with Red Cod the least overall. There was no decrease in rostral lengths of squid beaks in this study. We applied the DC's constructed from the 1994 experiments to a series of partially digested otoliths from known size prey, the result of a pilot study on the same three seals in 1993. After application of the DC's and the regressions shown in Table 2, a range of fish lengths were derived. This estimated range overlapped the size range of fish fed to the seals in 1993.

DISCUSSION

The results of this study, in combination with those reported by others, point to the importance of experimental feeding studies being conducted on different seal species. The value for mean percentage recovery of otoliths from the major fish species in this study (63%) is comparable to the highest found by others for captive seal feeding studies, and falls between the values recorded by Harvey (1989) of 58% (from Harbour Seals) and the recovery rates of Prime and Hammond (1986) of 78% and 88% (using Grey Seals). Cephalopod recovery was also higher at 57.98% recovery, than the 37% reported by Harvey (1989) or the 46.3% by Tollit *et al.* (1994). Both measures also contrast strikingly with very low values reported by da Silva and Neilson (1985) who only recovered 4% from a Harbour Seal, and the distinct differences between fish otoliths and cephalopod beaks reported by Gales and Cheal (1992) with the Australian Sea Lion. It may be that some variation between the above-mentioned studies was due to experimental design, particularly whether the animals had access to a swimming area. Even so, it is clear that the utility of generalizing results from any one species of seal, are limited.

However, the *raison d'etre* for this series of experiments was to examine the utility of faecal and regurgitate analysis as a means of determining diet composition of natural populations of New Zealand Fur Seals. From our results, it appears that, provided caution is used in extrapolations to other prey species, it may be possible to use faecal and regurgitate analysis as a means of looking at variation in numerical representation, either inter-annually or seasonally. For four teleost prey species, that were distinctly different in regard to size and otolith robustness, comparable recovery rates were achieved. This is despite the fact that the degree of erosion was much higher for the smaller otoliths. However, for a fifth species, English Mackerel, with very fragile otoliths, a tentative result of poor recovery was attained. Thus, although it may be possible to look at how some species vary in proportion in the diet with time, the method is poor for looking at complete dietary representation by species.

Differential passage rates of prey hard parts may also bias evaluation of prey importance. We cannot rule out the possibility that otoliths and beaks not recovered in this study were not lost due to erosion, but were retained for longer in the digestive tracts of the seals. Our finding that 100% of otoliths scattered in the pool were recovered in the outlet net also supports this. Jobling and Breiby (1986) stated that retention of large indigestible particles may occur with frequent feeding in mammals. However, they, also suggested that marine mammals may be meal eaters rather than frequent feeders, and that hard parts should therefore be voided rapidly. Bigg and Fawcett (1985) and Gales and Cheal (1992) found that in Northern Fur Seals and Australian Sea Lions, respectively, squid beaks tended to cluster near the pyloric sphincter, and so were either retained for longer than otoliths, or regurgitated later. Remains of cephalopods are often regurgitated by fur seals, (along with other large hard parts of prey) and the resulting detritus may represent accumulation of prey material from a number of feeds (Bigg and Fawcett 1985). Following Dellinger and Trillmich (1988), we used the passage half-time as an indication of the average time otoliths (and squid beaks) were exposed to digestion, and this was fairly consistent across prey species in this study. Most of the digested material appeared within 2–3 days of ingestion. For the New Zealand Fur Seal in the wild, this means that more otoliths and beaks are likely to be egested and therefore collected on land in the summer, when foraging trips are shorter in duration, than in the autumn (Harcourt and Davis 1997). They are also more likely to be more representative of what the animals are eating when foraging trips of adult seals are shorter. In the autumn, female fur seals may be at sea for up to 14 days and forage over 200 km from the rookery (Harcourt and Davis 1997). It is not unlikely that the prey encountered when foraging beyond the continental slope is different from that encountered in shallow water nearer shore. With passage half times of around 50 hours, these offshore prey items will probably be under represented in faeces and/or regurgitates. In addition, there are other factors which will affect the rate of passage for diet items such as size of meal, time of consecutive feeds, activity of the seal and perhaps the oil content of the fish, but these factors have not as yet been adequately studied.

Incomplete numerical representation of prey species will have profound importance on quantifying prey consumed. When sampling faeces of wild seals from haul-out sites, loss of otolith or beak material may occur at several stages. Seals have been observed consuming only parts of their prey, in particular discarding the heads of large prey (Rae and Shearer 1965) and as this study and others have shown, complete digestion of some otoliths may occur (Murie and Lavigne 1986). Calculations of prey biomass consumption by wild seals involves multiplying the number of otoliths in the sample by the mean weight of the fish represented by those otoliths. It assumes that all prey items contain otoliths or beaks and that the fish remains found in the faeces at the collection sites to be representative of the prey consumed locally by that seal colony (Prime and Hammond 1987). It therefore ignores the probable loss of a significant number of otoliths and beaks as described above, as well as any prey items which do not contain hard parts. However, for commercial prey species with robust otoliths such as those used in this study, at least some idea of their relative importance may be gained.

This may be further enhanced by measures of erosion. From measurements of otolith erosion we have constructed digestion coefficients which allow us to predict the size of fish consumed from their recovered otoliths. Many previous studies which used otolith size to calculate fish size (Murie and Lavigne 1988; Beck *et al.* 1993) considered erosion of otoliths to be a negligible factor. In line with other workers (e.g., Prime and Hammond 1990) we have illustrated how the size of the fish consumed may be considerably underestimated due to otolith digestion. Having measured otolith erosion for fish species fed to captive seals, we have estimated the degree of inaccuracy inherent in measurements of eroded otoliths. Considering the many factors that may affect passage rates and erosion of prey items (as noted above), applying digestion coefficients to otoliths retrieved from wild seals assumes that passage and erosion of these prey species has little variation, even within prey species at different stages of growth. Prime and Hammond (1987) suggested that a weighted mean DC calculated from several species could be used to correct for digestion. We argue that a single figure for combined fish species will not prove reliable because of variability between species.

ACKNOWLEDGEMENTS

This research was funded by the New Zealand Lottery Board and the University of Otago, and conducted under permit by the Department of Conservation, New Zealand. RH was supported by a University of Otago Post-Doctoral Fellowship. We thank Gary McDonald and the staff of the Marineland of New Zealand for all their help. Bruce Dix gave helpful suggestions. Chris Lalas kindly loaned us his reference collection for the back calculations and taught us how to measure and identify many different types of otoliths. The reviewers, Rosemary Gales and Ken Green provided numerous constructive comments which greatly improved the manuscript.

REFERENCES

Antonelis, G. A., Lowry, M. S., DeMaster, D. P. and Fiscus, C. H., 1987. Assessing northern elephant seal feeding habits by stomach lavage. *Mar. Mamm. Sci.* **3:** 308–22.

Aurioles, D., Fox, C., Sinsel, F. and Tanos, G., 1984. Prey of the California sea lion *(Zalophus californianus)* in the Bay of La Paz, Baja California Sur, Mexico. *J. Mamm.* **65:** 519–21.

Beck, G. G., Hamill, M. O. and Smith, T. G., 1993. Seasonal variation in the diet of Harp seals *(Phoca groenlandica)* from the Gulf of St Lawrence and Western Hudson Strait. *Can. J. Fish. Aquat. Sci.* **50:** 1363–371.

Bigg, M. A. and Fawcett I., 1985. Two biases in diet determination of northern fur seals *(Callorhinus ursinus)*. Pp. 284–91 *in* Marine Mammals and Fisheries ed by J. R. Beddington, R. J. H. Beverton and D. M. Lavigne. Allen and Unwin: London.

Carey, P., 1992. Fish prey species of the New Zealand fur seal *(Arctocephalus forsteri,* Lesson). *N.Z. J. Ecol.* **16:** 41–46.

da Silva, J. and Neilson, J. D., 1985. Limitations of using otoliths recovered in scats to estimate prey consumption in seals. *Can. J. Fish. Aquat. Sci.* **42:** 1439–442.

Dellinger, T. and Trillmich, F., 1988. Estimating diet composition from scat analysis in otariid seals (Otariidae): is it reliable? *Can. J. Zool.* **66:** 1865–870.

Fiscus, H. C. and Baines, G. A., 1966. Food and feeding behaviour of Steller and California sea lions. *J. Mamm.* **47:** 195–200.

Frost, K. J. and Lowry, L. F., 1981. Trophic importance of some marine gadids in northern Alaska and their body-size otolith relationships. *Fish. Bull.* **79:** 187–92.

Gales, N. J. and Cheal, A. J., 1992. Estimating diet composition of the Australian sea lion (*Neophoca cinerea*) from scat analysis: an unreliable technique. *Wildl. Res.* **19:** 447–56.

Green, K., Burton, H. R. and Williams, R., 1989. the diet of Antarctic fur seals *Arctocephalus gazella* during the breeding season at Heard Island. *Ant. Sci.* **1**:317–24.

Green, K., Williams, R., Handasyde, K. A., Burton, H. R. and Shaughnessy, P. D., 1990. Interspecific and intraspecific differences in the diets of fur seals, *Arctocephalus* species (Pinnipedia: Otariidae) at Macquarie Island. *Aust. Mamm.* **13:** 193–200.

Harcourt, R. and Davis, L., 1997. The use of satellite telemetry to determine fur seal foraging areas. Pp. 137–42 *in* Marine Mammal Research in the Southern Hemisphere Volume 1: Status, Ecology and Medicine ed by M. Hindell and C. Kemper. Surrey Beatty & Sons: Chipping Norton.

Harvey, J. T., 1989. Assessment of errors associated with Harbour seals *(Phoca vitulina)* faecal sampling. *J. Zool.* **219:** 101–11.

Harwood, J., 1992. Assessing the competitive effects of marine mammal predation on commercial fisheries. *S. Afr. J. Mar. Sci.* **12:** 689–93.

Harwood, J. and Croxall, J. P., 1988. The assessment of competition between seals and commercial fisheries in the North Sea and the Antarctic. *Mar. Mammal Sci.* **4:** 13–33.

Jobling, M. and Breiby, A., 1986. The use and abuse of fish otoliths in studies of feeding habits of marine piscivores. *Sarsia* **72:** 255–60.

Lipinski, M. R. and David, J. H. M., 1990. Cephalopods in the diet of the South African fur seal (*Arctocephalus pusillus pusillus*). *J. Zool. Lond.* **221:** 359–74

McConnell, B. J., Chambers, C., Nicholas, K. S. and Fedak, M. A., 1992. Satellite tracking of grey seals *(Halichoerus grypus)*. *J. Zool., Lond.* **226:** 271–82.

McConnell, B. J., Chambers, C. and Fedak, M. A., 1993. Foraging ecology of southern elephant seals in relation to the bathymetry and productivity of the Southern Ocean. *Ant. Sci.* **4:** 393–98.

Markussen, N. H., 1993. Transit time of digesta in captive harbour seals *(Phoca vitulina)*. *Can. J. Zool.* **71:** 1071–073.

Murie, D. J. and Lavigne, D. M., 1986. Interpretations of otoliths in stomach contents analyses of phocid seals: quantifying fish consumption. *Can. J. Zool.* **64:** 1152–157.

North, A. W., Croxall, J. P. and Doidge, D. W., 1983. Fish prey of the Antarctic fur seal *Arctocephalus gazella* at South Georgia. *Br. Antarct. Surv. Bull.* **61:** 27–37.

Prime, J. H. and Hammond, P. S., 1987. Quantitative assessment of grey seal diet from faecal analysis. Pp. 165–81 *in* Approaches to Marine Mammal Energetics ed by A. C. Huntley, D. P. Costa, G. A. J. Worthy and M. A. Castellini. Society for Marine Mammalogy, Special Publication No. 1.

Prime, J. H. and Hammond, P. S., 1990. The diet of grey seals from the South-Western North Sea assessed from analyses of hard parts found in faeces. *J. Appl. Ecol.* **27:** 435–47.

Rae, B. B. and Shearer, W. M., 1965. Seal damage to salmon fisheries. *Mar. Res. Dep. Agric. Fish. Scotl.* **2:** 1–39.

Street, R. J., 1964. Feeding habits of the New Zealand fur seal. *NZ Mar. Fish. Tech. Rep.* No. 9.

Tate, M. L., 1981. The autumn-winter diet of the New Zealand fur seal *Arctocephalus forsteri* (Lesson) with special reference to its cephalopod prey. Dip. Sci Thesis, University of Otago: New Zealand.

Tollit, D. J., Froome, M. A., Hall, A. J. and Donovan, T., 1994. Assessment of errors associated with current estimates of the diet composition of common seals. *In* The behaviour and ecology of common and grey seals in the Moray Firth ed by P. M. Thompson, D. J. Tollitt, D. Wood, A. MacKay and P. A. Racey. Final Contract Report to the Scottish Office Agriculture and Fisheries Department.

Treacy, S. D. and Crawford, T. W., 1981. Retrieval of otoliths and statoliths from gastrointestinal contents and scats of marine mammals. *J. Wildl. Manage.* **45:** 990–93.

Diet and diving behaviour of sympatric fur seals *Arctocephalus gazella* and *A. tropicalis* at Macquarie Island

S. D. GOLDSWORTHY,[1,2] M. A. HINDELL[3] and H. M. CROWLEY[4,5]

The diet and diving behaviour of lactating Antarctic and Subantarctic Fur Seals *Arctocephalus gazella* and *A. tropicalis*, was studied at Macquarie Island during the 1990–91 summer-autumn period. There was no marked difference in the diet of each species, which was dominated by fish in both cases. Twelve fish species were identified in the seal's diet, all from the family Myctophidae. One species, *Electrona subaspera*, dominated, accounting for 92% numerical abundance of identifiable otoliths for both seal species. Data recovered from time-depth recorders (TDR's) indicated no difference in the diving behaviour of each species. Diving behaviour exhibited a strong nocturnal pattern, with almost all diving occurring at night and at very shallow depths (median 8 m). Bouts of deeper diving (15–20 m) occurred at sunset and sunrise, most other, presumably foraging dives, were about 5–6 m. Foraging trips of females wearing TDR's lasted 2.5 and 5.0 days for *A. gazella* and *A. tropicalis*, respectively. In addition to these, half of the females of each species performed short overnight foraging trips lasting less than 10 hours. Comparisons of foraging suggests that prey may be more accessible to fur seals at Macquarie Island than South Georgia, but further longer term studies will be required to confirm this.

INTRODUCTION

RECENT studies examining maternal care in fur seals (Pinnipedia, Otariidae) highlight environmental factors influencing the seasonal and predictable availability of prey as major factors which have shaped the evolution of pup rearing (or maternal strategies) in these species (Gentry *et al.* 1986; Oftedal *et al.* 1987; Trillmich 1990; Trillmich and Ono 1991). Perhaps the clearest example of this, is the marked difference in lactation length between species. Species breeding in highly seasonal yet productive high latitudes have short lactation periods lasting about four months (Gentry *et al.* 1986; Trillmich 1990). In contrast, species breeding in less productive and less seasonal seas near the equator have extended lactation periods which may last up to three years (Gentry *et al.* 1986; Trillmich 1990; Gales and Costa 1997). Species breeding in more temperate latitudes are intermediate with respect to lactation length (8–12 months), and there is no clear relationship between lactation length and latitude (Boness 1991; Trillmich 1990).

In recent years there have been a number of studies which have examined how fur seals have adapted to different environmental regimes (Gentry and Kooyman 1986; Oftedal *et al.* 1987; Trillmich 1990; Trillmich and Ono 1991). Apart from lactation length, these studies have examined a suite of traits which are involved in the care of young from birth to weaning, both on-shore (such as pup growth rates and weaning mass, shore attendance, nursing behaviour, milk quality) and at-sea (diet, distance to foraging areas, diving behaviour, foraging energetics). Comparative studies seeking to determine the proximate causes for the evolution of different lactation strategies have been somewhat constrained by the inability to control for different environmental conditions across the range of the species studied. However, this is possible at some locations where species with different lactation lengths breed sympatrically. Such sites provide natural experiments where species with different maternal strategies can be studied under the same environmental conditions.

At Macquarie Island (54°30′S, 158°57′E) Antarctic *Arctocephalus gazella* and Subantarctic *A. tropicalis* fur seals breed sympatrically (Shaughnessy *et al.* 1988). Macquarie Island is one of three locations where these species breed sympatrically, the other two being Marion Island and Iles Crozet (Kerley 1985; Jouventin *et al.* 1982). These islands all lie just north of the Antarctic Polar Front (APF). All other breeding populations of *A. gazella* are on islands south of the APF, with major populations centered in the Scotia Arc (South Georgia and South Shetland Islands) and the Southern Indian Ocean (Iles Kerguelen and Heard Island) (Boyd 1993; Jouventin and Stonehouse 1985; Shaughnessy

[1]Department of Ecology and Evolutionary Biology, Monash University, Clayton, Victoria, Australia 3168.
[2]Current address: Department of Zoology, University of Tasmania, GPO Box 252C, Hobart, Tasmania, Australia 7001.
[3]Department of Zoology, University of Tasmania, GPO Box 252C, Hobart, Tasmania, Australia 7001.
[4]Division of Botany and Zoology, Australian National University, Canberra, Australian Capital Territory, Australia 2601.
[5]Current address: Reserve de Berenty, Tolognaro (Fort Dauphin), BP 54, Madagascar.

Pages 151–63 *in* MARINE MAMMAL RESEARCH IN THE SOUTHERN HEMISPHERE VOLUME 1: STATUS, ECOLOGY AND MEDICINE ed by M. Hindell and C. Kemper. Surrey Beatty & Sons, Chipping Norton. 1997.

and Goldsworthy 1990). Apart from a single breeding record at Heard Island (Goldsworthy and Shaughnessy 1989), all *A. tropicalis* breeding populations occur north of the APF, with major concentrations at Gough, the Prince Edward and Amsterdam Islands (Hes and Roux 1983; Bester 1990; Wilkinson and Bester 1990). Macquarie Island is the most isolated breeding population of *A. gazella* and *A. tropicalis*, some 5 500 km east of the nearest breeding colonies in the Southern Indian Ocean.

The timing of the pupping season is similar for both *A. gazella* and *A. tropicalis* (Bester and Bartlett 1990; Kerley 1983), however, the duration of lactation is vastly different. *A. gazella* wean at approximately four months of age, while *A. tropicalis* pups wean between 8 and 10 months of age (Lunn *et al.* 1993; Kerley 1985; Goldsworthy and Crowley, unpubl. data). These differences remain fixed even where species breed sympatrically (Kerley 1985; Goldsworthy and Crowley, unpubl. data). The diet of *A. gazella* has been studied at several sites across the specie's range (Bonner 1968; Daneri and Coria 1992; Daneri and Coria 1993; Doidge and Croxall 1985; Green *et al.* 1989; Green *et al.* 1990; Green *et al.* 1991; North *et al.* 1983), while the diet of *A. tropicalis* has been examined at Gough, Marion and Amsterdam Islands (Bester and Laycock 1985; Condy 1981; Paulian 1964). The diet of *A. gazella* and *A. tropicalis* has been examined previously at Macquarie Island by Green *et al.* (1990), however, they were unable to examine interspecific differences in the diets of each species because of non-specific collection of scats. There have been a number of studies conducted on the diving behaviour and foraging energetics of *A. gazella*, but almost all of these have been restricted to the population at South Georgia (Kooyman *et al.* 1986; Costa *et al.* 1989; Boyd and Croxall 1991; Boyd *et al.* 1994). The exception to this is one study conducted on diving behaviour at Heard Island (Green 1997). As yet there are no published data on the diving behaviour or foraging energetics of *A. tropicalis*.

In this study, we aim to investigate inter-specific differences in the at sea components of maternal care in *A. gazella* and *A. tropicalis* breeding sympatrically at Macquarie Island. Specifically we wished to examine whether the different lactational strategies of these species reflected differences in how they utilized marine resources in terms of diet and diving behaviour.

METHODS

Study site and species identification

Fur seals were studied at North Head, Macquarie Island, where *A. gazella* and *A. tropicalis* breed sympatrically. The fur seal population is small, but currently increasing at about 12.5% per annum. In the 1990/91 breeding season, 67 pups were born (Goldsworthy and Shaughnessy, unpubl. data). Adult female *A. gazella* and *A. tropicalis* were identified on the basis of external characteristics and vocalizations as described in Shaughnessy *et al.* (1988).

Diet

Fur seal scats were collected at Secluded Beach and Goat Bay (North Head), Macquarie Island between December and March 1990–91. Scats were collected opportunistically on a daily basis, and where possible, were collected fresh from individuals of known species and age-class (adult male, adult female, sub-adult male and juvenile). If the age-class and species origin of scats was not known, they were classed as "unknown origin". Scats were frozen, and later broken apart with water in a series of two sieves with mesh diameter of 1 mm and 0.5 mm mesh. Fish otoliths, squid beaks and other remains were removed from the 1 mm sieve, and the residual in the 0.5 mm sieve was then sorted in a glass petri-dish over a black surface to aid extraction of small otoliths less than 1 mm in diameter. Otoliths were identified using Williams and McEldowney (1990). Where possible otoliths were identified to species, but when otolith erosion prevented this, they where identified to genus. If the extent of erosion prevented identification at any taxonomic level, they were classed as being too eroded to identify. The standard lengths and mass of fish prey were estimated from measurements of otoliths showing very little or no sign of erosion in conjunction with regression equations presented in Williams and McEldowney (1990).

Diving behaviour

Time-depth recorders (TDR's) (MK 3+, Wildlife Computers) containing 256 Kbytes of memory were deployed on ten lactating *A. gazella* and four *A. tropicalis* during February and March 1991. Study animals of each species were selected at random, although care was taken to select those with healthy pups. Females were captured using a hoop-net, physically restrained, then fitted with a TDR (secured with two hose-clamps looped through a small perspex mount) which was glued to the seal with fast setting epoxy glue (Ceiba Geigy, Araldyte K268 road marker adhesive). TDR's were deployed on the seals for between 2 and 12 days. Seals were recaptured after returning to shore, when their TDR and epoxy mount were removed. TDR's were programmed to sample depth every 5 s, water temperature every 60 s, and light levels every 60 s. Dive records recovered from TDR's were analysed using Dive Analysis software (Wildlife Computers). Foraging locations were

Table 1. Percentage frequency of occurrence of remains in faecal samples of *A. gazella* and *A. tropicalis* at Macquarie Island, December 1990 to March 1991.

	December	January	February	March	Whole period
Fish	100.0	100.0	100.0	100.0	100.0
Squid	8.3	1.9	11.1	22.7	8.7
Mollusc	0.0	1.9	0.0	0.0	0.7
Amphipod	0.0	0.0	3.7	4.5	0.7
Nematode	5.6	22.6	29.6	50.0	23.9
Plastic	0.0	0.0	14.8	4.5	3.6
Total Scats	36	53	27	22	138

estimated using data on light levels obtained from TDR records and Geolocation software (Wildlife Computers).

Arrival and departure times were estimated from the mid-point between observation periods (2–3 times daily) in two females, and using small radio transmitters (150–151 MHz, Sirtrack, Havelock North, New Zealand) attached to flipper tags of two other females. Arrival and departure times for these individuals were monitored with an Advance Telemetry Systems (Minnesota, USA) 150 and 151 MHz model 2000B scanning receiver in conjunction with an ATS model 5040 data collection computer, which were housed in a weather-proof box on a steep slope approximately 60 m above the study area.

RESULTS

Diet

A total of 138 faecal samples were collected between December 1990 and March 1991. Fish remains in the form of otoliths, bones and eye-lenses occurred in every faecal sample, while squid (in the form of beaks and eye-lenses) were a minor component of the diet occuring in 8.7% of the samples collected (Table 1). Fish otoliths (6 048 in total) occured in all but two samples (average of 44.4 otoliths per sample, range 1–319). Twelve species of fish were identified from sagital otoliths (Table 2). In addition to these, small (<1 mm) non-sagital (asteriscus or lapillus) otoliths (identified as being from Myctophid species (D. Williams, pers. comm)), were also identified. These accounted for 29.2% of all othoilths recovered from scats (Table 2), and because they were non-diagnostic with respect to species, were excluded from further analysis. The following analyses were conducted on the remaining 4 314 diagnostic sagital otoliths (mean 31.7 per sample, range 0–282).

Six per cent of all otoliths were too eroded to be identified (Table 2). Otoliths from the genera *Electrona* and *Gymnoscopelus* were most numerous with respect to frequency of occurrence and percentage by number (Table 2). Otoliths from seven other species only occured incidentally (<6% frequency of occurrence and <1% by numbers). *E. subaspera* dominated almost all samples in their frequency of occurrance (91%) and percentage by numbers (67%). Of the 51.7% of otoliths which could be identified to species level, 91.0% were *E. subaspera*, 4.3% were *E. carlsbergi* and 2.4% were *G. piabilus* (Table 2). All other species occurred incidentally (<0.26% frequency of occurrence).

Of the 4 314 sagital otoliths recovered from scats, 8.1% (351) were considered uneroded. These were measured to estimate fish standard length and mass. The mean standard length (± s.d.) and mass of the three main species of fish taken by fur seals was: *E. subaspera* length, 83.7 ± 16.3 mm (range, 16.9–125.9 mm, n = 287), mass, 12.3 ± 7.01 g (range, 0.1–39.9 mm, n = 287), *E. carlsbergi*, length, 58.7 ± 3.5 mm (range, 48.5–67.9 mm, n = 30), mass, 3.7 ± 0.6 g (range, 2.2–5.5 g, n = 30), *G. piabilus*, length, 89.4 ± 30.4 mm (range, 67.9–110.9 mm, n = 2).

Interspecific comparisons

For both *A. gazella* and *A. tropicalis*, otoliths from the genera *Electrona* (numerical abundance, NA: 90.1 and 83.1%, respectively) and *Gymnoscopelus* (NA: 3.2 and 9.2%, respectively) were most numerous (Table 2). For both species, otoliths from *E. subaspera* dominated (NA: 91.5 and 91.6%, respectively). The main difference between otoliths recovered from scats of *A. gazella* and *A. tropicalis*, was in the numerical abundance of *E. carlsbergia* (5.5% and 0.6%, respectively) and *G. piabilus* (0.7% and 7.3%, respectively) (Table 2). The estimated standard length of *E. subaspera* taken by *A. gazella* (85.2 ± 15.9 mm, n = 103) and *A. tropicalis* (82.1 ± 14.8 mm, n = 58), did not differ significantly ($F_{1,159}$ = 1.49, P = 0.225). Similarly, the estimated mass of *E. subaspera* taken by *A. gazella* (12.8 ± 6.5 g, n = 103) and *A. tropicalis* (11.4 ± 6.4 g, n = 58), did not differ significantly ($F_{1,159}$ = 1.80, P = 0.187).

Age/sex-class differences

Enough scats were collected from adult females (AF's), sub-adulat males (SAM's) and juveniles (JUV's) to enable comparison across these three age/sex-classes. Non-sagital otoliths 'able 2. Fish prey species identified from otoliths collected from *A. gazella* and *A. tropicalis* at Macquarie Island December 1990–March 1991.

	All samples					*A. gazella*					*A. tropicalis*				
	Freq. of occurrence		Numerical abundance			Freq. of occurrence		Numerical abundance			Freq. of occurrence		Numerical abundance		
ish species	(n)	(%)	(n)	(%)^A	(%)^B	(n)	(%)	(n)	(%)^A	(%)^B	(n)	(%)	(n)	(%)^A	(%)^B
subaspera	125	91.0	2 849	66.6	91.1	56	93.0	886	66.7	91.5	17	89.0	323	65.3	91.6
carlsbergi	12	8.7	136	3.2	4.3	4	6.7	53	4.0	5.5	1	5.3	2	0.4	0.6
antarctica	1	0.7	8	0.2	0.3	0	0.0	0	0.0	0.0	0	0.0	0	0.0	0.0
nindent. *Electrona* spp.	133	96.0	765	17.9				60	100.0	258	19.4	18	95.0	87	17.[illegible]
'ectrona spp.	271	96.4	3 758	87.8		60	100.0	1 197	90.1			18	95.0	417	83.1
ymnoscopelus nicholsi	5	3.6	6	0.1	0.2	3	5.0	8	0.6	0.8	0	0.0	0	0.0	0.0
piabilis	25	18.9	74	1.7	2.4	7	12.0	7	0.5	0.7	4	21.0	26	5.2	7.[illegible]
nindent. *Gymnoscopelus*	37	27.0	131	3.1	13	22.0	28		2.1		4	21.0	20	4.0	
ymnoscopelus spp.	67	27.0	211	4.9		13	22.0	43	3.2		4	2.1	46	9.2	
uthylagus sp.	1	0.7	1	0.02	0.03	0	0.0	0	0.0	0.0	0	0.0	0	0.0	0.0
'refftichthys andersoni	7	5.1	24	0.6	0.8	2	3.3	3	0.2	0.3	1	5.3	1	0.2	0.[illegible]
'agnisudis prionosa	6	4.3	11	0.3	0.4	3	5.0	7	0.5	0.7	0	0.0	0	0.0	0.0
'otomyctophum bolini	3	2.2	6	0.1	0.2	0	0.0	0	0.0	0.0	0	0.0	0	0.0	0.0
normani	3	2.2	5	0.1	0.2	2	3.3	4	0.3	0.4	1	5.3	1	0.2	0.[illegible]
tennisonii	2	1.4	8	0.2	0.3	0	0.0	0	0.0	0.0	0	0.0	0	0.0	0.0
uranotothenia magellanica	1	0.7	1	0.02	0.03	0	0.0	0	0.0	0.0	0	0.0	0	0.0	0.0
oo eroded to identify	49	35.5	255	6.0	18	30.0	75	5.6	10	52.6	37	7.4			
on-sagital otoliths	124	90.0	1 768	29.2*	58	97	756	36.6*	14	74.0	257	33.9*			
umber of otoliths			6 048				2 085				759				

'ercentage numerical abundance excluding non-sagital otoliths. *Exception.
'ercentage numerical abundance using otoliths which could be indentified to species.

were most common in the scats of AF's, compared with SAM's and JUV's (NA: 46.4%, 23.7% and 23.0%, respectively). There also appeared to be minor differences in the numerical abundance of some prey species in each age/sex-class. *E. subaspera* was most abundant in samples collected from JUV's (NA: 95.1%), compared with AF's and SAM's (NA: 88.8% and 89.2%, respectively). In addition, *E. carlsbergi* was more abundant in the scats of AF's (NA: 8.4%) compared with SAM's and Juv's (NA: 0.0% and 0.9%, respectively), and *G. piabilus* was most abundant in the scats of SAM's (NA: 9.8%) compared with AF's and JUV's (1.4% and 3.6%, respectively). The estimated mean standard length of *E. subaspera* taken by JUV's (78.8 ± 14.7 mm, n = 64), was significantly smaller than those taken AF's (90.7 ± 14.2 mm, n = 119) ($F_{1,181}$ = 28.57, P<0.0001). The size of *E. subaspera* taken by SAM's (93.7 ± 19.9 mm, n = 8) was similar to that of AF's.

Temporal differences in diet

PREY OCCURRENCE

There was little temporal fluctuation in the occurance and abundance of the major prey genera (*Electrona* and *Gymnoscopelus*), between December and March (1). The numerical abundance (using only otoliths identifiable to species level) of *E. carlsbergi* otoliths (which came from mostly *A. gazella* samples) declined steadily between December and March from 12.9% to 0.0%. Concommitent with this, the numerical abundnce of *E. subaspera* otoliths increased from 83.5% to 96.6% (Fig. 1).

Within fur seal species the pattern was similar. In *A. gazella*, the numerical abundance of *E. carlsbergi* otoliths declined significantly between December and March from 23.0% to 0.0%, con-commitent with an increase in *E. subaspera* otoliths from 76.4 to 96.4%. In *A. tropicalis*, the numerical abundance of *G. piabilus* fluctuated between 9.7 and 7.0%, for 3 of 4 months, and were absent in January. This absence may reflect the small number of scats collected from *A. tropicalis* in that month. In general there was less evidence for temporal fluctuation in the prey of *A. tropicalis*.

PREY SIZE

This analysis is restricted to *E. subaspera* due to sample size constraints for other prey species. The standard length of *E. subaspera* (estimated from otolith size), decreased significantly between December and March ($F_{3,283}$ = 71.2, P < 0.0001), as did estimated mass ($F_{3,283}$ = 74.2, P < 0.0001). The changing length and mass of *E. subaspera* in the diet of fur seals is most apparent when the frequency distributions of fish length and mass are compared for each month (Fig. 2).

The average number of *E. subaspera* (maximum number of left or right otoliths)/scat/month increased significantly between December and March from 7.8 to 19.4 ($F_{3,132}$ = 4.82, P = 0.0032) (Fig. 3), concomittent with a reduction in the estimated size of *E. subaspera* taken (Figs 2 and 3).

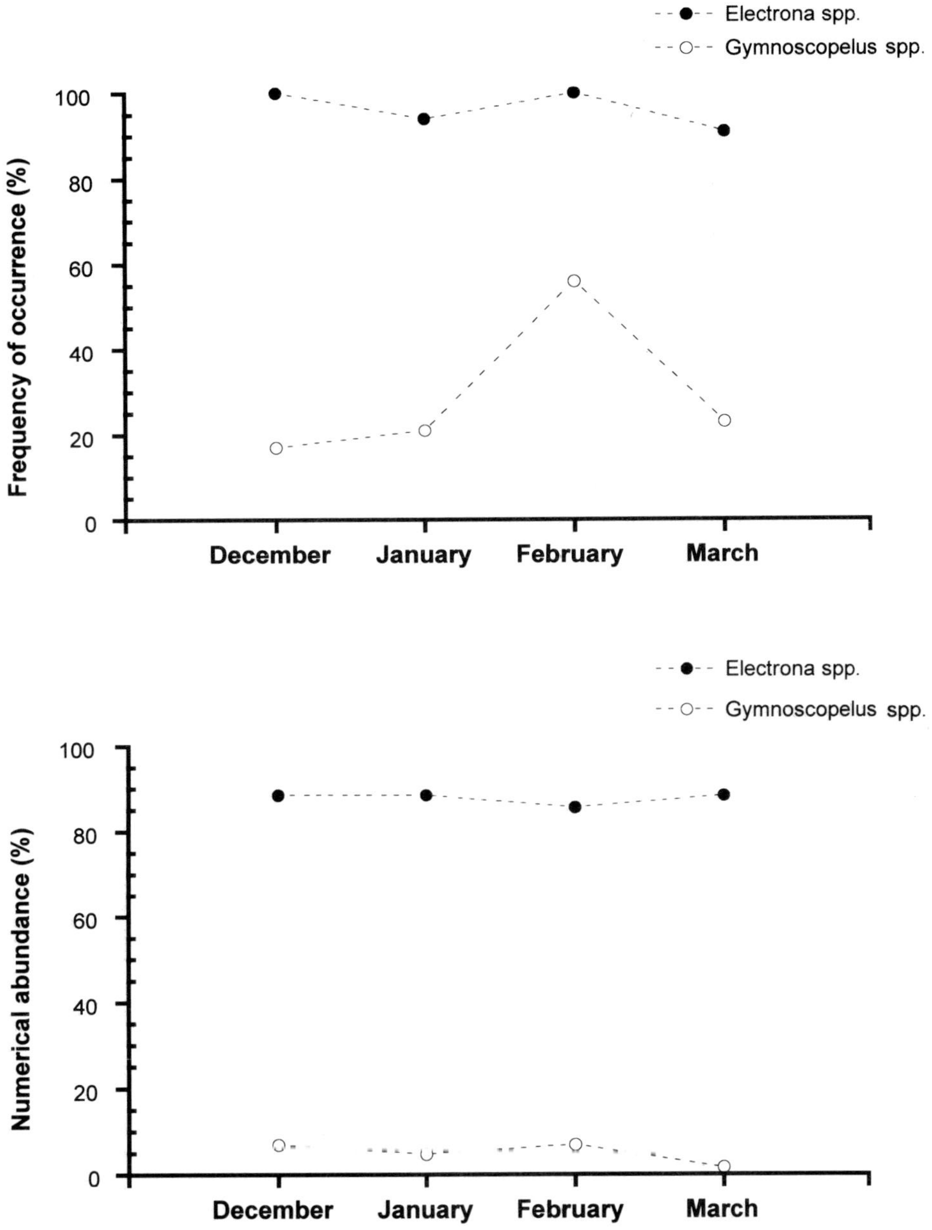

Fig. 1. Temporal changes in the frequency of occurrence and numerical abundance of the two most common fish genera taken (*Electrona* spp. and *Gymnoscopelus* spp.) by *A. gazella* and *A. tropicalis* at Macquarie Island between December 1990 and March 1991.

Diving behaviour

Fourteen of the 15 TDR's deployed returned usable data, 10 for *A. gazella* and four for *A. tropicalis*. Data were collected from a total of 26 foraging trips, 20 from *A. gazella* and six from *A. tropicalis* (Table 3). The foraging trips of *A. gazella* were on average 59.7 ± 53.9 h and those of *A. tropicalis* were 119.1 ± 104.8. Despite the apparent disparity between the two means a nested ANOVA on log transformed data revealed no significant difference in the foraging trip duration between the species ($F_{1,12}$ = 0.258, P = 0.620). The considerable variance about the means is due to several of the seals making very short, overnight foraging trips of less than 10 hours duration. Exactly half of the seals from each species performed these short trips (Table 3). The mean duration of these short trips were 8.6 ± 2.3 and 7.0 ± 4.2 h for the two species respectively.

A total of 15 770 dives of greater than 2 m depth were recorded: 10 718 from *A. gazella* and 5 052 from *A. tropicalis*. As the data on depth, frequency and surface interval are all highly skewed to the left, medians are the most appropriate measure of central tendancy,

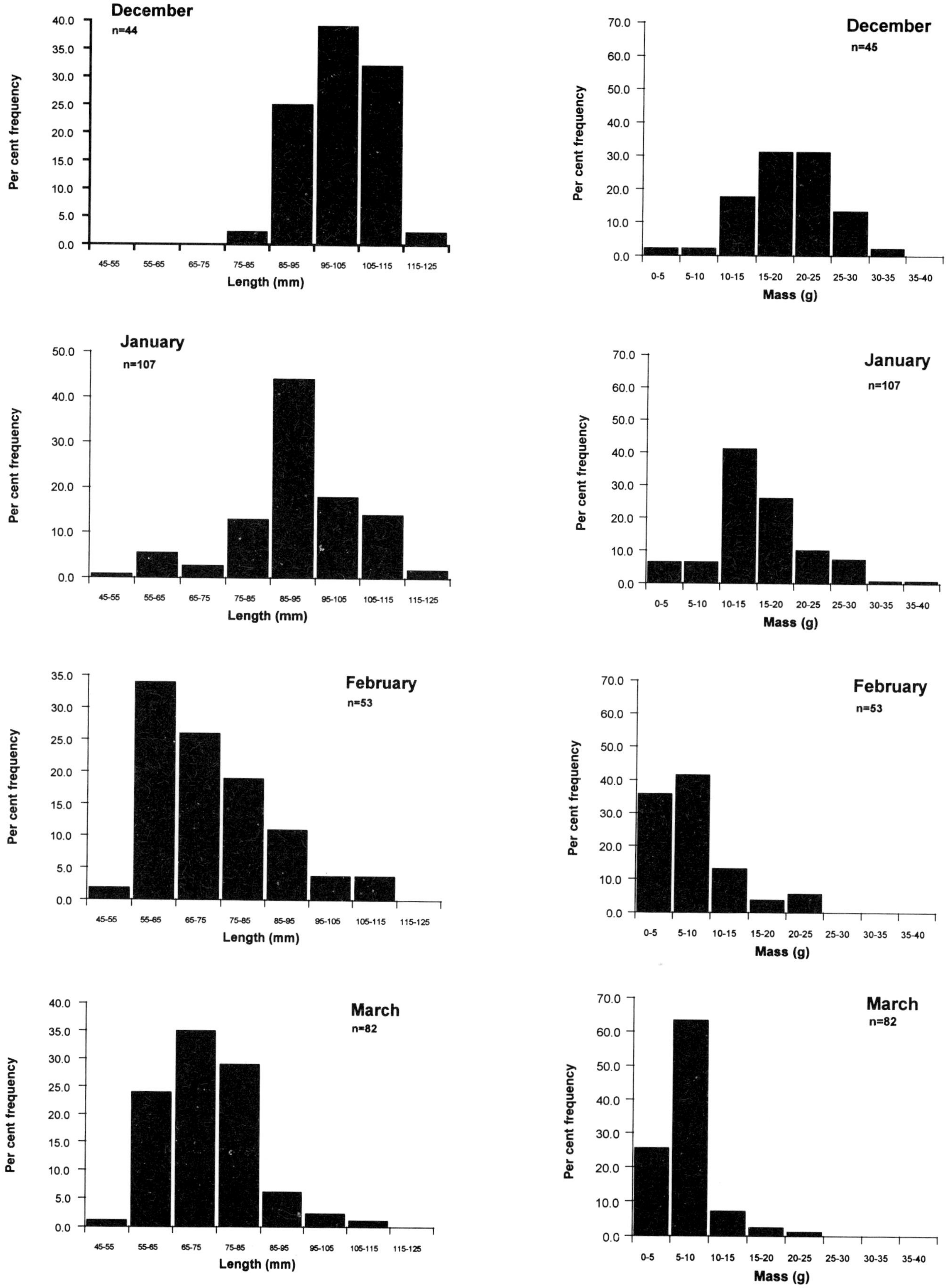

Fig. 2. Temporal changes in the frequency distribution of estimated standard length and mass for *E. subaspera* taken by *A. gazella* and *A. tropicalis* at Macquarie Island between December 1990 and March 1991.

Table 3. Basic summary statistics for the 10 *A. gazella* and four *A. tropicalis* females studied between late January and early March 1991 at Macquarie Island. These data summarize all dives greater than 2 m in depth, and are expressed as the mean ± s.d. with the median given in italics. *Denotes those seals that made one or more short, overnight foraging trips.

Seal	Species	No. of forage trips	Foraging trip length (days)	No. of dives	Depth (m)	Max depth (m)	Duration (min.)	Max duration (min.)	Surface interval (min.)	Max surface interval (min.)
MINP19*	*A. gazella*	3	42.9 ± 27.0 *57*	1 567	12 ± 10 *9*	72	0.7 ± 0.5 *0.6*	3.7	2.7 ± 31.5 *0.2*	864.3
TDRNP3*	*A. gazella*	2	39.4 ± 1.1	459	11 ± 11 *8*	76	0.6 ± 0.6 *0.5*	3.4	6.2 ± 39.9 *0.2*	950.9
MISC1	*A. gazella*	1	186.0	1 973	9 ± 10 *6*	84	0.6 ± 0.6 *0.4*	4.3	1.6 ± 27.7 *0.2*	966.2
TDRNP2*	*A. gazella*	2	34.2 ± 37.7	489	15 ± 13 *10*	90	0.8 ± 0.6 *0.7*	3.7	6.6 ± 57.9 *0.5*	971.3
MINC22	*A. gazella*	1	40.8	185	7 ± 5 *8*	46	0.5 ± 0.4 *0.3*	2.5	7.5 ± 49.9 *0.5*	645.3
TDRNP11*	*A. gazella*	4	37.6 ± 59.4 *8.8*	2 060	8 ± 8 *6*	95	0.6 ± 0.5 *0.4*	3.7	1.0 ± 12.3 *0.2*	555.9
MINC2*	*A. gazella*	3	28.4 ± 23.4 *14.8*	265	18 ± 19 *10*	84	0.9 ± 64.0 *0.8*	3.3	10.2 ± 64.0 *0.8*	603.9
TDRNP5	*A. gazella*	1	115.7	1 650	12 ± 11 *9*	89	0.7 ± 0.5 *0.5*	3.3	1.1 ± 20.6 *0.2*	689.3
TDRNP10	*A. gazella*	1	63.8	349	12 ± 12 *8*	82	0.6 ± 0.6 *0.3*	3.3	0.3 ± 0.5 *0.2*	8.9
TDRSE22	*A. gazella*	2	137.9 ± 1.8	1 811	16 ± 14 *10*	88	0.8 ± 0.6 *0.6*	3.0	4.2 ± 56.1 *0.4*	926.6
MIC10*	*A. tropicalis*	2	119.2 ± 162.8	1 333	11 ± 12 *7*	83	0.7 ± 0.6 *0.4*	3.6	1.6 ± 20.5 *0.3*	588.0
MINP13*	*A. tropicalis*	2	107.4 ± 137.8	1 495	14 ± 15 *8*	109	0.7 ± 0.6 *0.5*	3.8	4.6 ± 57.7 *0.3*	1 001.7
TDRSE21	*A. tropicalis*	1	197.4	1 705	12 ± 9 *10*	65	0.7 ± 0.5 *0.6*	3.3	1.7 ± 27.3 *0.3*	947.5
TDRC8	*A. tropicalis*	1	64.0	519	11 ± 10 *6*	52	0.6 ± 0.5 *0.3*	3.1	2.3 ± 42.8 *0.3*	976.1
TOTAL	*A. gazella*	20	59.7 ± 53.9 *48.2*	10 718	12 ± 11 *8*	95	0.7 ± 0.6 *0.5*	4.3	2.6 ± 35.4 *0.3*	971.3
TOTAL	*A. tropicalis*	6	119.1 ± 104.8 *131*	*5 052*	12 ± 12 *9*	109	0.7 ± 0.6 *0.5*	3.8	2.6 ± 39.2 *0.3*	1 011.7

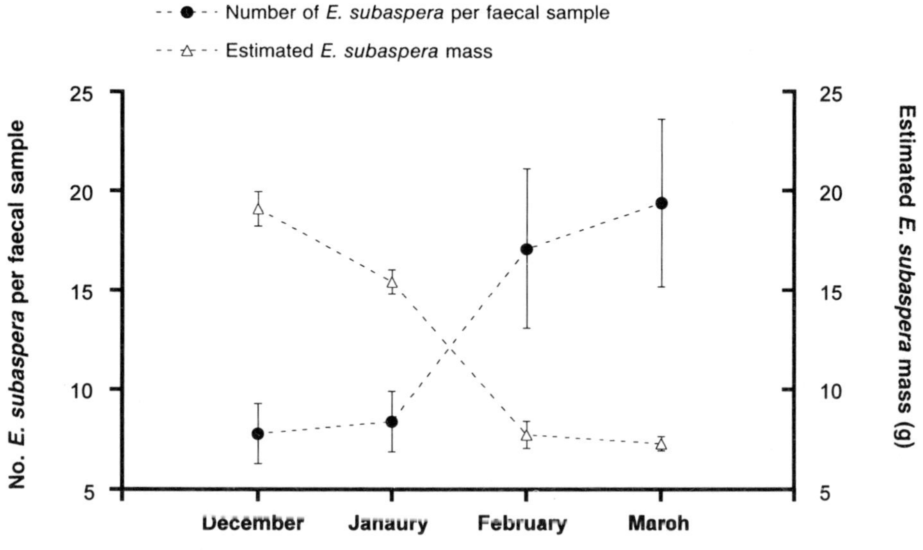

Fig. 3. The average number of *E. subaspera* otoliths obtained from each fur seal faecal sample plotted within the mean size (mass –kg) of *E. subaspera* taken by *A. gazella* and *A. tropicalis* at Macquarie Island between December 1990 and March 1991.

although means and standard deviations have also been included for comaprisons with other studies (Table 3). Median dive depths for all seals were shallow, between 6 and 10 m, and the durations were correspondingly short, between 0.3 and 0.8 min. Surface intervals were also short with median values of 0.2–0.8 min. The maximum dive depth recorded in this study was 109 m for an *A. tropicalis*, while the longest dive was 4.3 min for an *A. gazella*.

Nested ANOVAs (seals nested within species) on log transformed data indicated no significant difference between the two species with respect to either depth of dive ($F_{1,12} = 0.393$, $P = 0.542$) or duration of dive ($F_{1,12} = 0.029$, $P = 0.867$).

The dives were subsequently divided into two groups, "travelling dives" with durations of less than 0.2 min and less than 6 m depth, and "foraging" dives, which constituted all of the remaining dives. The time from when the seal first entered the water until it made it is first foraging dive was calculated for each foraing trip (Table 4). The mean values ranged from 1.0 to 6.9 h, but there was no significant difference between the two species in the time that it took to begin foraging dives (Nested ANOVA, $F_{1,2} = 1.53$, $P = 0.239$), giving an overall average of 2.6 ± 1.7 h. The time from the last foraging dive until the seal returned to land was also calculated (Table 4). Again there was no significant difference between the species

Table 4. The amount of time spent in various activities by female *A. gazella* and *A. tropicalis* during trips to sea in February and March. For seals with records for more than one foraging trip values are expressed as mean ± s.d.

Seal	Species	Total time at Sea (hrs)	Mean time to first foraging dive (hrs)	Mean time since last foraging foraging dive (hrs)	Time spent submerged on foraging dives (%)	Dives per hour
MINP19	*A. gazella*	128.8	1.5 ± 0.3	3.2 ± 0.9	15.0	9.31
TDRNP3	*A. gazella*	78.8	3.9 ± 4.3	3.9 ± 4.3	6.4	3.78
MISC1	*A. gazella*	185.9	4.1	6.9	10.9	11.11
TDRNP2	*A. gazella*	63.8	3.0 ± 2.3	2.4 ± 3.3	10.5	6.88
MINC22	*A. gazella*	40.8	4.4	1.5	3.6	4.64
TDRNP11	*A. gazella*	150.4	0.9 ± 0.7	0.8 ± 1.2	13.5	14.66
MINC2	*A. gazella*	85.1	2.5 ± 2.2	5.5 ± 2.0	4.6	3.09
TDRNP5	*A. gazella*	115.7	6.9	4.9	16.0	14.56
TDRNP10	*A. gazella*	63.8	3.3	3.9	5.6	12.7
TDRSE22	*A. gazella*	275.8	3.3	8.2	8.8	8.29
MIC10	*A. tropicalis*	238.3	1.0	0.7	6.7	6.75
MINP13	*A. tropicalis*	214.8	2.3 ± 0.7	2.3 ± 1.9	7.7	7.05
TDRSE21	*A. tropicalis*	64.0	3.1	8.4	8.5	9.31
TDRC8	*A. tropicalis*	197.4	1.9	6.4	10.8	8.61
TOTAL	*A. gazella*	1 193.3	2.7 ± 1.8	3.5 ± 2.7	10.4	8.9 ± 4.3
TOTAL	*A. tropicalis*	714.8	2.1 ± 0.8	4.0 ± 3.3	8.3	7.9 ± 1.2

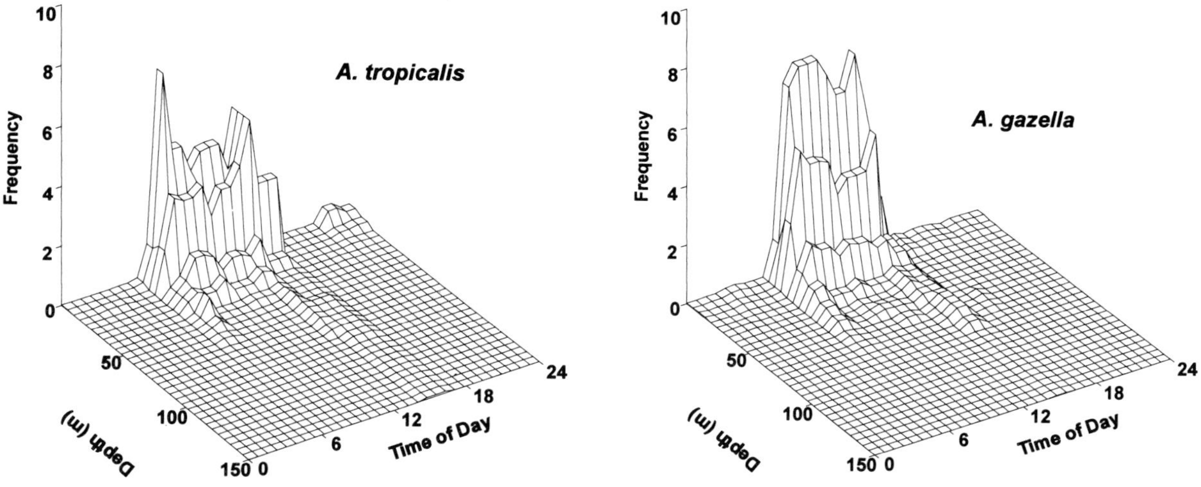

Fig. 4. Frequency distributions of foraging dives of *A. gazella* and *A. tropicalis* in relation time of day and depth.

(Nested ANOVA, $F_{1,2} = 0.036$, $P = 0.853$). The overall mean duration of these return trips was 3.6 ± 2.8 h. There was also no significant difference between the time taken to begin foraging dives and the time taken to return to the colony after the last foraging dive (Paired *t*-test, $t_{22} = 0.194$, $P = 0.848$), using data from both species combined.

The seals spent between 3.6% and 16.0% of their total time at sea doing foraging dives (Table 4). There was no difference between the species in this regard (*t*-test, $t_{23} = 1.244$, $P = 0.250$). Pooling the data from both species indicates that the female fur seals from Macquarie Island spent on average 9.2 ± 2.8% of their total time at sea performing foraging dives. The mean rate of diving for *A. gazella* (8.9 ± 4.3 foraging dives per hour) was not significantly different to that for *A. tropicalis* (7.9 ± 1.2, $t_{12} = 0.434$, $P = 0.672$).

There were marked diel patterns in the diving behaviour (Fig. 4). Both species dived almost exclusively at night. *Arctocephalus gazella* performed 94.1% of all foraging dives during the hours of darkness, and *A. tropicalis* made 95.8% of all foraging dives at night. There was a similar pattern with respect to the depth of dives, with the very few foraging dives made during the day being very shallow (<5 m), and all of the deeper foraging dives being restricted to night time. (Fig. 4). The deepest dives were made around sunrise and sunset, with mean values in excess of 20 m. Dives at other times of the night had mean values between 8 and 17 m.

Approximate foraging locations identified using light levels measured by TDR's and analysed by geoloction software, could be determined for 9 of 14 deployments (Fig. 5). Most foraging was concentrated to the west and north of Macquarie Island, with very little activity

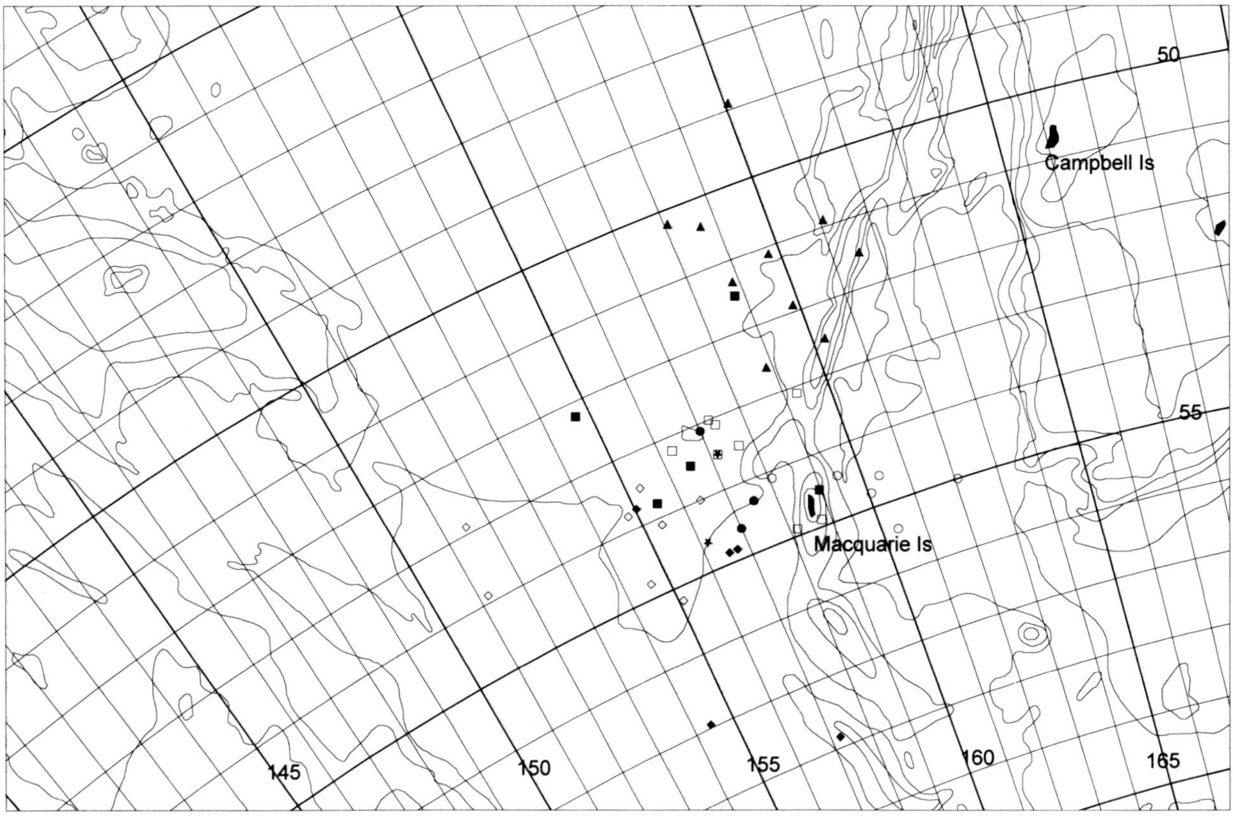

Fig. 5. Geolocation estimates of the foraging locations of nine fur seals (*A. gazella* closed symbols, *A. tropicalis* open symbols) from Macquarie Island during February and March 1991. Batymatry contours at every 1 000 m.

south of the island. The mean distance of each geolocation point from Macquarie Island for each female was 215.0 ± 105.6 km (range 94.1–386.1 km, n = 9). There was no obvious distinction betwen areas used by *A. gazella* and *A. tropicalis* (Fig. 5), nor was there any difference in the mean distance of foraging areas from Macquarie Island (*A. gazella*: mean = 227.3 ± 106.3 km, range 103.7–386.1 km, n = 6, *A. tropicalis* mean = 190.3 ± 122.5 km, range 94.1–328.2, n = 3, $t_7 = 0.470$, $P = 0.653$). These locations are very approximate, as the longitudinal error associated with geolocation points was 2° (which corresponds to an error of 118.2–149.2 km between 47.9°S and 57.9°S), while the latitudinal error was approximately 2.5 ± 0.88° (n = 58), which is equivalent to 279.1 ± 97.9 km (range 166.9–645.2, n = 58).

DISCUSSION

Diet

Interspecific differences

No major differences were detected in the diets of *A. gazella* and *A. tropicalis* over the course of the study. Both species fed predominantly on the pelegic myctophid, *Electrona subaspera*. The only detectable difference was that *E. carlsbegi* occured more frequently in samples from *A. gazella*, and *Gymnoscepelus* spp., were more prevalent in samples from *A. tropicalis*. Similarly, Green *et al.* (1990), found myctophid fish (especially *E. subaspera*) to dominate in their combined samples from *A. gazella* and *A. tropicalis* at Macquarie Island at the same time of year.

Fish also dominated in the diet of *A. gazella* at Heard Island, although here, a mixture of pelagic and benthic species were taken, most likely due to the large area of the Kerguelen Plateau surrounding the Island (Green *et al.* 1989; Green *et al.* 1991; Green *et al.* 1997). The main pelagic species taken include the icefish, *Chamsocephalus gunnari, E. antarctica, E. subaspera* and *G. nicholsi*, while benthic species include skate, *Bathyraja* spp., *Notothenia squamifrons* and *Channichthys rhinoceratus* (Green *et al.* 1989; Green *et al.* 1991; Green *et al.* 1997). In addition, squid from the genus *Mastigoteuthis*, were major prey in autumn and winter months (Green *et al.* 1991).

Whereas the diet of *A. gazella* at Macquarie Island and in the Southern Indian Ocean consists mostly of fish, in the Southern Atlantic Ocean at South Georgia, the species prey primarily on Antarctic Krill *Euphausia superba* (Bonner 1968; North *et al.* 1983). Fish are also taken, but mostly by sub adult and adult males (Doidge and Croxall 1985; Prince in North *et al.* 1983). Pelagic Icefish *C. gunnari* predominate, while notothenids comprise the remainder of the fish diet (North *et al.* 1983). Further south in the South Orkneys, moulting male *A. gazella* prey on a mixture of krill and fish (Daneri and Coria 1992; Daneri and Coria 1993). The fish prey are mostly pelagic myctophids, principally *E. antarctica* and *G. nicholsi* (Daneri and Coria 1992; Daneri and Coria 1993).

The diet of *A. tropicalis* has been less intensively studied. At Gough Island, they prey on mostly cephalopods (50%) and fish (45%) and to a less extent euphausids (5%) (Condy 1981; Bester and Laycock 1985). At Amsterdam Island, *A. tropicalis* have been reported to take Rockhopper Penguins *Eudyptes chrysocome* (Paulian 1964). Although penguins were reported in the diets of *A. gazella*/*A. tropicalis* at Macquarie Island in Green *et al.* (1990), they were not recorded in this study.

Temporal differences

There was little evidence for temporal changes in the occurance of any of the main prey species over the period of this study. Although Green *et al.* (1990) reported an increase in the proportion of *Gymnoscopelus* spp. with a conconmittent decrease in *Electrona* spp. over the summer season of 1988/89, this pattern was not found in this study. There were, however, temporal change in the size prey taken. As the season progressed, there was a steady decline in the the size of *E. subaspera* taken by seals, a pattern also detected by Green *et al.* (1990). Little is known of the biology of *E. subaspera*, but the declining size may reflect changes in age structure associated with spawning, or movements of adult cohorts further from the Island. Interestingly, as the size of *E. subaspera* in the diet of fur seals declined over the season, the number of *E. subaspera* otoliths found in each faecal sample increased. This may reflect greater number of prey being eaten to compensate for their smaller size. Greater numbers of otoliths per faecal sample may also reflect other changes in foraging behaviour, such as feeding nearer to the island, or a reduction in transit times from foraging areas to shore resulting in less otoliths being voided at sea.

Green *et al.* (1990) found that although there were a greater number of otoliths per faecal sample at Macquarie Island than at Heard Island (Green *et al.* 1989), and that a greater percentage of the Macquarie Island otoliths were eroded (98.3% and 89.7%, respectively). They suggested this reflected a slower gut passage rate (resulting from a lower food intake), or voiding of feaces at sea due to greater transit times from foraging areas. They concluded that these explanations were consistent with food availabily being lower at Macquarie Island compared with Heard Island, either because it is less abundant, more dispersed or further away.

During the four month period when faecal samples were collected in this study, the duration of foraging trips made by lactating *A. gazella* and *A. tropicalis* increased from about 2 to 5.5 days, and then declined to about 3.5 days by the end of March (Goldsworthy 1992). Over the same period when faecal samples where collected at Heard Island in Green *et al.*'s (1990) study (1987/88 summer), the foraging trip length of lactating *A. gazella* increased from 5 to 7 days (Goldsworthy 1995). In addition, the growth rates of *A. gazella* pups at Heard Island (in 1987/88) were lower than those reported for *A. gazella* pups at Macquarie Island (1990/91) (Goldsworthy 1992; Goldsworthy 1995). Contrary to the arguments of Green *et al.* (1990), these data, plus the wide use of overnight foraging trips by fur seals at Macquarie Island (this study; Goldsworthy 1992), suggest that foraging areas are closer to Macquarie Island than they are to Heard Island.

Diving behaviour

Inter-species comparisons

There were no discernable difference in the aspects diving behaviour of *A. gazella* and *A. tropicalis*. Both species exhibited a strong nocturnal pattern of diving, with deeper dives occuring at dusk and dawn. Most of the feeding dives were very shallow, with a median value of less than 10 m. This behaviour is consistent with what is known about the distribution of their major prey *Electrona,* which display pronounced vertical migration, coming close to the surface during the hours of darkness (Perissinotto and McQuaid 1992).

There was also no obvious separation of foraging areas, with individuals from both species travelling for about 2.5 hours before foraging behaviour commenced. As the shelf break is within 3 km of the breeding sites on North Head, the seals would have easy access to offshore and slope waters, which are areas of potenially high productivity, within this period of travelling. This is supported by data derived from geolocation, which also indicated little difference in the foraging areas utilized by females from both species at the time of this study. However, given the large errors associated with approximating foraging areas using geolocation techniques, these data can only be regarded as an indication of foraging range. Better resolution of spatial use in these species must await the use of satellite telemetry.

Intra-species comparisons

There are no other published data on the diving behaviour of *A. tropicalis*, but the diving behaviour of lactating *A. gazella* at Macquarie Island are similar to those found at both South Georgia (Boyd and Croxall 1992; Boyd *et al.* 1991; Kooyman *et al.* 1986) and Heard Island (Green 1997), which are also characterized by shallow nocturnal diving. Despite the considerable difference in the diet of lactating fur seals at Macquarie Island (predominantly myctophiids) and South Georgia (almost exlusively krill), *A. gazella* from both areas spend the same proportion of their time underwater on foraging dives, 10.7 ± 2.4% ($t_{23} = 1.244, P = 0.250$). The depth and durations of each dive are also similar between these two populations, although the median depth of 8 m for Maquarie Island animals was at the low end of the range described for South Georgia, suggesting that the myctophid prey were closer to the surface in the present study.

The most significant difference between the two populations is the number of seals from Macquarie Island that performed short, over-night foraging trips. Half of the 14 animals equipped with TDR's in this study made these trips, while such trips were not reported at all from either South Georgia or Heard Island. This may be a consquence of the close proximity of the deep, oceanic water to Macquarie Island.

CONCLUSION

There was no evidence that the diet or foraging behaviour of lactating *A. gazella* and *A. tropicalis* differed, despite the substantial difference in the species' lactational strategies. This, however, may not be surprising as during the first three to four months of lactation, the relative growth rates of pups from each species are very similar (Goldsworthy and Crowley, unpubl. data; Kerley 1985). This pattern appears true for most fur seal species (Gentry *et al.* 1986).

The lack of interspecific differences may also indicate that present circumstances at Macquarie Island (in terms of food availabilty) are so favourable that there is little pressure for species to partition resources. This situation may change with increasing population density and competition. The apparently flexible foraging strategies and diet of these species which appear to have genetically fixed lactation lengths (or weaning age), may indicate that they evolved in isolation from congeners, as strong congeneric competition may be expected to have resulted in the evolution of more specialized foraging strategies. What is still unclear, however, is whether one lactation strategy is more appropriate or successful at Macquarie Island than the other.

The endemic population of fur seals at Macquarie Island became extinct as a result of 18th century sealing, and we do not know which species bred on the island prior to this time (Shaughnessy and Fletcher 1987). Since the first breeding record in 1955, the annual pup production increased slowly, until the early 1980s when it began to increase in a rapid, exponential fashion (Shaughnessy and Goldsworthy 1993). This increase in pup production has been largely due to increasing numbers of *A. gazella*, which account for approximately 80% of the total pup production in the 1994/95 season (Goldsworthy and Shaughnessy, unpubl. data). In contrast, the numbers of *A. tropicalis* pups being born annually has changed little in the last five years (Goldsworthy and Shaughnessy, unpubl. data). This differs markedly with the other two sympatric popultions at Iles Crozet and Marion Islands, where the larger and rapidly increasing populations are *A. tropicalis*, while the *A. gazella* populations are small and are increasing relatively slowly (Wilkinson and Bester 1990; Jouventin *et al.* 1982). These differences may be due to how the animals utilize their marine environment over winter, whether there is marked seasonal fluctuations in food availability between summer and winter months, and how energy resources are partitioned into lactation and gestation.

Both *A. gazella* and *A. tropicalis* mate about a week post-partum. This is followed by a four month delayed implantation period and then eight months active gestation (Boyd 1991). In *A. gazella*, this results in a lactation period which is free from the additional energy demands of gestation, as their pups wean at four months of age. During the eight months of gestation which follow (April to November/December), females remain at sea foraging until returning ashore to pup. This contrasts dramatically with *A. tropicalis* females which have a lactation period 2–2.5 times longer than *A. gazella*, but which also have the additional energy demands of gestation through much of their lactation period. The only respite for *A. tropicalis* is a 2 to 2.5 month period between weaning one pup and giving birth to the next. The ability of *A. gazella* females to exploit food resources away from Macquarie Island during their non-lactating period may place them at an advantage over *A. tropicalis* females, which have to depend on food resources near the Island for most of the year, and for three quarters of their gestation period.

ACKNOWLEDGEMENTS

We thank those people who helped with field work; Paul Cooper, Rupe Woods, Helen Cooley, Phil Barnaart, Kieran Lawton, Guy Bedford, Noel Carmichael and especially Dave Slip and Old Toby. We also thank Doreen Marchesan who provided much assistance with otolith identification. The Antarctic Science Advisory Committee, CSIRO Monash Collaborative Research Fund, and the Winnifred Violet Scott Trust provided funding for the project.

REFERENCES

Bester, N. M., 1990. Population trends of Subantarctic fur seals and southern elephant seals at Gough Island. *S. Afr. J. Antarct. Res.* **20:** 9–12.

Bester, M. N. and Bartlett, P. A., 1990. Attendance behaviour of Antarctic and subantarctic fur seals at Marion Island. *Antarct. Sci.* **2:** 309–12.

Bester, N. M. and Laycock, P. A., 1985. Cephalopod prey of the sub-Antarctic fur seal, *Arctocephalus tropicalis*, at Gough Island. Pp. 551–54 *in* Antarctic nutrient cycles and food webs ed by W. R. Siegfried, P. R. Condy and R. M. Laws. Proceedings of the 4th SCAR Symposium on Antarctic Biology. Springer-Verlag: Berlin.

Boness, D. J., 1991. Determinants of mating systems in the Otariidae (Pinnipedia). Pp. 1–44 *in* The Behaviour of Pinnipeds ed by D. Renouf. Chapman and Hall: Cambridge.

Bonner, W. N., 1968. The fur seal of South Georgia. *Br. Antarct. Surv. Sci. Rep.* **56:** 81 pp.

Boyd, I. L., 1991. Environmental and physiological factors controlling the reproductive cycles of pinnipeds. *Can. J. Zool.* **69:** 1135–148.

Boyd, I. L., Lunn, N. J. and Barton, T., 1991. Time budgets and foraging characteristics of lactating Antarctic fur seals. *J. Anim. Ecol.* **60:** 577–92.

Boyd, I. L., 1993. Pup production and distribution of breeding Antarctic fur seals *(Arctocephalus gazella)* at South Georgia. *Antarct. Sci.* **5:** 17–24.

Boyd, I. L. and Croxall, J. P., 1992. Diving behaviour of lactating Antarctic fur seals. *Can. J. Zool.* **70:** 919–28.

Boyd, I. L., Arnould, J. P. Y., Barton, T. and Croxall, J. P., 1994. Foraging behaviour of Antarctic fur seals during periods of contrasting prey abundance. *J. Anim. Ecol.* **63:** 703–13.

Condy, P. R., 1981. Anual food consumption and seasonal fluctuations in biomass of seals at Marion Island. *Mammalia* **45:** 21–30.

Costa, D. P., Croxall, J. P. and Duck, C. D., 1989. Foraging energetics of Antarctic fur seals in relation to chanes in prey availability. *Ecology* **70:** 596–606.

Daneri, G. A. and Coria, N. R., 1992. The diet of Antarctic fur seals, *Arctocephalus gazella*, during the summer-autumn period at Mossman Peninsula, Laurie Island (South Orkneys). *Polar Biol.* **11:** 565–66.

Daneri, G. A. and Coria, N. R., 1993. Fish prey of Antarctic fur seals, *Arctocephalus gazella*, during the summer-autumn period at Laurie Island, South Orkney Islands. *Polar Biol.* **13:** 287–89.

Doidge, D. W. and Croxall, J. P., 1985. Diet and energy budget of the Antarctic fur seal *Arctocephalus gazella* at South Georgia. Pp. 543–50 *in* Antarctic nutrient cycles and food webs ed by W. R. Siegfried, P. R. Condy and R. M. Laws. Proceedings of the 4th SCAR Symposium on Antarctic Biology. Springer-Verlag: Berlin.

Gales, N. J. and Costa, D. P., 1997. The Australian Sea Lion: a review of an unusual life history. Pp. 78–87 *in* Marine Mammal Research in the Southern Hemisphere Vol. 1: Status, Ecology and Medicine ed by M. Hindell and C. Kemper. Surrey Beatty & Sons: Chipping Norton.

Gentry, R. L., Costa, D. P., Croxall, J. P., David, J. H. M., Davis, R. W., Kooyman, G. L., Majluf, P., McCann, T. S. and Trillmich, F., 1986. Synthesis and conclusions. Pp. 220–64 *in* Fur Seals: Maternal Strategies on Land and at Sea ed by R. L. Gentry and G. L. Kooyman. Princeton University Press: Princeton.

Gentry, R. L. and Kooyman, G. L., 1986. Fur seals: Maternal Strategies on Land and at Sea. Princeton University Press: Princeton.

Goldsworthy, S. D., 1992. Maternal care in three species of southern fur seal (*Arctocephalus* spp.). Ph.D. Thesis, Monash University.

Goldsworthy, S. D., 1995. Differential maternal expenditure of maternal resources in Antarctic fur seals, *Arctocephalus gazella*, at Heard Island, southern Indian Ocean. *Behav. Ecol.* **6:** 218–28.

Goldsworthy, S. D. and Shaughnessy, P. D., 1989. Subantarctic Fur Seals *Arctocephalus tropicalis* at Heard Island. *Polar Biol.* **9:** 337–39.

Green, K., 1997. Diving behaviour of Antarctic Fur Seals *Arctocephalus gazella* Peters around Heard Island. Pp. 97–104 *in* Marine Mammal Research in the Southern Hemisphere Volume 1: Status, Ecology and Medicine ed by M. Hindell and C. Kemper. Surrey Beatty & Sons: Chipping Norton.

Green, K., Burton, H. R. and Williams, R., 1989. Diet of Antarctic fur seals *Arctocephalus gazella* during the breeding season at Heard Island. *Antarct. Sci.* **1:** 317–24.

Green, K., Burton, H. R. and Williams, R., 1991. The diet of Antarctic fur seals during the late autumn and early winter around Heard Island. *Antarct. Sci.* **3:** 359–61.

Green, K., Williams, R. and Burton, H. R., 1997. Foraging ecology of Antarctic Fur Seals *Arctocephalus gazella* Peters around Heard Island. Pp. 105–13 *in* Marine Mammal Research in the Southern Hemisphere Vol. 1: Status, Ecology and Medicine ed by M. Hindell and C. Kemper. Surrey Beatty & Sons: Chipping Norton.

Green, K., Williams, R., Handasyde, K. A., Burton, H. R. and Shaughnessy, P. D., 1990. Interspecific and intraspecific diets of fur seals, *Arctocephalus* species (Pinnipedia: Otariidae), at Macquarie Island. *Aust. Mammal.* **13:** 193–200.

Hes, A. D. and Roux, J. P., 1983. Population increase in the sub-Antarctic fur seal *Arctocephalus tropicalis* at Amsterdam island. *S. Afr. J. Antarct. Res.* **13:** 29–34.

Jouventin, P. and Stonehouse, B., 1985. Biological survey of Ile de Croy, Iles Kerguelen, 1984. *Polar Rec.* **22:** 688–91.

Jouventin, P., Stahl, J. C. and Weimerskirch, H., 1982. La recolinsation des iles Crozet par les otaries (*Arctocephalus tropicalis* et *A. gazella*). *Mammalia* **46:** 505–14.

Kerley, G. I. H., 1983. Comparison of seasonal haul-out patterns of fur seals *Arctocephalus tropicalis* and *A. gazella* on subantarctic Marion Island. *S. Afr. J. Wild. Res.* **13:** 71–77.

Kerley, G. I. H., 1985. Pup growth in the fur seals Arctocephalus tropicalis and *A. gazella* on Marion Island. *J. Zool., Lond. (A)* **205:** 315–24.

Kooyman, G. L., Davis, R. W. and Croxall, J .P., 1986. Diving behavior of Antarctic fur seals. Pp. 115–25 *in* Fur Seals: Maternal Strategies on Land and at Sea ed R. L. Gentry and G. L. Kooyman. Princeton University Press: Princeton.

Lunn, N. J and Boyd, I. L., 1993. Effects of maternal age and condition on parturition and the perinatal period of Antarctic fur seals. *J. Zool. Lond.* **229:** 55– 67.

North, A. W., Croxall, J. P. and Doidge, D. W., 1983. Fish prey of the Antarctic fur seal *Arctocephalus gazella* at South Georgia. *Br. Antarct. Surv. Bull.* **61:** 27–37.

Oftedal, O. T., Iverson, S. J. and Boness, D. J., 1987. Milk and energy intakes of suckling Californian sea lions *Zalophus californianus* pups in relation to sex, growth, and predicted maintenance requirements. *Physiol. Zool.* **60:** 560–75.

Paulian, P., 1964. Contribution a l'etude de l'otarie de l'le Amsterdam. *Mammalia* 28, suppl. **1:** 1–146.

Perissinotto, R. and McQuaide, C. D., 1992 Land-based predator impact on virtically migrating zooplankton and micronecton advected to a Southern Ocean archipelago. *Mar. Ecol. Prog. Ser.* **80:** 15–27.

Shaughnessy, P. D., Shaughnessy, G. L. and Fletcher, L., 1988. Recovery of the fur seal population at Macquarie Island. *Pap. Proc. Roy. Soc. Tas.* **122:** 177–87.

Shaughnessy, P. D. and Goldsworthy, S. D., 1990. Population size and breeding season of the Antarctic fur seal Artocephalus gazella at Heard Island — 1987/88. *Mar. Mamm. Sci.* **6:** 292–304.

Shaughnessy, P. D. and Goldsworthy, S. D., 1993. Feeding ecology of fur seals and their management at Heard and Macquarie Islands. *Proc. NIPR Symp. Polar Biol.* **6:** 173–75.

Shaughnessy, P. D. and Fletcher, L., 1987. Fur seals, *Arctocephalus* spp., at Macquarie Island. Pp. 177–88 *in* Status, biology and ecology of fur seals ed by J. P. Croxall and R. L. Gentry. *NOAA Technical Report. NMFS* **51**.

Trillmich, F., 1990. The behavioural ecology of maternal effort in fur seals and sea lions. *Behaviour* **114:** 3–20.

Trillmich, F. and Ono, K. A., 1991. Pinnipeds and El Nino: Responses to environmental stress. Springer-Verlag: Berlin.

Wilkinson, I. S. and Bester, M. A., 1990 Continued population increase in fur seals, *Arctocpehalus tropicalis* and *A. gazella*, at Prince Edward Islands. *S. Afr. J. Antarct. Res.* **20:** 58–63.

Williams, R. and McEldowney, A., 1990. A guide to the fish otoliths from waters off the Ausralian Antarctic Territory, Heard and Macquarie Islands. *ANARE Res. Notes* **75.**

The role of stones in the seal lion stomach: investigations using contrast radiography and fluoroscopy

D. J. NEEDHAM[1]

The presence of stones in the stomach of pinnipeds has been reported on numerous occasions. The function of these stones is not known but speculations vary from buoyancy control to grinding of food. This study aimed to examine the function of stones in the stomachs of two juvenile Australian Sea Lions *Neophoca cinerea* by fluoroscopy. The animals were fluoroscoped in sternal and lateral recumbency using dorso-ventral and lateral projections. The size of the stones varied from 1.5 to 7 cm, but 2 to 5 cm were the most common. The stones were observed in the body of the stomach at the level of the angular notch. With gastric contractions the smaller stones moved into the pyloric antrum and back again, appearing to collide with the larger stones. Fish skeletons remained in the fundus during the period of observation (up to 5 hours), suggesting that the stones act as a sieve, retaining larger food particles, and perhaps aid in mixing or trituration of the food.

INTRODUCTION

AUSTRALIAN Sea Lions *Neophoca cinerea* are found in small colonies along the southern and western Australian coast. Their range extends from the Pages (35°46′S, 138°18′E) near Kangaroo Island in the east, to the Houtman Abrolhos (28°51′S, 114°03′E) off Western Australia. The latest population estimate for Australian Sea Lions is in the order of 10 000 animals, of which about 6 900 are in South Australia (Gales 1990).

There is little published information on the morphology and function of the alimentary tract in this species (Richardson and Gales 1987; Needham 1981; Ling 1992). The presence of stones and pebbles (Fig. 1) has been recorded in the stomachs of many pinnipeds (King 1983). Wood Jones (1925) reported that Australian Sea Lions swallowed faceted granite pebbles, with a maximum of 40 stones and a weight of 2 kg being recorded. However, the function of the stones is not known. Speculation has included: grinding of stomach parasites, breaking up ingested food, assisting in vomiting of fish bones, facilitating ballast and buoyancy control and providing bulk in the stomach during the enforced fasting of the breeding season (King 1983).

In this study the stomachs of two juvenile Australian Sea Lions were examined by radiography and fluoroscopy in an attempt to understand the function of stones in this species. Barium studies were then performed using either liquid or solid food on different occasions with each animal. The relative movements of the stomach and the stones were compared with each type of food.

METHODS

Animals

The two juvenile sea lions used in this study were rescued animals. The male was a suckling pup (weighing 9.0 kg) which was presented at approximately 10–14 days of age (for age data mentioned in the text it was assumed that this animal was 10 days old) and in good condition. It was subsequently hand raised and bottle fed using a formula primarily composed of ground whole fish and cream. Weekly radiographs were taken as part of another study. The barium studies were conducted over the period when the sea lion pup was between 80 and 200 days of age. During this period its weight increased from 22.8 kg to 42.8 kg. The female sea lion was about 8–9 months old and weighed 15.1–17.2 kg during the study. Presented as a thin stranded animal (weight 12.2 kg) she was rehabilitated for about ten weeks prior to the commencement of this project. Her diet was mainly Ruff *Arripis georgianus* weighing approximately 20–30 g each. When possible the sea lions were fasted overnight before a barium experiment was conducted.

---

[1]294 Greenhill Road, Glenside, South Australia, Australia 5065.

Pages 164–69 *in* MARINE MAMMAL RESEARCH IN THE SOUTHERN HEMISPHERE VOLUME 1: STATUS, ECOLOGY AND MEDICINE ed by M. Hindell and C. Kemper. Surrey Beatty & Sons, Chipping Norton. 1997.

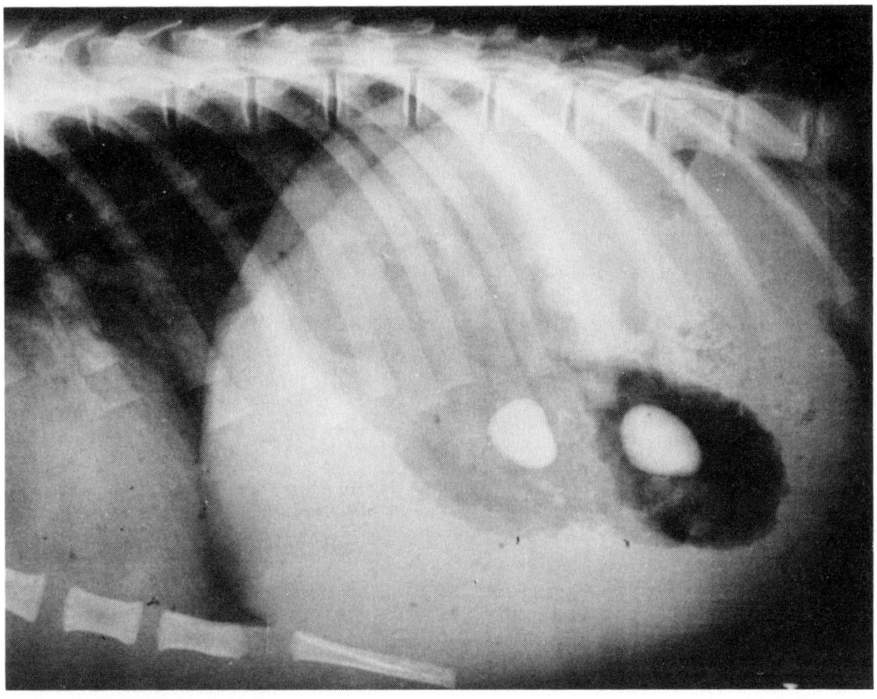

Fig. 1. Left lateral radiograph of sea lion stomach showing the position of stones. The stomach has been filled with air to aid contrast.

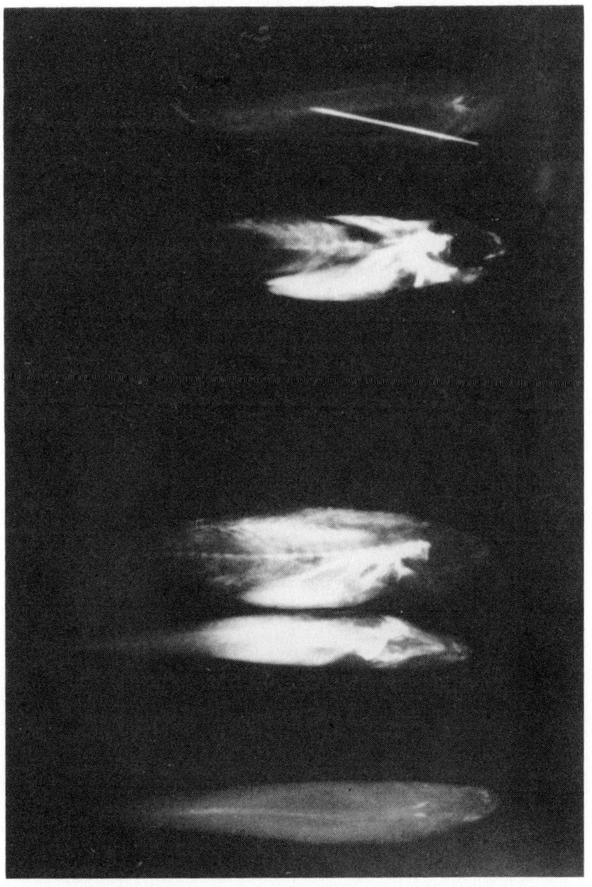

Fig. 2. Radiographs of solid food fish showing (from top): location of needle for injection of contrast; barium injected to one side of fish (lateral view); barium injected on both sides (lateral and dorso-ventral views); plain fish (dorso-ventral view).

Contrast media

Barium sulphate suspension ("Medebar"®, Medicon) was mixed with the formula for the male pup (200 ml "Medebar" with 300 ml formula) and fed by bottle and teat. In the case of the female, the barium was mixed in 300 ml of water and administered by stomach tube. One liquid barium experiment was carried out on the female and several were done on the male as he generally could not be persuaded to eat fish.

For the studies utilizing whole fish, two contrast mixtures were used; (i) 7.5 g of barium sulphate powder (BP) was suspended in 8 ml of water and (ii) 1.5–3 ml of "Medebar" suspension. The contrast media was injected intramuscularly into the fish using a 21 G needle directed in a postero-dorsal manner through the gill region (Fig. 2). About 1.5–2.0 ml of barium suspension was placed on each side of the vertebral column.

The body of the fish was therefore much easier to see and the removal of flesh from the skeleton during digestion could be visualized leaving the bones and skin with a light coating of contrast. On some occasions plain fish (without barium) were fed to maintain the animal's interest, but these fish were not visible in the stomach during fluoroscopy.

Selecting appropriate fish for solid food experiments was important. Sea lions must swallow fish head first. If the fish were too large or not presented to the animal in a suitable orientation, they were bitten and turned around in the mouth prior to swallowing or being spat out. The fish weighed about 20–30 g each.

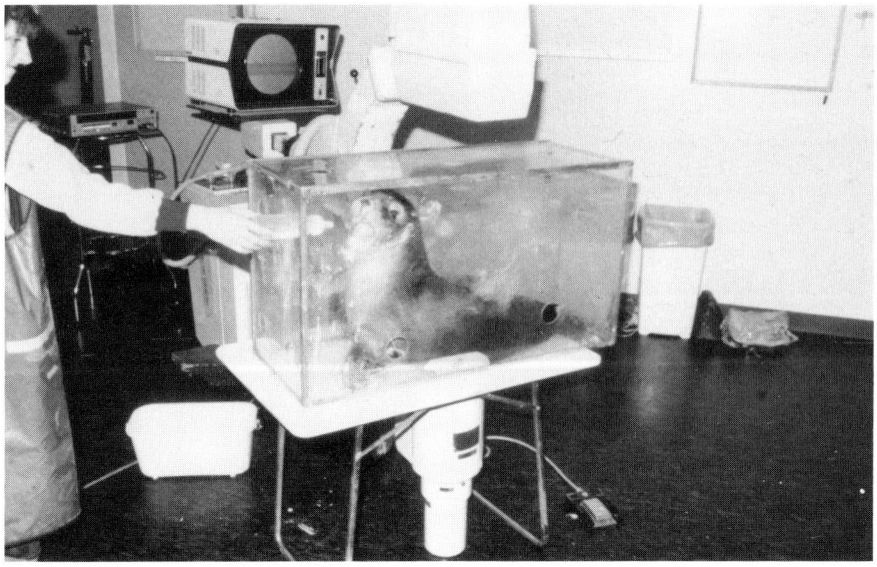

Fig. 3. Sea lion in perspex restraining box. Shows orientation of fluoroscope and method of bottle feeding.

Solid food was fed to the female on three occasions. Two, five or ten, barium impregnated fish were fed and followed through the stomach with the fluoroscope. The male sea lion could only be persuaded to take fish on one occasion, when he swallowed 12 out of 14 offered fish.

Fluoroscopy

The fluoroscope used was a Toshiba mobile X-Ray, TV unit, Model #: SXT6–6B–5, fitted with an image intensifier Model No. IG-96G/MCI–11–KD. The X-ray tube and the camera were situated at each end of a C-shaped arm that was mounted in the centre and could rotate about this axis for 90° to either right or left. This permitted both dorso-ventral and lateral views. Unfortunately, due to the configuration of the equipment, it was not possible to move quickly from one view to the other. The kilovoltage (kV) employed was generally between 60–90 kV and adjusted to obtain the most suitable image. The duration of recordings were chosen in an attempt to minimize overheating of the fluoroscope. Observations and results reported in this study are timed from the first oral administration of the contrast medium.

Animal restraint

It was necessary to be able to keep the abdomen of the sea lion in the line of the x-ray beam. This was achieved by constructing a rectangular perspex box which permitted the sea lion to lie prone or maintain a normal sitting position in comfort (Fig. 3). A series of 75 mm diameter holes was cut in the sides and top of the box to provide ventilation and allow the insertion of the feeding bottle. As the sea lions were free to move around within the confines of the box, there were frequent changes in the view obtained through the fluoroscope. The box was placed on a mobile trolley that could be moved to keep the animal's abdomen in the line of the X-ray beam to minimize these problems. The height of the trolley was just sufficient to allow the camera end of the fluoroscope to fit beneath the perspex box. This close apposition of the camera to the animal minimized the distortion of the recorded image.

RESULTS AND DISCUSSION

Stones are frequently found in the stomachs of Australian Sea Lions at necropsy. They are more commonly found in wild animals than in captive specimens. The maximum number recovered by the author is 14 stones that collectively weighed 1.5 kg. Generally the stones are smooth and rounded often with one flattened facet. The size varies from 1.5 to 7 cm diameter, with smaller stones being found in younger and smaller animals.

Fluoroscopy is frequently used to evaluate the anatomy and function of the gastro-intestinal tract. It has the advantage of providing moving, real time images which can be enhanced by the addition of contrast media to the ingesta.

Liquid barium series

Female sea lion

Plain films showed the presence of two stones in the stomach. When the 300 ml of liquid barium was introduced, the body and pyloric regions of the stomach were the first to fill. Contractile movements began almost immediately, with a contraction designated as a consistent wave in the stomach wall, or when the stones were noted to move relative to each other. The dorso ventral view showed the gastric contractions well. On lateral projections, the cardia and body could be clearly distinguished but the pyloric region was very difficult to visualize and contractions were not easily detected. This is possibly due to the relatively small volume of barium contained in the pyloric region, the presence of contrast in the long, convoluted small intestine and the thickness of other overlying tissue. The barium hid the stones from view in both projections.

In the distended stomach, contractions originated at about the junction of the cardia and body regions (Richardson and Gales 1987). The contraction wave was more readily visible on the greater curvature of the stomach. It began as a small wave that appeared to increase in magnitude as it passed the angular notch heading into the pyloric region.

Strong contractions in the pyloric region were observed, particularly as waves emanating from the body near the angular notch and moving towards the pylorus. As each wave approached the pylorus, there appeared to be a rapid opening and closing of the pyloric canal which released a small amount of barium into the duodenum. It is possible that this valve-like action is due to the operation of the pyloric torus (Richardson and Gales 1987). Following the closure of the pyloric canal, a portion of the barium returned to the main part of the pyloric region and joined the next wave of contraction. Barium was not seen to enter the duodenum with every wave of stomach contraction, although it was not possible to determine the frequency of passage. Within five minutes some of the contrast medium had entered the duodenum. This is considerably quicker than that observed by Needham (1981), when the barium took at least 12 minutes to enter the duodenum.

After 13 minutes a stone was visible in the area of the angular notch during a contraction wave and was not seen again until 25 minutes had elapsed. Both stones were seen for up to 60 minutes. They appeared to move with the contraction wave from the distal portion of the stomach body to the pyloric region and back again. After an hour, both stones were visible due to their different radio density to the surrounding barium residue. The large intestine contained contrast medium by 94 minutes, was nearly full at 120 minutes and the sea lion urinated and defecated at 128 minutes. The stomach was empty at 138 minutes and the stones appeared to be in the cranial section of the pyloric region.

Male sea lion

This animal had been radiographed on a weekly basis for several months as part of another study. No stones were noted in the stomach on any occasion up to the second barium series. The sequence of events and their timing were very similar to those of the female sea lion. Stomach contractions were observed to occur at intervals of about 6.5 seconds when the stomach was distended at the beginning of the fluoroscopy. After 25 minutes the average interval between contractions was 6.7 seconds. The contraction interval slowed to an average of 9.2 seconds after 55 minutes when it was estimated that the stomach was about half full.

On the third occasion that the male sea lion was given a barium meal (age 107 days) two small stones were observed in his stomach. These pebbles were not present one week earlier, but were observed on a radiograph two days prior to the third barium series. The stones were occasionally visible and were located one on either side of the angular notch. Because the barium suspension was mixed with formula it was diluted and therefore less radio dense through the fluoroscope. The stones were therefore more readily observed in the male sea lion and appeared as radio-dense highlights in the contrast medium. Movements of the pebbles from the body to the pyloric region and back again were similar to those of the female sea lion, giving the impression of a mixing action in the distal stomach contents.

On day 112 the seal was provided with a selection of stones that had been previously collected from the stomachs of other sea lions at necropsy. Sixteen of these stones were photographed, radiographed and numbered. Eight pebbles (assigned odd numbers) were placed in the pool where he had day-time access and the other eight (with even numbers) were located in his evening accommodation. Over the next three days, three of the even numbered pebbles disappeared along with one odd numbered stone. Plain radiographs taken on day 116 revealed the presence of four stones in the animal's stomach.

The fourth time the male sea lion was fluoroscoped (day 128), he had five stones in his stomach. There was one large stone located at the level of the angular notch with two stones immediately cranial in the body and the other two (smaller) pebbles in the pyloric region. On this occasion he was fed 500 ml of formula followed by three 10 ml boluses of barium. Each swallow appeared as a ball of contrast on the fluoroscope and did not immediately mix with the stomach contents. The contrast media remained on the surface of the stomach contents in the cardiac region for a minute or so, then slowly sank to the ventral area of the stomach. The balls of contrast remained cranial to the group of stones and took nearly five minutes to disperse into the stomach contents. The barium began to surround the stones 15 minutes after administration of the second bolus. There was insufficient contrast present to adequately delineate the stomach wall.

Contractions of the stomach occurred from the time the barium was given but proved difficult to record as the sea lion was very restless. The stones remained in the same relative positions with the two smallest pebbles being subjected to the greatest movements and periodic changes of position. Barium first entered the duodenum after 26 minutes.

After 67 minutes, the four smaller pebbles were located in the pyloric region with the largest stone at the level of the angular notch and aligned along the greater curvature. Contractions continued but most of the contrast medium had been expelled from the stomach at this time. Another two 10 ml boluses of barium were administered at 70 minutes. The contrast had surrounded the large stone by 75 minutes and contractions of a smaller magnitude continued at about 7 second intervals. Most of the stone movement occurred in the caudal most stone. Occasionally, two or three of the pyloric pebbles moved in unison without changing their relative position. This second bolus of barium began entering the duodenum at 77 minutes.

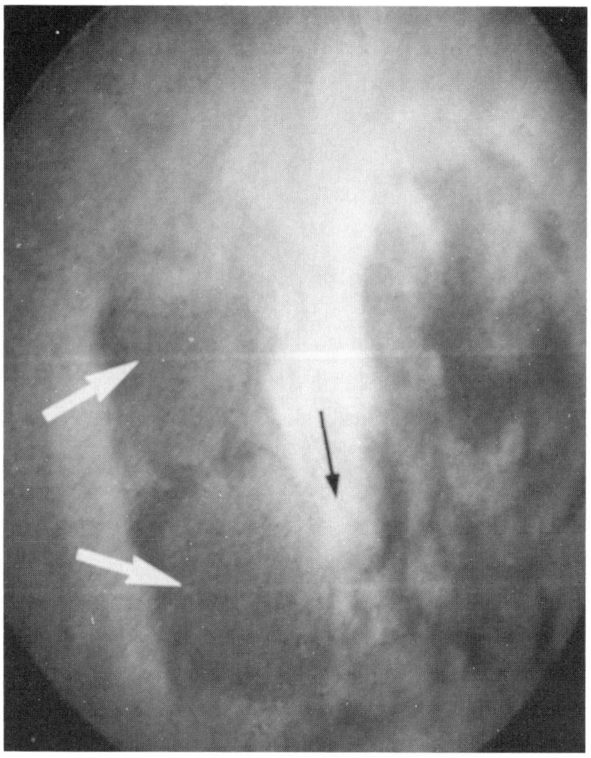

Fig. 4. Fluoroscope view of seal lion stomach showing 10 fish in the body and the two stones (white arrows) forced into the pyloric region beyond the angular notch (black arrow). Some barium surrounds the stones.

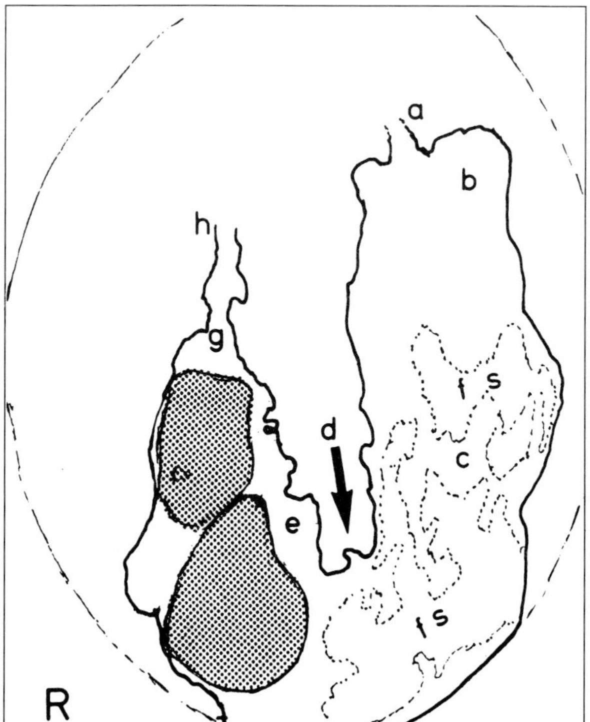

Fig. 5. Schematic diagram of Figure 4 showing the anatomical relations of the stomach and the stones (shaded). Legend: a, oesophagus; b, cardia; c, body; d, angular notch; e, pyloric region; fs, fish skeletons; g, pyloric canal; h, duodenum.

Solid food barium series

Again, the results for the two sea lions were essentially similar. When fewer than five fish were ingested, very little activity was recorded in the stomach. However, regardless of the amount of solid food consumed, the fish skeletons were always located cranial to the stones. When 10 fish were swallowed the body of the stomach was greatly distended and the stones were positioned beyond the angular notch and into the pyloric region˙ (Figs 4 and 5). The stones may be important in delaying the passage of large food particles from the stomach while permitting the rapid transit of liquids and small particles. This is consistent with the findings of Richardson and Gales (1987) who found that 46% of small markers (<1.2 mm diameter) passed through the alimentary tract within 6 h and 40% of the large markers (3 mm diameter.) were retained for periods longer than six days.

The stomach appears to need to be distended for contractions to begin. As the soft tissues separated from the skeletons during digestion, the barium impregnated material enveloped the stones as it entered the pylorus. This occurred between four and five minutes after the meal. Due to the smaller volume of contrast present contraction waves were not visible in the body of the stomach but were sometimes seen in the pyloric wall. The relative movements of the stones was a useful indicator of contraction waves but there were many waves that did not result in movement of the pebbles. Frequently

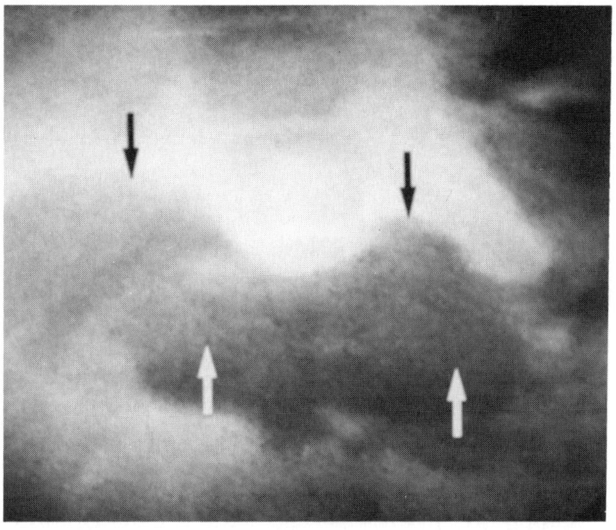

Fig. 6. Contraction waves (black arrows) in the pyloric region passing the stones (white arrows).

there would be stone movement with two or three consecutive contractions but the pattern was irregular. The interval between contractions involving stone movement was generally about 2.5 seconds during the first 90 minutes, in the third hour the time had lengthened to 7.5 seconds.

As each wave travelled along the pylorus, the liquid ingesta was forced around the more cranially located stones and between the other pebbles (Fig. 6). Once the wave has passed, the stones moved back towards the angular notch. The .fluid contents continued around and beyond all the stones to the anterior end of the pyloric canal. If the canal is open when the wave arrives, a small amount of digesta passes into the duodenum. When the pyloric canal closes (or if it does not open) there is a reflux of the wave into the next approaching contraction increasing its amplitude. The most caudal stones in the pyloric region undergo considerable movement during this procedure. The returning wave may continue to about the middle of the pylorus. This combined action of wave and stone movement mixes the liquid stomach contents before it enters the narrow duodenum (circumference about 4 cm, Richardson and Gales (1987)).

CONCLUSIONS

During the digestion of a meal which distended the stomach, the stones moved backwards and forwards in the pyloric region as the contraction waves passed distally. The stones were generally located in the body of the stomach at about the level of the angular notch. When solid food was consumed the stones remained between the ingested fish and the pylorus, and seemed to act as a physical barrier or valve that kept the larger food particles in the body of the stomach and prevented them from entering the narrow small intestine. For an animal whose food is mainly swallowed whole without mastication this could be a very useful strategy. Not all contraction waves released gastric contents to the duodenum. More often than not the pyloric canal was closed and the returning wave proceeded cranially in the pylorus moving the stones again. This action facilitated further trituration of the liquid stomach contents.

ACKNOWLEDGEMENTS

Special thanks are due to Mr Ray Yates who kindly arranged for us to use the facilities of the Flinders University of South Australia animal house and assisted with the setting up of the fluoroscope. Leslie McDonald and Judy Rand cared for the animals and provided assistance with the procedures. Dr R. A. Tedman read several drafts of the paper and made many helpful suggestions for improving the manuscript.

REFERENCES

Gales, N. J., 1990. Abundance of Australian Sea Lions *Neophoca cinerea* along the southern Australian coast and related research. Report to the Western Australian Department of Conservation and Land Management, South Australian National Parks and Wildlife Service and the South Australian National Wildlife Conservation Fund. (Unpublished).

King, J. E., 1983. Seals of the World. University of Queensland Press: St. Lucia, Queensland.

Ling, J. K., 1992. *Neophoca cinerea.* Mammalian Species No. 392, Pp. 1–7. The American Society of Mammalogists.

Needham, D. J., 1981. Barium meal studies in Australian Sea Lions *(Neophoca cinerea).* Pp. 181–85 *in* Wildlife Diseases of the Pacific Basin and other Countries ed by M. E. Fowler. Fruitridge Printing: Sacramento.

Richardson, K. C. and Gales, N. J., 1987. Functional morphology of the alimentary tract of the Australian Sea Lion, *Neophoca cinerea. Aust. J. Zool.* **35:** 219–26.

Wood Jones, F., 1925. The Mammals of South Australia. Part III. The Monodelphia. Government Printer: Adelaide.

Report on the healing of a large wound in a Bottlenose Dolphin *Tursiops truncatus*

M. B. ORAMS[1] and R. B. DEAKIN[2]

INTRODUCTION

THERE are several reports that dolphins (Delphinidae) sustain injuries from shark bites (Corkeron *et al.* 1987a) and that these wounds heal rapidly (Corkeron *et al.* 1987b; Bruce-Allen and Geraci 1985). However, there have been few opportunities of observing the reaction of a dolphin to a major injury and monitoring its rate of healing. At Tangalooma Resort on Moreton Island in south-east Queensland, Australia, a programme has been established which allows tourists to regularly hand feed a group of wild Bottlenose Dolphins *Tursiops truncatus* (Orams 1994). The programme also permits researchers to observe and record behaviour and incidents involving the habituated dolphins. This paper reports on the occurrence and healing of a large wound, probably caused by a shark bite, on a mature male Bottlenose Dolphin, regularly visiting the feeding sessions at Tangalooma.

Tangalooma dolphin care programme

Routine feeding of groups of wild Bottlenose Dolphins by humans close inshore is rare in Australia. At Monkey Mia in Western Australia this practice has been occurring for over 30 years and is well reported in the literature (e.g., Doak 1981; Connor and Smolker 1985; Edwards 1988; Morris 1988). A more recent programme which allows people to hand feed a group of Bottlenose Dolphins in Moreton Bay, Queensland has been in operation since 1992 (Orams 1994). However, dolphin interaction with human beings in this area dates back many decades (Fairholme 1854).

There are currently nine Bottlenose Dolphins regularly visiting the beach area next to the jetty at Tangalooma. These dolphins are hand fed from the shallows, mostly with Silver Biddys *Gerres ovatus*, a small, soft flesh bait fish commonly found in the area.

Each of the nine visiting dolphins is now well known to researchers based at Tangalooma. Individual animals are identified by the unique shapes of their dorsal fins (Caldwell and Caldwell 1972). They are fed in an area which is easily viewed from elevated positions on the jetty, consequently unobstructed observation of the dolphins is available to researchers. The water is shallow (<2 m) and usually clear, so continuous observation is possible before, during and after feeding sessions. Records are kept of each feeding session, and each session is video taped from an elevated position to permit subsequent review and analysis of animal behaviour. Therefore, the Tangalooma Dolphin Care and Research Programme provides a valuable opportunity for studying Bottlenose Dolphin behaviour in the wild.

Appearance of, and reaction to, a large wound

On 20 May 1993 a mature male Bottlenose Dolphin, "Freddy", arrived at the feeding with a large open wound located on his left side, above and approximately level with the flipper (see Fig. 1). This injury must have occurred during the previous 24 hours since the dolphin had attended the feeding session on 19 May 1993 and did not have the wound then. The lesion was roughly semi-circular in shape and approximately 45 cm in diameter. It penetrated approximately 3 cm into the dolphin's blubber layer at the deepest point, and in two places muscle tissue was visible. The wound was deepest on the dorsal side, and several punctures resembling teeth marks were present on its ventral aspect. The shape and type of injury was consistent with a shark bite. Previous research indicates that over a third of all dolphins in Moreton Bay have evidence of shark bites (Corkeron *et al.* 1987b).

The wound caused concern amongst staff controlling the feeding of the dolphins at Tangalooma. Consequently, "Freddy's" behaviour was monitored closely during the following weeks. (As part of the regular record keeping for the dolphin feeding, the behaviour of all dolphins was recorded throughout this period.) From 20 May, the day when the injury occurred, "Freddy" continued to attend the feeding sessions. His fish intake at these sessions did not vary from normal (3–5 kg day^{-1}) nor was there any noticeable reaction by the dolphin to the injury. He neither favoured nor sought to protect the area of the wound and there was no

[1]Department of Management Systems, Massey University, Albany, Private Bag 102 904, North Shore MSC, New Zealand.
[2]Sea World Enterprises, P.O. Box 190, Surfers Paradise, Queensland, Australia 4217.
Pages 170–73 *in* MARINE MAMMAL RESEARCH IN THE SOUTHERN HEMISPHERE VOLUME 1: STATUS, ECOLOGY AND MEDICINE ed by M. Hindell and C. Kemper. Surrey Beatty & Sons, Chipping Norton. 1997.

Fig. 1. Wound on adult male Bottlenose Dolphin *Tursiops truncatus* thought to have been caused by a shark. Dorsal view of the injury within 48 hours of its occurrence.

Fig. 2. Visible unpigmented scar, six months after the injury occurred.

Fig. 3. The scar nine months after the injury occurred.

Fig. 4. Twelve months after the injury the scar is scarcely visible.

detectable change in his movements. These observations are interesting since it is the first detailed record of behavioural reactions to a large shark inflicted injury on a Bottlenose Dolphin. Other research has shown that dolphins appear to survive these injuries and they heal remarkably quickly (Bruce-Allen and Geraci 1987).

Treatment

In response to concerns over the possible trauma this injury caused, staff at the Tangalooma Dolphin Care and Research Programme administered a treatment commonly used on injured captive dolphins (Dierauf 1990). A portion of "Freddy's" fish (all *G. ovatus*) were hydrated by injecting small quantities of fresh water (approximately 30 ml) with a syringe into the gut of about 2 kg of fish each day. In addition, two 250 mg multi-vitamin tablets were slipped inside a fish each day in order to assist in the animal's recovery. This treatment was administered on 12 occasions during the two weeks following the injury. It is unknown whether this treatment contributed to the animal's recovery but it seems likely the wound would have healed irrespective of any therapy.

Healing rate

There appeared to be no bleeding from the wound, although small amounts would have been difficult to detect due to dilution in surrounding water. Changes in the colour and width of the wound were noticeable within seven days. It did not appear as bright, as deep, nor as open. Complete closure occurred within 30 days, and after 45 days all that remained was an obvious indented white unpigmented scar. This scar faded over time, and six months after the wound appeared was less clear but still noticeable (Fig. 2). After nine months the scar was difficult to detect unless one was close to the animal (Fig. 3); and 12 months later the scar was scarcely visible at all (Fig. 4).

Observations made by Corkeron *et al.* (1987a) showed that approximately 36% of Moreton Bay dolphins had evidence of shark attack. It is possible that many more dolphins in the bay have been wounded by sharks, but the injuries have healed so that indications of attacks are no longer detectable. This record of rapid healing from a major wound resulting in minimal visible scarring, suggests that 36% may be an underestimate.

CONCLUSION

The occurrence of shark attacks on Bottlenose Dolphins in Moreton Bay, Queensland, Australia is quite common. Many dolphins appear to survive these attacks, even when the injuries seem quite severe. The case of the mature male Bottlenose known as "Freddy". at Tangalooma produced a photographic record which shows the healing from a large shark bite at regular intervals. It provides further evidence that the healing rate of Bottlenose Dolphins from shark bites is rapid. In addition, this case provides the first record that dolphins do not appear to be visibly traumatized by severe wounds such as the one described. Furthermore, the barely visible scar left from this wound suggests that shark attack on Moreton Bay dolphins may be even more common than previously thought.

ACKNOWLEDGEMENTS

Financial support for this research was provided by a University of Queensland Postgraduate Research Scholarship, The Royal Geographical Society of Queensland and Tangalooma Moreton Island Resort. The authors also wish to thank the management and staff of Tangalooma Moreton Island Resort for their assistance with this project and in particular Philip Robbins for the photography. Thanks also to the referees who provided useful suggestions which have improved this paper.

REFERENCES

Bruce-Allen, L. J. and Geraci, J. R., 1985. Wound healing in the bottlenose dolphin (Tursiops truncatus). *Can. J. Fish. Aqua. Sci.* **42:** 216–18.

Caldwell, D. K. and Caldwell, M. C., 1972. The world of the Bottlenose Dolphin. Lippincott: New York.

Connor, R. C. and Smolker, R. S., 1985. Habituated dolphins (*Tursiops* sp.) in Western Australia. *J. Mammal.* **66**(2): 398–400.

Corkeron, P. J., Morris, R. J. and Bryden, M. M., 1987a. Interactions between Bottlenose Dolphins and sharks in Moreton Bay, Queensland. *Aqua. Mammal.* **13**(3): 111–14.

Corkeron, P. J., Morris, R. J. and Bryden, M. M., 1987b. A note on healing of large wounds in Bottlenose Dolphins, *Tursiops truncatus. Aqua. Mammal.* **13**(3): 96–98.

Dierauf, L. A. (ed), 1990. Handbook of marine mammal medicine: Health, disease and rehabilitation. CRC Press: Boca Raton, Florida.

Doak, W., 1981. Dolphin, dolphin. Hodder and Stoughton: Auckland, New Zealand.

Edwards, H., 1988. The Monkey Mia dolphins. Pp. 200–13 *in* Whales, dolphins and porpoises ed by R. Harrison and M. M. Bryden. Facts of File Publications: New York.

Fairholme, J. K. E., 1856. The blacks of Moreton Bay and the porpoises. *Proc. Zool. Soc. Lond.* **24:** 353–54.

Morris, R., 1988. Human contact. Pp. 204–07 *in* Whales, dolphins and porpoises ed by R. Harrison and M. M. Bryden. Facts on File Publishing: New York.

Orams, M. B., 1994. Tourism and marine wildlife: The wild dolphins of Tangalooma, Australia. *Anthrozoos* **7**(3): 195–201.

Preliminary investigations of techniques for killing whales

D. K. BLACKMORE,[1] A. NUTMAN,[2] G. R. G. BARNES,[3] P. MADIE,[4] A. S. DAVIES,[2] M. C. BOWLING,[3] M. DONOGHUE[5] and E. J. KIRK[2]

Attempts have been made, based on the use of projectiles and explosives, to develop better methods for the euthanasia of stranded whales *in extremis*. The physiological effects of electrical lancing, as used in commercial whaling, has been investigated. These investigations were preceded by anatomical studies based on CT scans of the heads and skulls of 10 different species of cetaceans, acrylic casting of the arterial supply to the brain of Pilot Whales, and EEG studies of sheep shot through the occipito-atlantal junction. The value of using a shotgun loaded with buckshot for destroying smaller whales when shot from the lateral side of the head has been demonstrated. Acrylic casts have confirmed that the common carotid arteries do not supply blood to the brain. Electrical lancing is unlikely to affect the brain or heart in terms of causing either insensibility or death.

Key words: Whales, Euthanasia, Explosives, Projectiles, Blood supply to the brain, Electric lance.

INTRODUCTION

In a paper presented to the Commission in 1992 (Blackmore 1992) by one of the authors of this paper, an essentially theoretical approach was taken to assess the problems associated with the humane slaughter of whales. It was concluded that none of the methods used could be considered humane when compared with methods of slaughter used for domestic stock. Based on a desire to test some of the hypotheses posed and a wish to comply with the recommendation of the International Whaling Commission (IWC) for further research to be carried out in countries where natural strandings of whales frequently occur, work on stranded animals was initiated in November 1993. Prior to these investigations of stranded animals, a team of scientists from different disciplines agreed to form a working group without financial recompense (see list of authors). Experts in other fields gave their advice and assistance on an honorary and *ad hoc* basis. (See acknowledgements). Funds for equipment and other costs were obtained from organizations, institutions and individuals in the United Kingdom and New Zealand.

The investigations so far have been focused on the anatomy of the head of different species of cetacean, ballistical problems related to the euthanasia of stranded animals, and the likelihood of inducing brain and cardiac dysfunction by applying electrical currents to whales. The findings presented in this paper are of a preliminary nature and it is hoped to present more comprehensive data at the IWC humane killing workshop in 1995. However, it is believed that these preliminary findings are of considerable importance and already indicate some practical points which are of immediate relevance to both the euthanasia of stranded cetaceans and the slaughter of whales for commercial or other purposes.

The oral presentation of this paper will include numerous illustrations to assist in the explanation of procedures and results.

ANATOMICAL STUDIES

Methods of slaughter finally depend on inducing permanent dysfunction of the brain by either direct physical effects, or by inducing cerebral anoxia associated with disruption of the blood supply to the brain. Therefore it was decided that detailed studies of the blood supply to the brain should be carried out. Also, it was considered appropriate to study the comparative anatomy of the heads of various species of cetacean, in order to determine the most appropriate sites for further ballistic trials involving projectiles and explosive charges.

Blood supply to the brain

The studies of the blood supply to the brain were based on injection of the arterial system with liquid acrylic from the level of the descending aorta, just distal to the heart, in a cranial direction. Once the acrylic had polymerised, acrylic replica of the arteries were obtained by one of two

[1]371 Albert Street, Palmerston North, New Zealand.
[2]Department of Physiology and Anatomy, Massey University, Palmerston North, New Zealand.
[3]Department of Physics and Biophysics, Massey University, Palmerston North, New Zealand.
[4]Department of Veterinary Pathology and Public Health, Massey University, Palmerston North, New Zealand.
[5]Department of Conservation, Auckland, New Zealand.
Pages 174–78 *in* MARINE MAMMAL RESEARCH IN THE SOUTHERN HEMISPHERE VOLUME 1: STATUS, ECOLOGY AND MEDICINE ed by M. Hindell and C. Kemper. Surrey Beatty & Sons, Chipping Norton. 1997.

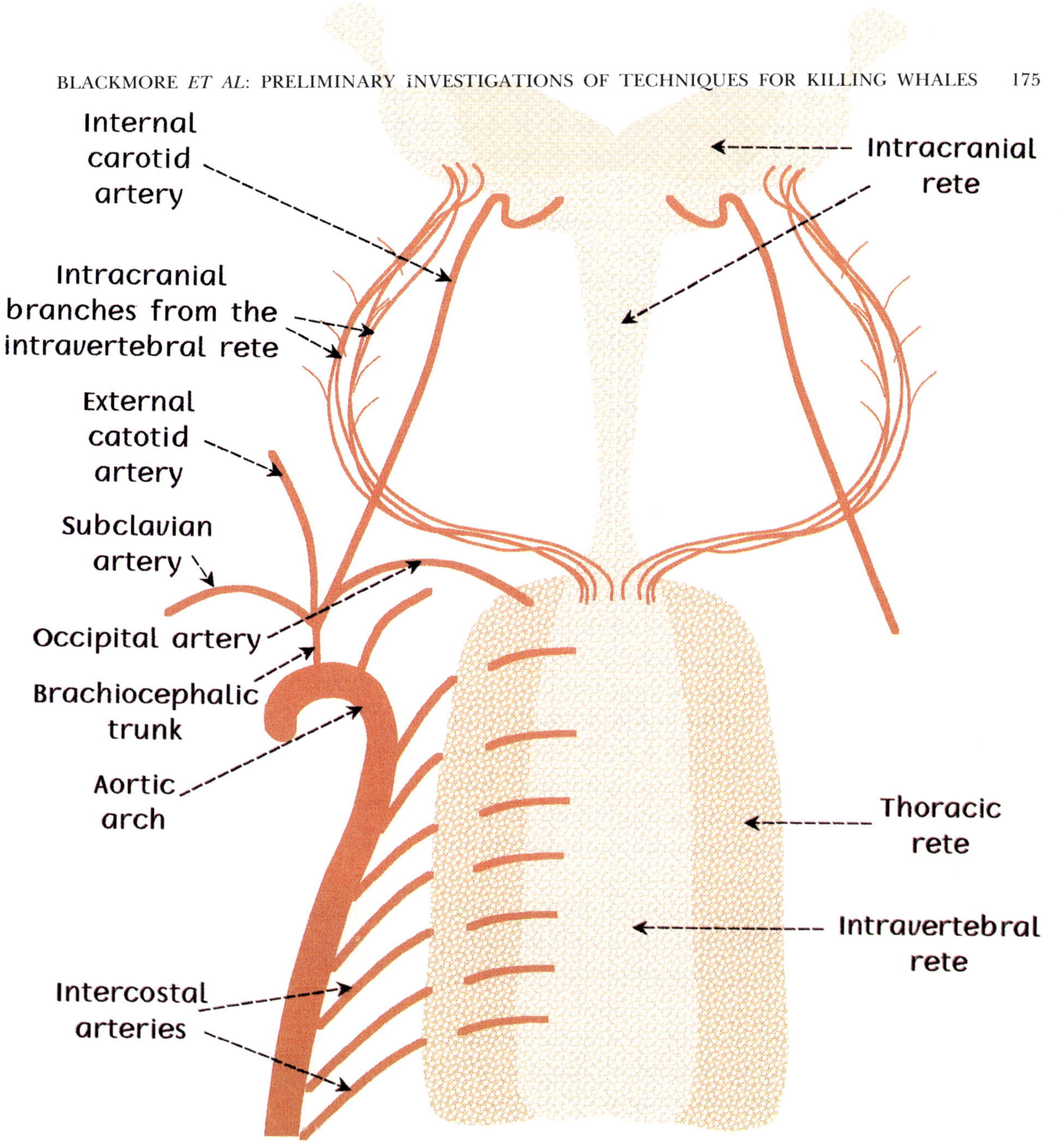

Fig. 1. Schematic diagram of the blood supply to the brain of an immature long-finned Pilot Whale *Globicephala melas*.

methods. One method involved the immersion of the specimen in concentrated hydrochloric acid for three to four months. This resulted in a complete corrosion cast of all the arteries rostral to (in front of) the level of the diaphragm. The other specimen was subjected to maceration of all soft tissues by a colony of Hide Beetles *Dermestides maculatus* over a period of six months. This produced a cast with bone retained, so that the topographical features of the skeletal structures could be studied in association with the arterial system.

These techniques were developed and carried out by one of the authors (A.N.). Preliminary studies of these casts have revealed some unique features of the arterial supply to the brain of the two immature long-finned Pilot Whales *Globicephala melas* so far examined (Fig. 1).

The pattern of the arterial connections between the aorta and the brain is undeniably different from that observed in terrestrial mammals, and many of the features of this vasculature appear to be similar to those observed in dolphins and porpoises (MacKay 1886; Morgane *et al.* 1966). Such features include an absence of a circle of Willis-basilar artery complex, the presence of an extensive thoraco-spinal retial system and lack of vertebral arteries. However, the aorta and its primary divisions bear some resemblance to the general arrangement found in terrestrial mammals, but the common carotid arteries do not supply blood to the brain.

In the Pilot Whale, the major pathway for arterial blood is via the thoraco-spinal retial system. From the aorta, branches of the dorsal intercostal arteries enter a massive thoracic rete, lying retroperitoneally directly beneath the bodies of the thoracic vertebrae. Efferent vessels from this pass into the vertebral canal via the intervertebral foraminae and enter the spinal rete, which is extradural and encircles the spinal cord. Large efferent arteries from the spinal rete pass rostrally and enter the cranial cavity with the spinal cord, through the foramen magnum. These vessels divide to form arteries which course laterally around each side of the brain, giving off branches as they go (cerebral arteries). Each group of lateral arteries terminates at a rostral retial complex. This rete extends to each orbit, forming an ophthalmic rete on each side and also passes caudally in the midline to join the spinal rete. A pair of large internal carotid arteries was present, though one of these vessels was not complete, possibly due to the presence of a blood clot in the vessel during infusion. The internal carotid arteries arise from a commom stem with the external carotid and the occipital artery, close to the origin of the subclavian vessels. The subclavian arteries in turn are derived from the ipsilateral brachycephalic trunk. The occipital arteries also supply the rostral part of the thoracic rete, which itself appears to extend beneath the ventral surface of the floor of the cranial cavity. These internal carotid arteries regress with maturity, as is the case in certain mammals, including dolphins and porpoises (MacKay 1886 and Morgane *et al.* 1966).

In relation to the euthanasia and slaughter of whales the failure of the common carotid arteries to supply blood to the brain is of vital significance

Anatomy of the head

The complete heads of a Risso Dolphin *Grampus griseus* and a Gray's Beaked Whale *Mesoplodon grayi*, and the skulls of Cuvier's Beaked Whale *Ziphius cavirostris*, False Killer Whale *Pseudorca crassidens*, Gray's Beaked Whale, Pygmy Right Whale *Caperea marginata*, Pygmy Sperm Whale *Kogia breviceps* long-finned Pilot Whale *Globicephala melas* and a common dolphin *Delphinus delphis* were examined. These were subject to radiographical computerized tomography (CT) scans. From these images, and by the use of a computer, one of the authors of this paper (A.S.D.) prepared lateral view diagrams showing the structures surrounding, or near to, the brain. These included the airways and pharynx (Figs 2 and 3).

Ballistic investigations

This work was primarily designed to develop better methods for the euthanasia of stranded cetaceans *in extremis*. Based on previous work on the use of a multiple projectile for killing domestic stock (Blackmore 1985) and the advice of one of the authors of this paper (M.C.B.) a 12 gauge shot gun was used. Projectiles were either a single lead slug with a mass of 28 g or buckshot consisting of nine individual pellets with a combined mass of 28 g. The isolated heads of cattle, pigs and a common dolphin, and several complete dead long-finned Pilot Whales were shot at different sites in the head from distances of up to 0.6 m. Detailed results of this work is published elsewhere (Blackmore *et al.* 1995a; Blackmore *et al.* 1995c). Arising from this work a method was developed for the emergency slaughter of large mature pigs (Blackmore *et al.* 1995c) and it has proved effective and safe for operators when used to kill both dolphins and medium-sized Pilot Whales. The projectile caused gross trauma of the brain without penetration of the opposite side of the head. When a rifle is used a bullet penetrating both sides of the head can create a danger to other animals and people in the vicinity and is less likely to destroy a vital centre in the brain compared with a multiple projectile or one which disintegrates in the cranial cavity.

Electroencephalographic (EEG) studies on sheep shot in the vicinity of the occipito-atlantal junction have been completed (Blackmore *et al.* 1994). This work shows that a bullet shot through the upper spinal cord does not cause insensibility. A similar technique has been used by others in an attempt to kill stranded whales. If the bullet hits the occipital condyles of sheep, immediate insensibility occurs. A technique to study the EEG of stranded Pilot Whales is being developed. Such a technique will provide unequivocal evidence on the effectiveness of many techniques of slaughter.

For large cetaceans, a technique is being developed for placing a small explosive charge within a "stomach tube", beneath the skull at the level of the pharynx. At this site, the distance between a charge and the brain is considerably less than on the dorsal surface of the head. Preliminary results are encouraging, and it is hoped that further work will lead to a method of euthanasia which is humane, safe and aesthetically acceptable. Such a method is necessary for the euthanasia of animals such as sperm whales *(Physeter macrocephalus)* which often strand in New Zealand and cannot be refloated.

Electrical investigations

This work has been designed to assess the effects of the electric lance as used by the Japanese as a secondary method of killing, in terms of effectiveness and welfare. Trials were carried out on dead stranded whales and current densities were measured in the brain and heart after applications of electrical currents similar to those used in Japanese whaling operations.

When this paper was presented to the Marine Mammal symposium in Tasmania, this section was essentially a precis of a paper presented to the International Whaling Commission in 1994 (Blackmore *et al.* 1994). Since that time considerably more data have been collected and amendments made to estimates of current densities required to induce the brain and cardiac dysfunction. These new data and revised interpretations have since been published or accepted for publication

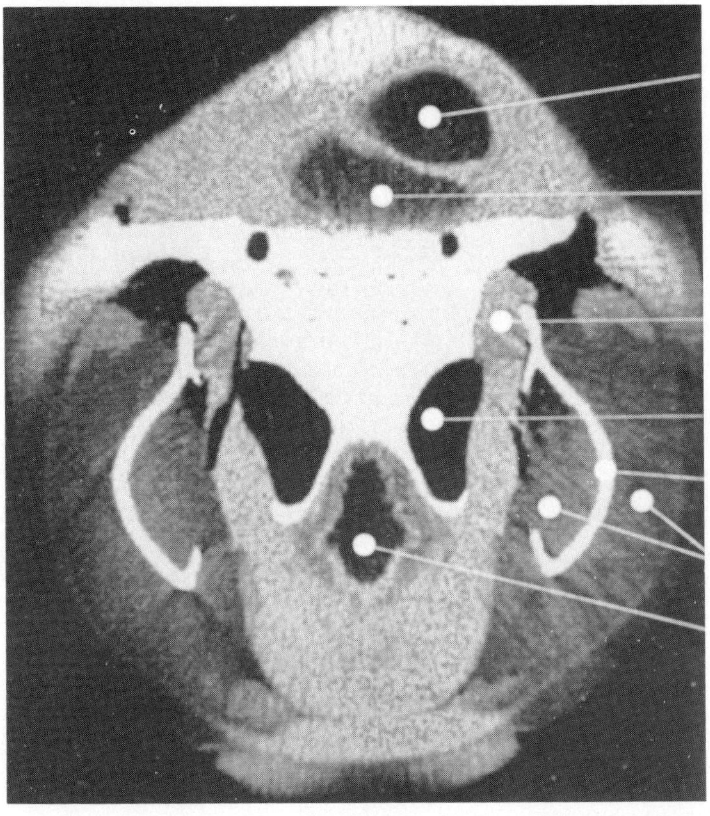

Fig. 2. A transverse CT scan of the head of a Gray's Beaked Whale *Mesoplodon grayi* at the level shown in Figure 3. The specialized fat depots form two melons dorsally and a huge deposit surrounding each mandible ventrally. Observations of such sections suggest that these fat deposits, because they have interfaces with gas filled sinuses, are involved in sound transmission and reception.

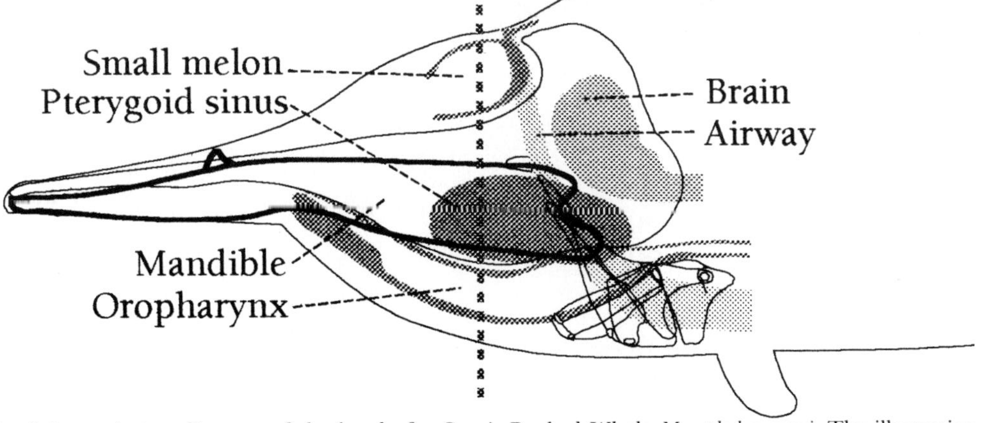

Fig. 3. Lateral view diagram of the head of a Gray's Beaked Whale *Mesoplodon grayi*. The illustration shows the close association of the pharynx with the base of the skull.

(Barnes *et al.* 1995, 1996; Blackmore *et al.* 1996). The last publication contains a comprehensive review and critical appraisal of the electric lance.

In summary, it has now been shown that minimal current densities required to cause depolarization of neurones are 10mA/cm^2, and 0.5mA/cm^2 to cause ventricular fibrillation of the heart. It has been shown that current densities in the heart and brain increase with time post mortem (Blackmore *et al.* 1996). When electrodes were placed in the body of dead whales, similar to those recorded as employed in whaling operations, no recorded current densities, which would have been over estimates of results from a live animal, were sufficient to have caused brain or cardiac dysfunction. Furthermore the low current densities recorded could have been associated with pain. The obvious conclusion of this work is that the electric lance is an ineffective method of secondary killing and can induce extra unnecessary pain and suffering to an already distressed animal.

DISCUSSION

As the work reported in this contribution is of a preliminary nature, it is inappropriate to discuss the findings at length. However, there are several important points to be made.

In spite of the morphological diversity amongst different species of mammal, there is a striking conformity in the way arterial blood reaches the brain of terrestrial mammals. All blood delivered to cerebral tissues is derived from the circle of Willis-basilar artery complex lying against the ventral surface of the brain. The only afferent connections possible with this structure are a pair of cerebral carotids and the ventral spinal artery. The vessels in the neck which conduct blood from the proximal divisions of the ascending aorta to the level of the head invariably involve the common carotid and the vertebral arteries. The only major differences between species in the gross pattern of the cerebral blood supply involves the manner and relative degree of connection between these vessels and the cerebral carotids and the ventral spinal artery, and whether or not retia mirabilia are present.

On the other hand, the relatively little information available on the cerebral blood supply of cetaceans suggests that the arrangement of the arteries appears to differ markedly from the standard terrestrial mammalian pattern (Ask-Upmark 1935; Slijper 1936). For instance, in dolphins and porpoises, as opposed to terrestial mammals, the entire blood supply is derived from a massive retial complex in the thorax and vertebral canal (MacKay 1886; Morgane *et al.* 1966).

In relation to the blood supply to the cetacean brain, it can now be conclusively stated that there is no circle of Willis-basilar artery complex, and that the common carotid arteries do not supply blood to the brain of pilot whales. Thus any method of killing whales based on cutting the carotid arteries must be considered inhumane, because it would not cause rapid cerebral anoxia, and it should be discontinued immediately. Conversely, the concentration of highly arborized vessels both within, or close to, the vertebral canal and some very large blood vessels just rostral to the brain, might provide the opportunity to disrupt the blood flow to the brain by hitherto unconsidered techniques.

The work on the anatomy of the head adds weight to the previous assumption (Blackmore 1992) that shooting whales with a rifle is a difficult and hazardous procedure, especially if they are shot in the sea from a boat. However, the use of a multiple projectile fired from a shot gun shows considerable promise as a safe and effective means for the euthanasia of smaller cetaceans. Greater details of the physics and ballistics of such a technique are published elsewhere (Blackmore 1985; Blackmore *et al.* 1995c). These anatomical studies of the heads of various species of cetacean should be useful for others interested in where best to place explosive chardes to kill such animals.

The work on current densities in the carcasses of whales demonstrates that the use of the electric lance as used in Japanese whaling operations is ineffective and likely to cause extra pain and suffering to an already distressed animal.

ACKNOWLEDGEMENTS

We wish to thank the World Society for the Protection of Animals, the RSPCA (UK) the Whale and Dolphin Conservation Society (UK), Massey University Research Fund, the New Zealand Pork Industry Board and Mrs Innes Murchie for financial assistance. We are also indebted to many others for their advice and technical expertise. These include Mr D. Gibbs of Triple R Engineering, Mr Fred King and Mr Peter Gallagher of the Department of Labour, Dr Peter Dickson and staff of Palmerston North Hospital, Anton van Helden of the Museum of New Zealand, Dr Clyde Daly and Dr Christian Cook of the Meat Research Institute of New Zealand, Sue Jenkins of the Department of Veterinary Clinical Studies and Dr Peter Davie of the Department of Physiology and Anatomy at Massey University and all the field officers of the Department of Conservation who assisted us, with special thanks to Mr Hans Rook. Last but not least we wish to thank Mrs S. Crawford and June Blackmore for typing innumerable letters, memoranda and documents.

REFERENCES

Ask-Upmark, E., 1935. The carotid sinus and the cerebral circulation — an anatomical, experimental and clinical investigation, including some observations on the rete mirabile caroticum. *Acta Psyciatrica and Neurologica Supplement* **6:** 1–374.

Barnes, G. R. G., Madie, P. and Blackmore, D. K., 1996. An assessment of the humane aspects of electric lancing of whales by measurement of current densities in the brain and heart of dead whales. *Medical and Biological Engineering and Computing* **37:** 1–11.

Blackmore, D. K., 1985. Energy requirements for the penetration of heads of domestic stock and the development of a multiple projectile. *Veterinary Record* **116:** 36–40.

Blackmore, D. K., 1992. The slaughter of cetaceans: A comparative review IWC/44/HKW 11.

Blackmore, D. K., Bowling, M. C., Madie, P, Nutman, A., Barnes, G. R. G., Davies, A. S., Donaghue, M. and Kirk, E. J., 1995a. The use of a shot gun for the emergency slaughter of large mature pigs. *New Zealand Veterinary Journal* **43:** 134–37.

Blackmore, D. K., Daly, C. and Cook, C., 1995b. Electroencephalographic studies of the nape shooting of sheep. *New Zealand Veterinary Journal* **43:** 160–63.

Blackmore, D. K., Madie, P., Bowling, M. C., Nutman, A., Davies, A. S., McLeod, W. R., Taylor, J. and Degen, M., 1995c. The use of a shot gun for the euthanasia of stranded cetaceans. *New Zealand Veterinary Journal* **43:** 158–59.

Blackmore, D. K., Madie, P. and Barnes, G. R. G., in press. Observations on the electric lance and the welfare of whales: a critical appraisal. *Animal Welfare* **5.**

Blackmore, D. K., Nutman, A., Barnes, G. R. G., Madie, P., Davies, A. S., Bowling, M. C., Donoghue, M. and Kirk, E. J., 1994. Preliminary investigations of techniques for killing whales. *Proceedings of the International Whaling Commission*. IWC/46/HK 4.

MacKay, J. J., 1886. The arteries of the head and neck and the rete mirabile of the porpoise *(Phocoena communis) Proc. Philosophy Soc. Glasgow* **17:** 366–76.

Morgane, P. J., MacFarland, W. L., Nagel, E. L., Viamonte, M. and Galliano, R. E., 1966. Arterial supply to the brain of the dolphin *(Tursiops truncatus) Physiologist* **9:** 249.

Slijper, E. J., 1936. Die Cetacean Vergleichendanatomisch und systematisch Capita Zoologica. Martinus Nijhoff. The Hague.

On capture, anaesthesia and branding of adult male New Zealand Fur Seals *Arctocephalus forsteri*

S. TROY,[1] D. MIDDLETON[2] and J. PHELAN[3]

Nineteen territorial adult male New Zealand Fur Seals *Arctocephalus forsteri* were captured and anaesthetized at Cape Gantheaume, Kangaroo Island, South Australia. Capture and initial physical restraint were effected using hand-held hoop nets. Halothane and isoflurane were used as inhalation anaesthetic agents. The period of induction and transport for each seal was not significantly different between the two drugs, suggesting similar anaesthetic induction times. The time until each seal was released following anaesthesia was shorter with isoflurane (mean = 39 minutes) than with halothane (mean = 240 minutes); the maximum duration of anaesthesia was 85 minutes and was similar with both drugs. Although there were no serious adverse effects and no mortalities with either drug, respiratory irregularities occurred with halothane on at least three occasions. These problems were not evident with isoflurane. In 1992, all anaesthetized animals were freeze-branded and in 1993 they were hot iron-branded. Although freeze brands were not clearly visible in the months following the first moult, they became more obvious with time and were quite clear following the second moult. Hot iron-brands were obvious immediately following the first moult, but two animals had infected brands six months following application. The majority of bulls subjected to these procedures regained their territories after release; those that were unsuccessful were significantly lower in weight.

INTRODUCTION

THE combination of large size, physical strength and an aggressive nature makes adult seals dangerous to handle without some form of restraint. Chemical restraint is often used to minimize the risks of handling for both the animals and the researchers (Laws 1993). Injectable drugs have been used successfully with pinnipeds, but fatalities have occurred due to difficulties with estimation of dosage, variations in drug metabolism and unpredictable physiological responses (Gales 1989; Erickson and Bester 1993). Injecting drugs remotely, for example with a projectile syringe, reduces the risks to handlers, but can be dangerous for the seal if it responds to the injection by moving into the water before the drug takes effect (e.g., David *et al.* 1988). Inhalation agents are among the safest drugs for producing anaesthesia in seals (Gales 1989), provided adequate physical restraint can be achieved during induction.

A study of the behaviour of territorial adult male (bull) New Zealand Fur Seals *Arctocephalus forsteri* began in November 1992 at Cape Gantheaume, Kangaroo Island, South Australia. The fur seal colony at Cape Gantheaume is on a narrow rocky shore platform, characterized by irregularly shaped boulders. Initially, physical restraint was used to tag, weigh, measure, brand, and collect tooth and tissue samples from breeding bulls, procedures which took up to 60 minutes per individual. In an attempt to reduce the time taken, minimize the risks to animals and handlers, and to maximize the welfare of the animals, chemical restraint was tested in these animals. The results of these trials are reported here, and although not comprehensive, they are indicative of the effect of two contemporary inhalation anaesthetic agents on adult male New Zealand Fur Seals.

METHODS

Physical restraint was used prior to the administration of any chemical agent to minimize the risks of drugged seals escaping to the water and drowning. The capture of fleeing bulls was difficult because of their size, the terrain and the short distance to the water. Consequently, bulls were captured at the beginning of the breeding season, when their reluctance to leave their territories made flight less likely.

Eleven adult male New Zealand Fur Seals were physically restrained and anaesthetized in late November 1992 and another 11 were similarly handled in late November 1993. Three of the animals anaesthetized in 1992 were anaesthetized again in 1993, in order to re-brand them and to replace tags.

[1]Department of Zoology, University of Melbourne, Parkville, Victoria, Australia 3052.
[2]Wild Vet, P.O. Box 626, Lilydale, Victoria, Australia 3140.
[3]Healesville Sanctuary, Healesville, Victoria, Australia 3777.

Pages 179–83 *in* MARINE MAMMAL RESEARCH IN THE SOUTHERN HEMISPHERE VOLUME 1: STATUS, ECOLOGY AND MEDICINE ed by M. Hindell and C. Kemper. Surrey Beatty & Sons, Chipping Norton. 1997.

Physical restraint

A maximum of five people (stalkers) with poles including one person with the net (described below) slowly surrounded the animal, blocking escape routes to the sea. The stalkers used large boulders and irregular terrain to conceal their presence as much as possible. Just before the emergence of the stalkers, the animal's attention was deliberately engaged by one person who was opposite the person with the net. Fur seal bulls that were surrounded in this way tended to stand their ground or move away from the stalkers, which assisted capture. Bulls sometimes moved toward the water (and incidentally, a stalker) but were often halted by threat vocalizations and arm waving from the stalkers. The bull was physically restrained in a hand-held, conical nylon mesh net which was reinforced with webbing and lashed to a hand-held aluminium hoop (diameter 82 cm) on a 150 cm long pole. The net was 1.3 m long, and consisted of 25 mm mesh with a 1 mm mesh inner lining for the first metre from the apex. There was a small reinforced hole in its apex through which the seal could put his nose, this held his jaws closed and enabled access to his nostrils for anaesthetic induction.

Anaesthesia

Once the seal was restrained within the net, an anaesthetic hood was placed over its head. The hood was a soft PVC cone about 1 m long, with many small holes punched at the apex and two small tapes sewn internally to hold a pad of cotton wool soaked with the anaesthetic agent. In November 1992 halothane was used (Fluothane; May and Baker) and in November 1993 isoflurane was trialed (Forthane; Abbot Laboratories).

The anaesthetic hood was held firmly around the seal's shoulders to increase the amount of drug being inhaled. The seal typically reacted to the net by vigorously thrashing his head. It was therefore easier and safer to administer the drug initially with the large hood than with an anaesthetic mask, which would require restraint of the animal's head.

Depth of anaesthesia was monitored by assessing the degree of muscular relaxation and testing for a pedal withdrawal reflex. When the withdrawal reflex was lost the seal was removed from his territory, in order to avoid the handlers and patients being attacked by neighbouring territorial males. Induction time was recorded as the time from placement of the hood over the animal's head until the pedal withdrawal reflex was lost. Induction/transport time was recorded as the time from placement of the hood over the seal's head until it was connected to the anaesthetic machine.

Following the removal of the animal from the colony and checking the depth of anaesthesia, the net and hood were removed and the seal was connected to a portable anaesthetic machine using a face mask placed over the nose and mouth. In 1992 the anaesthetic system incorporated a Drawover Fluotec (Cyprane, Yorkshire) halothane vaporizer using an open anaesthetic circuit with a five litre in-line re-breathing bag and an Ohmeda BOC Health Care Flowmeter. This system was modified, assembled and supplied by CIG Medisheild, Melbourne, Australia. This type of anaesthetic machine is designed for military use and does not require pressurized oxygen to function effectively provided the patient can produce a volume demand. In this instance the system was supplied with a variable flow oxygen source in order to enrich the inspired air and provide for the contingency of emergency oxygen administration. In 1993 the anaesthetic system used an out of circuit Penlon PPV flow-through isoflurane vaporizer connected to a standard re-breathing circuit using a soda-lime CO^2 absorption unit. A four litre re-breathing bag for positive pressure ventilation was fitted. The system used in 1993 was assembled and supplied by VMS Supplies, Melbourne, Australia.

The depth of anaesthesia was varied to suit the needs of the various procedures while maintaining as shallow a plane of anaesthesia as necessary. The level of anaesthesia was assessed regularly using reflexes (corneal, palpebral and pedal) and the degree of muscular relaxation (jaw tone). Vaporizer settings varied from 2 to 5%. Endotracheal tubes (9 mm–11 mm diameter) were placed in six seals in 1993. All equipment was disinfected prior to use. Rectal temperature, heart rate, respiration rate and the appearance of the mucous membranes of the mouth were monitored continually although recorded irregularly, and all subjects received a supportive drug regime according to the observed variation in their physiology. Additional drugs included atropine, antibiotics and doxapram (administered intramuscularly) and protective eye ointment (Brolene; see Table 1). Dopram–V was administered to five seals during 1992 to reverse apnoea or undesirably low respiratory rate (see Table 1). The subject was cooled or heated as necessary, using sea water on the flippers and a blanket, respectively.

Recovery

When all procedures were completed, each animal was placed in a 1 × 1 × 2 m wooden box with sliding doors. The doors were propped ajar to encourage air flow and the seal was monitored through small viewing holes. The animal was released once it was sitting up and showed co-ordinated movement, unless this took longer

Table 1. Additional drugs given to New Zealand Fur Seal bulls under anaesthesia.

Brand name (drug)	Manufacturer	Dosage	Year
Norocillin (Procaine penicillin and Benzine penicillin)	Norbrook/Heriot	15–30 mg/kg	1992
Alamycin LA (Oxytetracycline)	Norbrook/Heriot	20 mg/kg	1993
Atrosine Forte (Atropine sulphate)	Parnell Laboratories	0.1–0.4 mg/kg	both
Dopram–V (Doxapram hydrochloride)	Bomac Laboratories	1.3–2.7 mg/kg	1992
Brolene (Dibromopropamide isethionate)	Rhone–Poulenc Rorer	1–5 drops both eyes	both

than 30 minutes, in which case the seal was released after a minimum of three hours. Anaesthetic time was measured as the time from connection of the seal to the anaesthetic machine until it was disconnected. Recovery time was recorded as the time from the removal of the anaesthetic mask, or endotracheal tube, until the animal was either sitting or holding its head up. Release time was recorded as the time from the removal of the anaesthetic mask, or endotracheal tube, until the animal was released.

The location and behaviour of all tagged seals was noted during the subsequent breeding seasons to determine effects of the procedures on the territorial status of each bull.

Branding

All captured animals were freeze branded in 1992, and hot iron branded in 1993. For freeze branding, a 10 × 10 cm patch on either the right or left shoulder was closely shaved down to the skin, and was swabbed with methylated spirits to reduce the skin temperature and improve the conductivity of the skin. The first four animals were branded for 30 seconds with brands cooled in liquid nitrogen, the other seven animals were branded for 50 seconds with brands cooled in an alcohol/dry ice mixture. The alcohol/dry ice brands were cooled for at least 20 minutes prior to application. The brands were copper alloy heads 50 mm high and 6–8 mm wide. Each animal was individually marked with a single letter, in various orientations.

In hot iron branding, the heated brand was applied quickly to singe away the fur, then the area and the brand were wiped clean. The mild steel brand was then applied for four seconds. Each animal was individually marked with a two-digit brand on the flank.

The quality of brands was assessed when bulls were resighted during the breeding seasons (November–January) in 1992/3 and 1993/4, and during field trips in May and July 1993, and in May and August 1994.

RESULTS

Capture

Attempts to capture bulls began on November 17, 1992. Bulls generally became more aggressive and therefore easier to catch towards the day the first pup was born, which was on November 29 in both years. Four bulls were caught between November 17 and 22 despite daily attempts. Seven seals were caught between 27 and 30 November 1992. In 1993, 11 seals were caught from 27 November until 2 December. The weight of the captured animals ranged from 78 kg to 160 kg. Standard length ranged from 150 cm to 179 cm.

Anaesthesia

Table 2 describes the induction/transport time, induction time, anaesthetic time, time to recovery and time to release for the two anaesthetic agents used. Induction time was only recorded for five animals anaesthetized with halothane, and therefore this variable cannot be compared directly between anaesthetics. However, induction/transport time provides an indirect measure of induction time. The latter component (transport time) was similar in both years because the seals were caught in the same general area and the workstation was in the same place. Therefore, induction/transport time was analysed to provide an indication of anaesthetic induction time. Induction/transport time was not significantly different between drugs, nor did the length of time the seal was connected to the anaesthetic machine (anaesthesia) differ. Time to recovery and time to release were significantly shorter with isoflurane than with halothane. There was no significant difference between time to recovery from isoflurane-induced anaesthesia in intubated and non-intubated seals (unequal variances t-test = 0.38, d.f. = 4.7, p = 0.72).

The seals anaesthetized with isoflurane were released with a mean time of 39 minutes (Table 2) and all except one moved with co-ordination. The seals that were anaesthetized with halothane

Table 2. Anaesthetic induction and recovery times (mins) for New Zealand Fur Seal bulls under the two anaesthetic agents. Variables are defined in the text. †indicates differences are underestimated due to bias in data collection, see discussion. Times were not recorded for all trials because of distraction; hence the inconsistencies in sample size. All *t*-tests were calculated for equal variances with the exception of release time.

	Anaesthetic agent								
	Halothane 1992 n = 11			Insoflurane 1993 n = 11			Independent *t*-test		
	mean	s.e.	N	mean	s.e.	N	T	d.f.	P
Induction/transport	18.1	1.6	11	15.3	1.7	10	1.178	19	0.253
Induction	5.2	1.0	5						
Anaesthesia	47.0	5.8	11	58.4	4.0	8	−1.497	17	0.153
Recovery	12.6	3.4	5	6.4	1.3	11	2.077	14	0.057†
Release	242.1	48.0	11	44.6	9.8	8	4.031	10.8	0.002

Table 3. Physiological variables of New Zealand Fur Seal bulls under anaesthesia. Medians were calculated using the midpoint of the balues recorded for each animal, if more than one value was recorded. Ranges were taken from the highest and lowest value recorded for any animal.

	Halothane			Insoflurane		
Physiological variable	median	range	N	median	range	N
Heart rate (beats/min.)	107	60–126	9	59	50–80	10
Respiratory rate (breaths/min.)	10	3–22	9	10.5	5–24	9
Rectal temperature (°C)	35.2	35.0–36.7	4	36.2	35.3–38.4	11

were more variable in their co-ordination on release. Release times were recorded for all 11 seals, but evaluation of the recovery only made for seven seals. Four were released after a mean time of 254 ± 223 min and moved in an uncoordinated manner with a "floppy" belly indicating loose muscle tone; three were released after a mean time of 207 ± 114 min. and were well co-ordinated with better muscle tone. The other four seals were released after a mean time of 250 ± 44 min.

The anaesthetics performed with isoflurane were characterized by physiological stability and a rapid recovery relative to those conducted with halothane. Table 3 describes the physiological variables (recorded irregularly) as an indication of the variation observed. Respiratory irregularities occurred in at least three of the seals under halothane anaesthesia. These irregularities were accompanied by cyanosis and included periods of apnoea, variation in respiratory excursion (depth of breathing), inspiratory dyspnoea (associated with pharyngeal soft tissue and exudates partially obstructing the airway) and variations in the respiratory rate. There were no serious adverse effects and no mortalities with either anaesthetic agent. All animals were resighted in the colony, either on territories or in the non-breeding haul out areas, during the breeding season which followed their release.

Territoriality

Eight of the 11 seals that were anaesthetized with halothane regained their territories during the 24 hours following release; two of them did so immediately. The other three individuals did not hold a territory at all that breeding season. Similarly, nine of the 11 seals that were anaesthetized with isoflurane regained a territory in the days following release: three of them did so immediately and the other six began to hold a territory within 5 days (mean = 2.7 days, s.e. = 0.6, N = 6). Again, two seals did not hold territories at all during that breeding season. The five bulls that did not hold territories after release were significantly lighter than the successful bulls and were caught prior to the onset of pupping (Independent *t*-test = –4.33, d.f. = 17, $P < 0.001$; the three recaptured seals were included in the group for 1992 but excluded from the 1993 group to avoid over-representation in the data set).

Branding

The clarity and visibility of the brands were assessed once the wounds had healed after the January-April moulting period. Of the six freeze-branded animals resighted in May and in August 1993 after their first moult, two were apparently unbranded. However, after the subsequent breeding season and moult, all of these six seals had clear brands. The remaining five freeze-branded seals were sighted in November to January 1994, during the subsequent breeding season, and two of them had no discernible brands. These two "unbranded" seals were the first two that were captured and branded. In one case, the shaved brand-site had not been swabbed with methylated spirits prior to application of the brand. Although the freeze brands on the other nine seals were clear, they were small and not obvious to an inexperienced observer. Consequently, the three seals that were re-caught in 1993 for re-tagging were also hot iron branded to improve our ability to distinguish between individuals.

Ten hot iron-branded seals were resighted in the 14 months following their first moult and all had clearly visible brands. However, two of these animals had infected wounds from their brands, both of which had persisted for six months but were healed within 12 months.

DISCUSSION

The data collected in this study are largely qualitative and must be interpreted cautiously. Further studies using controlled drug dosages, application times, release times and more quantitative data would be of use in determining the full range of reactions of seals to these inhalation drugs.

This capture and anaesthetic technique is safe for territorial New Zealand Fur Seal bulls. The technique, using a heavy and unwieldy net, relies on the bulls standing their ground against intruders. A lighter, more manoeuvrable net would be required to enable animals to be caught in the non-breeding season, when they are not holding territories. The seal could then be transferred to the more restrictive, heavier net used in this study. This has been tried successfully with one small bull in the non-breeding season.

Isoflurane was the most satisfactory anaesthetic agent, producing more stable anaesthesia, shorter recovery times and dramatically shorter times to release. Seals that were anaesthetized with isoflurane were fit and capable of normal behaviour upon release, whereas halothane tended to produce lingering effects after release. It is important to minimize short-term after-effects of drugs used on territorial bulls, because animals that are not fully functional are highly susceptible to attack by other bulls. The longer recovery time from halothane anaesthesia is an underestimate because it was not recorded for six seals, five of which were still unconscious after 30 minutes. Similarity in anaesthetic induction times using the two drugs is suggested by the similar total times taken to initially restrain the seal and move him from his territory to the workstation.

Most bulls regained their territory within a few days of release. The low weights of the five bulls that did not regain a territory suggest they were likely to have been displaced by larger bulls before the onset of pupping. Although most bulls moved into the sea immediately after release, five went straight to their territories including one that successfully repulsed an intruder within 15 minutes of release.

Freeze-branding did not always produce an identifiable mark during the first year post-branding, although successful brands were obvious after 12 months, and hot-iron branding caused infection in some animals. The absence of a visible brand in the first two animals freeze-branded may have been because the branding irons were not cold enough. This may be due to insufficient cooling of the branding irons in the liquid nitrogen, rather than the change in the method of cooling them (from liquid nitrogen to dry ice/methylated spirits) and consequent change in brand application time, because liquid nitrogen is considerably colder. All wounds from freeze-brands healed cleanly without signs of infection. The danger of over-branding and consequent infection with hot-iron brands may make freeze-branding a better technique for permanently marking seals, particularly if the personnel are inexperienced. The results of the present study suggest that a minimum of two digits should be used in freeze-branding to make the brands easier to recognize. If identification of captured bulls in the first 12 months after freeze-branding is desired, we recommend also using temporary marking such as tags.

ACKNOWLEDGEMENTS

This work has been generously supported by the Holsworth Wildlife Research Fund, the Colman and Reckitt Down to Earth Research Fund and the M. A. Ingram Trust. We would also like to thank the many volunteers for their invaluable contribution, and South Australian National Parks and Wildlife Service (SANPWS) for their logistic support. This work was conducted under SANPWS permit numbers Z23211–01 and Z23211–02, and with University of Melbourne Animal Experimentation Ethics Committee Approval Number 654–082–0–92–1232. The manuscript was considerably improved by comments from Professor M. Bryden, Dr M. Elgar, Dr N. Gales, and Dr P. D. Shaughnessy, for which we are very grateful.

REFERENCES

David, J. H. M., Hofmeyr, J. M., Best, P. B., Meyer, M. A. and Shaughnessy, P. D., 1988. Chemical immobilization of free-ranging South African (Cape) fur seals. *S. Afr. J. Wildl. Res.* **18:** 154–56.

Erickson, A. W. and Bester, M. N., 1993. Immobilization and capture. Pp. 46–88 *in* Antarctic seals — research methods and techniques ed by R. M. Laws. Cambridge University Press: Cambridge, UK.

Gales, N. J., 1989. Chemical restraint and anaesthesia of pinnipeds: a review. *Mar. Mamm. Sci.* **5:** 228–56.

Laws, R. M., 1993. Antarctic seals — research methods and techniques. Cambridge University Press: Cambridge, UK.

Index

A

Abundance 78–79, 131
Aerial survey 3, 40–54
Aerobic Dive Limit (ADL) 115, 121–22
Age classes 57–58, 60
Ahuru 131–33, 141
Allele frequency 19
Amsterdam Island 88–90
Anaesthesia 97, 179–83
Anatomy 165, 168, 175–79
Antarctica 30–37
Antarctic Fur Seal 97–104, 105–13
  attendance patterns 103, 155, 158
  diet 105–13, 152–57, 159–61
  diel dive patterns 100, 111, 158, 160–61
  dive depth 98–99, 103, 109–10, 157–58
  dive duration 98–99, 110, 157–58
  dive types 98–100, 103, 157
  diving behaviour 97–104, 106, 151–61
  foraging ecology 105–13, 151–61
  moult 101, 111
  movements 99, 101, 103, 109–10, 158–61
  lactation 100, 112
  prey size 107, 109–10, 111–12, 152–53, 156
  site fidelity 101
Antarctic Whaling Management Areas 31, 37
Arctocephalus forsteri see New Zealand Fur Seal
Arctocephalus gazella see Antarctic Fur Seal
Arctocephalus pusillus see Australian Fur Seal
Arctocephalus tropicalis see Subantarctic Fur Seal
ARGOS 138–39
Arrow squid 133, 135, 145–47
Arteries 175–77
Atendance patterns 81–82, 103, 155, 157–58
Australian Fur Seal 56–61, 63–70
  anatomy 165, 168
  age classes 57–58, 60
  field identification 57–58, 63–70
  habitat 68
  haulouts 56
  distribution 57, 64
  Montagu Island 57–61
  locomotion 66, 68–69
  pelage 65
  taxonomy 70
  thigmotacism 68
  vocalizations 58, 66, 70
Australian Sea Lion 78–86, 70, 127, 148, 164–69
  abundance 78–79
  attendance cycles 81–82
  diet 84, 127, 148, 165
  digestion 165–69
  distribution 78–80
  duration of lactation 81–82
  duration of pupping cycle 79, 81
  embryonic diapause 82–83
  foraging energetics 84–85
  life history 78–86
  mating strategy 81
  milk composition 82–83
  population trends 78–79
  reproductive behaviour 79, 81–82
  reproductive physiology 82–84
  seasonality of pupping cycle 79, 81
  stomach stones 164–69

B

Balaenoptera acutorostrata see Minke Whale
Balaenoptera borealis see Sei Whale
Balleny Islands 30–31, 37
Ballistic investigations 176, 178
Birth mass 74, 93–95
Blood supply 174–79
Blubber/fat 73–74
Body composition 73–74
Body mass 74–76
Bottlenose Dolphin 1–8, 170–73
  age classes 2
  feeding 8, 170
  fisheries interactions 8
  group size 3–7
  healing rate 173
  inshore
    locations 3–6
  offshore
    photoidentification 2–3, 5
  predation 7–8, 170–73
  ranges 3–6
  sex classes 2
  shark attack 2, 7, 170–73
  site fidelity 2
  social structure 6, 8
  wound healing 170–73
Brain 174–79
Branding 181–83

C

Caperea marginata see Pygmy Right Whale
Capture/physical restraint 97, 166, 179–83
Census 30–37, 40–54, 57–61
Cephalopods 108, 111, 121–22, 126–27, 131, 133, 135, 145–49, 153–55
  beaks 108, 126–27, 131, 145, 147–49, 153
Champsocephalus gunnarii 108–09, 111–12
Common Dolphin 2, 31, 33, 36, 176
Computerized tomography (CT) 176–77
Conservation 24–25
Cuvier's Beaked Whale 176

D

Delphinus delphis see Common Dolphin
Dentition 93–96
Deuterated water (HDO) 73–76
Diel patterns 100, 120, 122, 158, 160
Diet 84, 105–13, 121–22, 125–28, 130–36, 143–49
  inter-annual variation 108
  inter-seasonal variation 108, 128, 132–35
  prey size 106, 109, 111, 126–27, 130–32, 133–35, 145–49
Digestion 165–69
Digestion coefficients 144, 147–49
Distribution 44–46, 53–54, 57
Dive classification 98, 117–19, 157
Dive depth 84, 99, 109–10, 114, 118–20 139, 157–58, 160–61
Dive duration 35, 84, 99, 110, 118–20, 139, 157
Diving behaviour 97–104, 109–10, 114–24, 151–61
Doubly labelled water (DLW) 93

E

Effect of instruments 138–39
Effective breeding ratio 15
Effective population size 15
Electric lance 176–79
Electroencephalographic (EEG) 176
Electrona sp. 108–09, 112, 153–60
Energetics 73–76
English Mackeral 145–46
Eubalaena australis see Southern Right Whale
Euphausia sp. *see* Krill
Euthanasia 174–78

F

False Killer Whale 3, 176
Feeding trials 125–27, 144–49
Female:calf ratios 48–49
Field identification 57–58, 152
Fisheries interactions 2, 8, 128
Fluoroscopy 166–69
Foraging areas 101–03, 109, 135, 137–41, 158–59
Foraging dives 119–20, 157
Foraging ecology 84–85, 96–104, 105–13, 114–24, 151–61
Foraging energetics 84–85
Freeze branding 181–83

G

Gene flow 20–23
Genetic bottleneck 85
Genetic diversity 12–13, 15–23
Genetic structure 17, 23–24
Genetics 11–25
Gestation 96
Globicephala melas see Long-finned Pilot Whale
Gough Island 88–90
Grampus griseus see Risso Dolphin
Gray's Beaked Whale 176–77
Group size 3–4, 31–36
Growth rate 74, 88–90, 160

H

Habitat 68
Hand feeding 170
Haplotype 19
Haulouts 56
Harbour Seal 14
Healing 170–73
Heard Island 13, 17–25, 96–104, 105–13, 114–24, 161
Heart rate 182
Hooker's Sea Lion 130–36
  diet 130–36
  foraging 135–36
  numbers 131
  prey size 131–32, 133–34
Hourglass Dolphin 31–32, 36
Humane killing 174–78
Humpback Whale 6, 15, 31–34, 37
Hydrurga leptonyx see Leopard Seal
Hyperoodon planifrons see Southern Bottlenose Whale

I

"Iniquity" 30–37
Inter-island comparisons 75, 89
International Whaling Commission (IWC) 37
IWC/IDCR cruises 36

J

Jack Mackeral 126–28, 145–47

K

Kerguelen Plateau 101, 103, 117–18, 121
Killer Whale 25, 31–33, 36
  behaviour 35
  group size 35
Kogia breviceps see Pygmy Sperm Whale
Koheru 145–47
Krefftichthys andersonii 108–09, 111–12
Krill 25, 103–04, 110–11, 159

L

Lactation 81–83, 73–75, 89, 93–94
Lagenorhynchus cruciger see Hourglass Dolphin
Leeuwin Current 85–86
Leptonychotes weddellii see Weddell Seal
Leopard Seal 14
Life history 78–86
Locomotion 65, 68–69
Long-finned Pilot Whale 175–76

M

Maatsuyker Island 125–28
Macquarie Island 13, 17–25, 30–31, 72–76, 92–96, 151–61
Management 24–25, 128
Marion Island 88–90
Maternal expenditure 72–76
Maternal size 73–76, 90
Mating strategy 81, 85
Megaptera novaeangliae see Humpback Whale
Mesoplodon grayi see Gray's Beaked Whale
Metabolic rate 73–75, 122
Milk composition 73, 82–83
Milk consumption/intake 73–75
Minke Whale 31–34, 36
Mirounga angustirostris see Northern Elephant Seal
Mirounga leonina see Southern Elephant Seal
Montagu Island 57–61
Moreton Bay 1–8
Moreton Island 1–8
Moult 23
Movements 48–51, 54, 101, 103, 109, 117–18, 121, 137–41, 158–59
Myctophids 103, 108–15, 122, 127–28, 141, 152–61

N

Neophoca cinerea see Australian Sea Lion
New Zealand Fur Seal 14, 58–59, 61, 63–70, 85–86, 125–28, 137–41, 143–49, 179–83
  anaesthesia 179–83
  attendance patterns 138
  branding 179–83
  diet 125–28, 143–49
  digestion coefficients 144, 147–49
  dive depth 138–39, 141
  dive duration 138–39
  effect of instruments 138–39, 142
  feeding trials 125–27, 145–48
  field identification 58, 63–70
  foraging areas 137–41
  habitat 68
  locomotion 65–66, 69
  movements 137–41
  otolith recovery rates 126–27, 145–47

New Zealand Fur Seal — *continued*
  pelage 64
  physical restraint 179–80
  prey size 126–27, 145–49
  rate of passage 145–49
  territoriality 182, 183
  thigmotacism 68–70
  vocalization 66, 70
Non-foraging dives 119–20, 157
North Stradbroke Island 1–8
Northern Elephant Seal 14–18, 75–76, 95–96, 122
Nursing period 94–95

O

Orcinus orca see Killer Whale
Otolith erosion 147–49, 153
Otolith recovery rates 126, 145–47
Otoliths 108–09, 125–27, 130–33, 145–49, 152–57

P

Pelage 65
Penguins 111
Phoca vitulina see Harbour Seal
Phocarctos hookeri see Hooker's Sea Lion
Phylogeographic analysis 16–18
Physeter macrocephalus see Sperm Whale
Photoidentification 2–3, 31, 43
  callosities 43
  matching 3, 43
Point Lookout 1–8
Population decline 24–25, 76
Population genetics 11–25
Population size 46–48, 78–79
Population trends 72, 76
Predation 7–8, 78–79, 170–73
Prey 90, 121–22, 126–28, 130–36, 141, 153–57
  abundance 90, 111–12, 128, 161
  size 108–09, 126–27, 131–33, 153, 155
Projectils 176
Pseudorca crassidens see False Killer Whale
Public reports 42–43, 53
Pygmy Right Whale 176
Pygmy Sperm Whale 176

Q

Quinnat Salmon 145–48

R

Radiography 165–69
Ranges 3–6
Rate of molecular evolution 13–15
Rate of passage 145–49, 166–69
Redbait 126–28
Red Cod 141, 145–47
Regurgitate analysis 125–26, 130–33
Reproduction 48–49
Reproductive behaviour 79–82
Reproductive physiology 82–84
Reproductive strategy 81–82
Respiratory rate 182
Rete 175
Risso's Dolphin 176

S

Satellite telemetry 104, 137–41
Scat analysis 125–26, 152–57
Sea-ice 37
Sealing 70, 77, 161
Sei Whale 31–33
Selection 20

Sharks 2, 7, 170–73
  attack 2, 7, 170–73
  Great White 7
  Tiger 7
Site fidelity 2, 24
Social structure 6, 8
South Georgia 13, 17–25, 74, 76, 111, 160–61
South-east Queensland 1
Southern Bottlenose Whale 31–32
Southern Elephant Seal 11–25, 72–76, 92–96, 114–23
  aerobic dive limit (ADL) 115, 121–22
  allele frequencies 19
  ancestral populations 22
  birth mass 74, 93–96
  body composition 73–75
  declining population 24–25, 76
  dentition 92–96
  development 94–96, 122
  diel patterns 120, 122
  diet 121–22
  dive classification 123
  diving behaviour 117–20, 122
  effective breeding ratio 15–16
  effective population size 15–16
  energetics 73–76
  gene flow 20–23
  genetics 11–25
  genetic diversity 12–13, 15–23
  genetic structure 17, 23–24
  growth 74, 93–96
  haplotype 19
  juveniles 114–23
  lactation 73–76, 93–95
  management 24–25
  milk consumption 73–74
  moult 23
  movements 22, 24, 114–18, 121
  phylogeographic analysis 16–18
  rate of molecular evolution 13–15
  sealing 25, 76
  site fidelity 24
  weaning mass 93–96
Southern Ocean 90, 101–04
Southern Ocean Whale Sanctuary 37
Southern Right Whale 40–54
  aerial surveys 40–54
  distribution 44–46, 53
  female calf ratios 48–49, 52
  movements 48–51, 54
  population size 46–48, 50–53
  public reports 42–43, 53
  reporting frequency 43–44
  reproduction 48–49
  seasonality 44–45
Sperm Whale 31–35
  dive durations 35
  respiration rates 35
Spotted Dolphin 3
Stenella attenuata see Spotted Dolphin
Stomach stones 127, 164–69
Subantarctic Fur Seal 88–90, 151–61
  age class 153–54
  attendance patterns 157–58
  birth mass 89
  diel patterns 158, 160–61
  diet 151–61
  diving behaviour 151–61
  dive depth 155, 157–58, 160–61
  dive duration 157–58
  field identification 152
  growth 88–90, 161
  inter-island comparisons 88–90
  lactation 89
  maternal size 90
  movements 158–60

Subantactic Fur Seal — *continued*
  prey abundance 90
  prey size 154, 156, 160
  weaning mass 89–90
Surface interval 120

T

Tangalooma 170–73
Teeth 92–96
Temperature depth profiles 115–18
Territoriality 182–83
Thigmotactism 66, 68
Thoraco-spinal reteal system 175–76
Time Depth Recorder (TDR) 98–99, 114–15, 138–39, 152–53
Trawling 7
Tritiated water (HTO) 75
Tursiops truncatus see Bottlenose Dolphin

V

Vocalizations 58, 64, 66

W

Water turnover 73–76
Weaning 94–95
Weaning mass 74, 89–90
Weddell Seal 14, 17
Whaling 40, 174
Wounds 2, 7, 170–73

Y

Yacht survey 30–37
Yellow-eyed Mullet 145–48

Z

Ziphius cavirostris see Cuvier's Beaked Whale